EVOLUTION OF THE CERCOPITHECOID FORELIMB

Theropithecus gelada in the Amhara highlands of central Ethiopia (photo courtesy of Robin Dunbar, University College London).

Evolution of the Cercopithecoid Forelimb

Phylogenetic and Functional Implications from Morphometric Analyses

Russell L. Ciochon

UNIVERSITY OF CALIFORNIA PRESS
Berkeley • Los Angeles • London

UNIVERSITY OF CALIFORNIA PUBLICATIONS IN GEOLOGICAL SCIENCES

Volume 138
Issue Date: October 1993

UNIVERSITY OF CALIFORNIA PRESS
BERKELEY AND LOS ANGELES, CALIFORNIA

UNIVERSITY OF CALIFORNIA PRESS, LTD.
LONDON, ENGLAND

PRINTED IN THE UNITED STATES OF AMERICA

Library of Congress Cataloging-in-Publication Data

Ciochon, Russell L.
Evolution of the cercopithecoid forelimb: phylogenetic and functional implications from morphometric analyses / Russell L. Ciochon.
p. cm. — (University of California publications in geological sciences; v. 138
Includes bibliographical references.
ISBN 0-520-09760-2 (alk. paper)
1. Cercopithecidae—Phylogeny. 2. Forelimb—Evolution.
I. Title. II. Series.
QL737.P93C55 1993
599.8'2—dc20 93-20876
CIP

The paper used in this publication meets the minimum requirements of American National Standard for Information Sciences—Permanence of Paper for Printed Library Materials, ANSI Z39.48-1984.

Contents

List of Illustrations

Figures

Plates

List of Tables

Acknowledgments

This monograph has been many years in the making. It is based on research conducted during my graduate studies at the University of California at Berkeley. Numerous colleagues and friends helped along the way. First and foremost I want to thank the members of my graduate committee—Professors F. Clark Howell (Chairman), W. A. Clemens, and D. E. Savage—for their expert guidance and patience. Second, I want to thank my good friend John Fleagle, who provided advice, moral support, and research facilities at a time when I was greatly in need of such help. Clearly, without his assistance and confidence this study would not have been completed. Additionally, John and his wife, Roberta Seidman, M.D., deserve many thanks for inviting me to stay in their home much longer than any of us had planned. I also gratefully acknowledge the faculty and staff of the Departments of Anatomical Sciences and Anthropology at the State University of New York at Stony Brook, especially the late Norman Creel, the late Maynard Dewey, Fred Grine, Bill Jungers, Dave Krause, Sue Larson, Lawrence Martin, Jack Stern, and Randy Susman, as well as John Fleagle, for making my stay at SUNY so stimulating and enjoyable.

During my graduate years at U.C. Berkeley I benefited greatly from discussions with Noel Boaz, Rob Corruccini, Eric Delson, Gerry Eck, Clark Howell, Eric Meikle, Geoff Pope, Yoel Rak, Vince Sarich, Mike Schwartz, Sherry Washburn, and Tim White. As graduate advisor, Clark Howell was always available for helpful chats, and provided facilities and seemingly endless research funds. Via telephone to New York I was often in contact with Eric Delson who was a limitless source of knowledge on fossil cercopithecoids. My good friend Rob Corruccini provided the computer programs and the technical advice on their use. In this same regard, Michael Schwartz deserves thanks for helping adapt several of the programs for my special needs and working closely with me to create the PCA Membership Quotient. Finally, I want to thank Sherry Washburn, who sparked my interest in physical anthropology and admitted me to the graduate program at U.C. Berkeley. Although we have not been in full agreement on how I should approach this project, his confidence in my research abilities has always been a source of inspiration.

I am very grateful to the curators, faculty, and staff of the following institutions for the courtesies and assistance they extended during the data-collection phase of this project: UNITED STATES: American Museum of Natural History, Department of Mammalogy, New York; Academy of Natural Sciences, Philadelphia; Cleveland Museum of Natural History; Field Museum of Natural History, Department of Mammalogy, Chicago; Museum of Comparative Zoology, Mammal Department, Harvard University, Cambridge, Mass.; Museums of Vertebrate Zoology and Paleontology, University of California, Berkeley; U.S. National Museum of Natural History, Smithsonian Institution, Division of Mammals (Primate Biology), Washington, D.C. BRITAIN: British Museum (Natural History), Departments of Palaeontology and Zoology, London. BELGIUM: Institut Royal des Sciences Naturelles de Belgique, Brussels. THE NETHERLANDS: Rijksmuseum van Natuurlijke Historie, Leiden. GERMANY: Anatomischen Institut der Universität (Dr. Senckenbergische Anatomie), Frankfurt; Zoologische Sammlung des Bayerischen Staates, Munich. FRANCE: Faculté des Sciences, Laboratoire des Sciences de la Terre, Université de Lyon; Museum d'Histoire Naturelle, Lyon; Museum d'Histoire Naturelle, Perpignan; Museum National d'Histoire Naturelle, Laboratoires d'Anatomie Comparée, Paleontologie, and Zoologie, Paris. SWITZERLAND: Anthropologisches Institut der Universität Zurich; Museum d'Histoire Naturelle, Geneva; Naturhistorisches Museum, Basel. ROMANIA: Center for Anthropological Research, Bucharest. EAST AFRICA: Ethiopian National Museum, Addis Ababa; Kenya National Museum, Osteology Department and Palaeontology Department, Nairobi. SOUTH AFRICA: Transvaal Museum, Pretoria; University of Witwatersrand Medical School, Anatomy Department, Johannesburg; Bernard Price Institute of Palaeontology, Johannesburg.

I would also like to acknowledge Professor Phillip Tobias, Head of Anatomy at the University of Witwatersrand Medical School; Dr. C. K. Brain, Director of the Transvaal Museum, Pretoria; Mr. Richard Leakey, Director, and Dr. Meave Leakey, Paleontologist, of the Kenya National Museums; and Dr. Donald Johanson, Director of the Institute of Human Origins, Berkeley, for providing access to the many unpublished fossil cercopithecoids in their respective charges.

Throughout the course of my graduate study at U.C. Berkeley I was generously supported by a National Institutes of Health (National Institute of General Medical Sciences GM 1224) Traineeship awarded to Professor George Foster. For support during the writing phase of this study at U.C. Berkeley I acknowledge funding from a Sigma Xi Research Grant, an Office of the Chancellor Dissertation Grant, the Department of Anthropology Subvention funds, the L.S.B. Leakey Foundation and the National Science Foundation. During the final preparation of the manuscript at the University of Iowa I acknowledge support from NIH Biomedical Research Seed Grant RR07035-22, the Carver Scientific Research Initiative Grants Program and a Faculty Scholar Award from the Office of Academic Affairs.

All of the original artwork in this study was drawn by Ms. Judith Ogden, except for two figures in Chapter 7 which were drafted by Hugh Nachamie. Robert Gill developed and printed the photographs. Dr. Eugene Hammel provided computer time for the analysis. The univariate and bivariate plots were produced at the U.C. Berkeley Computer Center using the Graphics Display System (GDS) and a Calcomp Plotter;

the final versions of these plots were then scanned and reformatted by Stephen Nash. My good friend and colleague, Ms. Joy Myers, proofread an early draft of the entire manuscript offering many useful suggestions; Dr. Steve Hartman carefully read and commented on Chapter 4; Dr. Eric Delson and two anonymous reviewers offered many useful comments regarding the revision and shortening of the manuscript; Mr. Richard Nisbett of Iowa State University read and edited later drafts of the manuscript; and Dr. Richard Cowen, who was chair of the editorial board for the University of California Publications in Geological Sciences when the manuscript was submitted, provided much needed editorial assistance and advice throughout the final review phase. For additional editorial assistance I also acknowledge Ms. Rose Anne White of the University of California Press.

During the publication phase at the University of Iowa I was assisted by Mary Allen, Alissa Bader, Jim Enloe, Tim Hare, Stephen Lensink, Richard Nisbett, Elizabeth Ramsey, Ted Roese, and Karen Wohlleben, as well as by numerous student volunteers working in the Laboratory of Biological Anthropology. I thank them all for their dedication to this project. Finally, I want to thank my dear wife, Noriko, who has always managed to make my life *away* from the laboratory such a delight.

Note: The author's present address is Department of Anthropology, 114 Macbride Hall, The University of Iowa, Iowa City, Iowa 52242-1322 where he is Associate Professor of Anthropology and Pediatric Dentistry.

Abstract

This study describes functional and systematic differences in the extant cercopithecoid primate forelimb, and uses this information to interpret cercopithecoid evolution in the African Plio-Pleistocene fossil record. During the past five decades a great many Old World monkey postcranial remains have been recovered from sites in East and South Africa. The great majority are isolated elements with no direct association with identifiable cranial remains. Though an extensive literature now exists on the evolutionary history of the Cercopithecoidea based on cranio-dental analyses, the postcranial skeleton has been regularly neglected. The lack of a rigorous methodological approach that can be applied in a systematic manner is largely responsible for this neglect.

In this monograph, both standard and novel multivariate statistical techniques are used to quantify patterns of variation in the forelimb skeleton of 20 species of living and extinct cercopithecoids. The forelimb is metrically described by 87 continuous linear variables which characterize the size and shape of joint surfaces and the areas of muscle origins and insertions, as well as the length, curvature, and robusticity of the long bones.

The comparative sample of 305 individuals comprises 13 extant and 7 extinct cercopithecoid species. This sample is first analyzed in detail and then used as a reference in a series of additional analyses of all 234 unidentified forelimb specimens from the Plio-Pleistocene of Africa and Eurasia. In this manner, the relationships between living and extinct cercopithecoid species of known taxonomic affinity can be studied and graphed before applying this knowledge to identification of the unknown fossil forelimb elements.

All species of the comparative sample are initially analyzed with univariate statistical techniques, and the functional anatomical implications of the variables are detailed. This is followed by principal components and stepwise discriminant multivariate techniques. Through computation of a newly developed PCA membership quotient, a PCA distance statistic, standard similarity coefficients, and Mahalanobis D^2 distances and posterior probabilities, it is possible to assign taxonomic allocations to virtually all of the previously unidentified forelimb specimens. The new classification techniques developed for this study should find numerous applications in zoology and paleontology.

The assignment, sometimes to species level, of hundreds of additional fossil cercopithecoids permits the range extension of various paleospecies including *Theropithecus brumpti*, *T. oswaldi*, *T. darti*, *Parapapio*, *Papio* sp., *Cercocebus* sp., *Paracolobus*, and *Rhinocolobus*. This information, and analysis of the extinct comparative sample allow better reconstructions of the forelimb skeleton of several large extinct colobines and three extinct species of *Theropithecus*, as well as new interpretations of the locomotor habits and ecological adaptations of these forms. Finally, this analysis of the forelimb results in the selection of a series of well-defined characters that separate the Colobinae and Cercopithecinae. Through outgroup comparisons with the earliest known Old World monkey, *Victoriapithecus*, and with the Oligocene catarrhine *Aegyptopithecus* and basal anthropoid *Apidium*, it is now possible to chart the primitive vs. derived nature of the Old World monkey forelimb. This knowledge advances our understanding of the order of development of character-complexes within the Cercopithecoidea, and the Catarrhini in general. In particular, this character analysis supports the growing body of literature demonstrating that the Colobinae is the most derived group in the Cercopithecoidea, and that this group's habitat preference for the tropical forest biome is a secondarily derived specialization.

Institutional Abbreviations

The institutional abbreviations used in this study are listed below in alphabetical order. They are used throughout the text, tables, and appendix to designate the location of both extinct and extant specimens analyzed in this study. Many of these same abbreviations are currently in use by the institutions listed below.

AIUF	Anatomischen Institut der Universität (Dr. Senckenbergische Anatomie), Frankfurt, Germany
AIUZ	Anthropologisches Institut der Universität Zurich, Switzerland
AMNH-M	American Museum of Natural History, Department of Mammalogy, New York
ANSP	Academy of Natural Sciences, Philadelphia, Pennsylvania
BMNH-P	British Museum (Natural History), Department of Palaeontology, London
BMNH-Z	British Museum (Natural History), Department of Zoology, London
BPIP	Bernard Price Institute of Palaeontology, Johannesburg, Republic of South Africa
CARB	Center for Anthropological Research, Bucharest, Romania
CMNH	Cleveland Museum of Natural History, Cleveland, Ohio
ENM	Ethiopian National Museum, Addis Ababa
FMNH-M	Field Museum of Natural History, Department of Mammalogy, Chicago, Illinois
FSUL	Faculté des Sciences, Laboratoire des Sciences de la Terre, Université de Lyon, France
IRSNB	Institut Royal des Sciences Naturelles de Belgique, Brussels
KNM-O	Kenya National Museum, Osteology Department, Nairobi
KNM-P	Kenya National Museum, Palaeontology Department, Nairobi

MCZ-M	Museum of Comparative Zoology, Mammal Department, Harvard University, Cambridge, Massachusetts
MHNG	Museum d'Histoire Naturelle, Geneva, Switzerland
MHNL	Museum d'Histoire Naturelle, Lyon, France
MHNP	Museum d'Histoire Naturelle, Perpignan, France
MNHN-A	Museum National d'Histoire Naturelle, Laboratoire d'Anatomie Comparée, Paris
MNHN-P	Museum National d'Histoire Naturelle, Laboratoire de Paléontologie, Paris
MNHN-Z	Museum National d'Histoire, Laboratoire de Zoologie, Section des Mammifères et Oiseaux, Paris
MVZ	Museum of Vertebrate Zoology, University of California, Berkeley
NMB	Naturhistorisches Museum, Basel, Switzerland
RNH	Rijksmuseum van Natuurlijke Historie, Leiden, The Netherlands
TMP	Transvaal Museum, Pretoria, Republic of South Africa
UCB-AH	University of California, Berkeley, Department of Anthropology; F. C. Howell, laboratory collection
UCB-AW	University of California, Berkeley, Department of Anthropology; S. L. Washburn, laboratory collection
UCMP	University of California Museum of Paleontology, Berkeley
USNM-M	United States National Museum of Natural History, Smithsonian Institution, Division of Mammals (Primate Biology), Washington, D.C.
UWMS-A	University of Witwatersrand Medical School, Anatomy Department, Johannesburg, Republic of South Africa
ZSBS	Zoologische Sammlung des Bayerischen Staates, Munich, Germany

Taxon Abbreviations

The following abbreviations for all living and extinct taxa in the comparative sample used in this study are listed below in alphabetical order. They appear in many of the figures and tables.

CA	*Cercocebus albigena*
CB	*Colobus badius*
CG	*Colobus guereza*
CM	*Cercopithecus mitis*
D	*Dolichopithecus ruscinensis*
M	*Mesopithecus pentelici*
MS	*Macaca sylvanus*
N	*Nasalis larvatus*
P	*Paracolobus chemeroni*
PA	*Papio anubis*
PC	*Papio cynocephalus*
PE	*Presbytis entellus*
PG	*Paradolichopithecus geticus*
PH	*Papio hamadryas*
PS	*Papio sphinx*
PU	*Papio ursinus*
R	*Rhinocolobus turkanaensis*
TB	*Theropithecus brumpti*
TG	*Theropithecus gelada*
TO	*Theropithecus oswaldi*

1

INTRODUCTION

This monograph is a comparative osteological study of the large bones of the forelimb (scapula, humerus, ulna, and radius) in extant and fossil cercopithecoid monkeys. The study is composed of several interrelated parts. First, a series of forelimb measurements are compared in large samples of selected extant Old World monkeys, using both univariate and multivariate statistical techniques. The aims of this part of the study are (1) to determine whether osteological measurements of the forelimb bones can be used to distinguish and identify different species of extant Old World monkeys; (2) to determine whether the two subfamilies of cercopithecoid monkeys (colobines and cercopithecines) are characterized by differences in the dimensions of forelimb bones; and (3) to examine whether these osteometric features that distinguish the species or subfamilies of extant Old World monkeys can be related to locomotor or ecological differences.

In the second part of the study, the forelimb bones of fossil cercopithecoid monkeys are examined and compared with the extant sample, using the same set of measurements and a variety of statistical techniques. The goals here are (1) to compare the fossil limb elements of known taxonomic allocation with the extant sample to determine likely phyletic and functional affinities of the extinct monkeys; and (2) to compare isolated, unassociated limb elements from Plio-Pleistocene deposits in East and South Africa (and elsewhere) with both the extant and extinct comparative samples, to determine the most likely taxonomic identification of the isolated elements and the probable locomotor habits of their owners.

The third part of the study, based on the preceding analyses of extant and fossil limb bones, attempts to reconstruct the broad outlines of the evolutionary history of the cercopithecoid forelimb as well as the locomotor and associated ecological diversity of extinct Old World monkeys.

Traditionally, studies of primate evolution have concentrated primarily on the dentition and skull, and with good reason. Teeth are the elements that are most commonly preserved in the fossil record and have been most thoroughly examined by past generations of paleontologists and comparative anatomists. Likewise, skulls play a major role in neontological studies of mammalian systematics, providing a firm basis for systematic determinations. In many respects, studies of primate

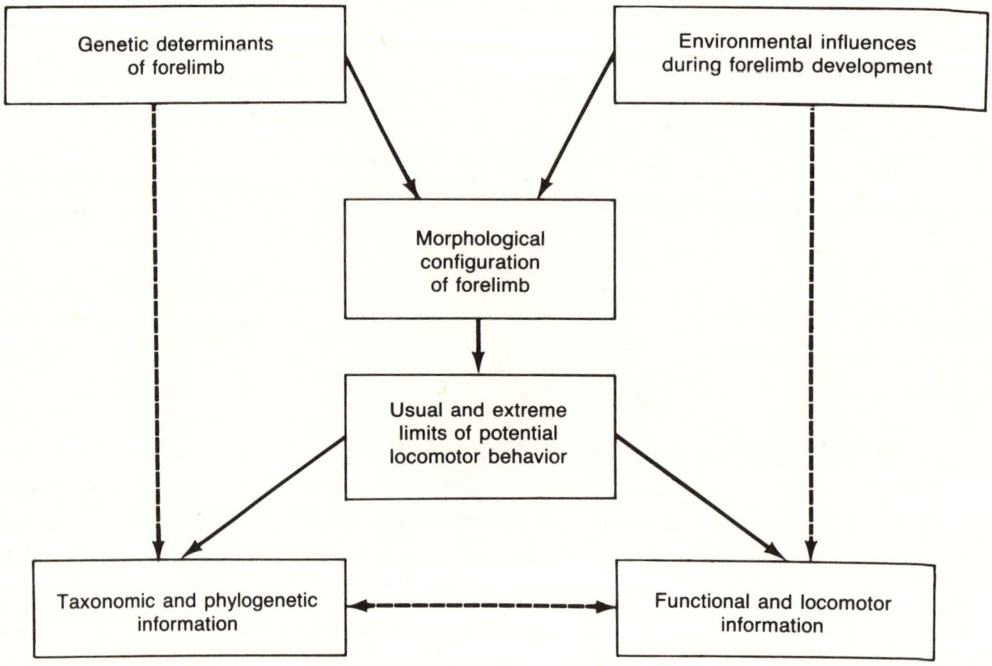

Figure 1. Flow chart illustrating the factors that determine forelimb morphology and the information that can be derived from their analysis.

evolution based on skulls and teeth have distinct advantages over studies based on limb bones, since the latter are relatively rare both in the fossil record and in museum collections. However, limb elements can potentially make very important and very different contributions to our knowledge of primate paleontology by providing information that can be used in conjunction with cranial and dental evidence to reconstruct both the phyletic affinities and the paleoecology of extinct primates. Conroy (1976), Szalay and Dagosto (1980), Fleagle and Simons (1982), Fleagle (1983), Oxnard (1984), Senut (1986a, 1986b), Rose (1988), Harrison (1989) and Krentz (1993), among others, have demonstrated that significant amounts of information about phylogeny and function can be derived from careful analysis of fossil postcranial data. The large bones of the cercopithecoid forelimb were chosen as an appropriate region of analysis because (1) past anatomical and morphometric analyses of primate forelimb elements (see Miller, 1932; Oxnard, 1967; McHenry and Corruccini, 1975; Corruccini and Ciochon, 1976, 1978; Feldesman, 1976; Birchette, 1982; Krentz, 1992) have yielded phylogenetic and functionally interpretable results; and (2) the bones of the forelimb, especially portions of the humerus and radius, are among the most prevalent postcranial elements in the fossil record because their compact shapes resist fragmentation during deposition.

A flow chart (see Fig. 1) illustrates the factors that determine the form of the forelimb elements and the information that can be derived from their analysis. Information encoded in the genome of a species is a major determining factor of forelimb morphology (see discussion below), but this information is subject to modification during ontogeny by a variety of environmental influences. Thus the final morphological configuration of the forelimb and the movement potentials across its joints determine the range of locomotor behaviors the species can perform. As Szalay and Dagosto (1980:7) have aptly stated, "joint surfaces, like all features which have evolved, are compromises; their form and function resulting from both conflicting *selection pressures*, as well as the further compromise necessarily superimposed by the *constraints of heritage*" (their italics). Since the morphology of the forelimb elements and their functional/locomotor correlates are in large part determined by genetic factors, a close study of their form should reveal both taxonomic and phylogenetic information as well as functional and locomotor information. In other words, if the morphology is analyzed at a fine enough level, features reflecting habitus and features reflecting heritage should both emerge.

The case for a strong genetic component in the determination of the form and structure of limb bones is well documented by Amtmann, who concludes (1979:350): "Since prenatal bone growth is under genetic control, and since the transformation of the bone is produced by the controlled activity of chondroclasts, osteoclasts, [and] chronoblasts, the modification of the bone by these cells can only proceed normally if the hereditary mechanism has not been disturbed." The prenatal phase in the development of the skeleton, primarily under genetic control, is termed "preadaptive morphogenesis" (ibid., p. 349). During the second phase, "postadaptive morphogenesis," ontogenetic mechanisms and direct adaptations to environmental stimuli play a greater role, but by this stage the shape and structural characteristics of the skeleton have already been determined (ibid.; see also experimental ontogenetic studies of joint mobility in patas monkeys by Turnquist, 1983, 1985).

A number of techniques can be employed in the study of postcranial specimens, with results that emphasize either heritage or habitus factors. Fig. 2 presents a flow chart modified from Day (1979) which illustrates this point. Any analysis of postcranial specimens, whether biomechanical, statistical, or otherwise, must be grounded on a firm understanding of anatomy. Chapter 3 presents the anatomical basis for this study. Fossil postcranial specimens can be compared with living and extinct primates along two pathways (see Fig. 2). In my opinion, the pathway leading through measurements and statistics is more likely to aid in useful taxonomic judgments, while the pathway through biomechanical analysis will be more useful in locomotor judgments. Obviously, the two methodologies and their end results are not mutually exclusive but interdependent, as the arrows in Fig. 2 indicate. Heritage and habitus are better regarded as two separate but overlapping components. Szalay (1981:164) defines heritage characters as all features "present in the ancestry of the taxon prior to the acquisition of other taxon-specific features of the lineage." Characters of habitus are features "newly acquired by the species in response to environmental selective forces" (ibid.). Szalay concludes that the heritage aspects of

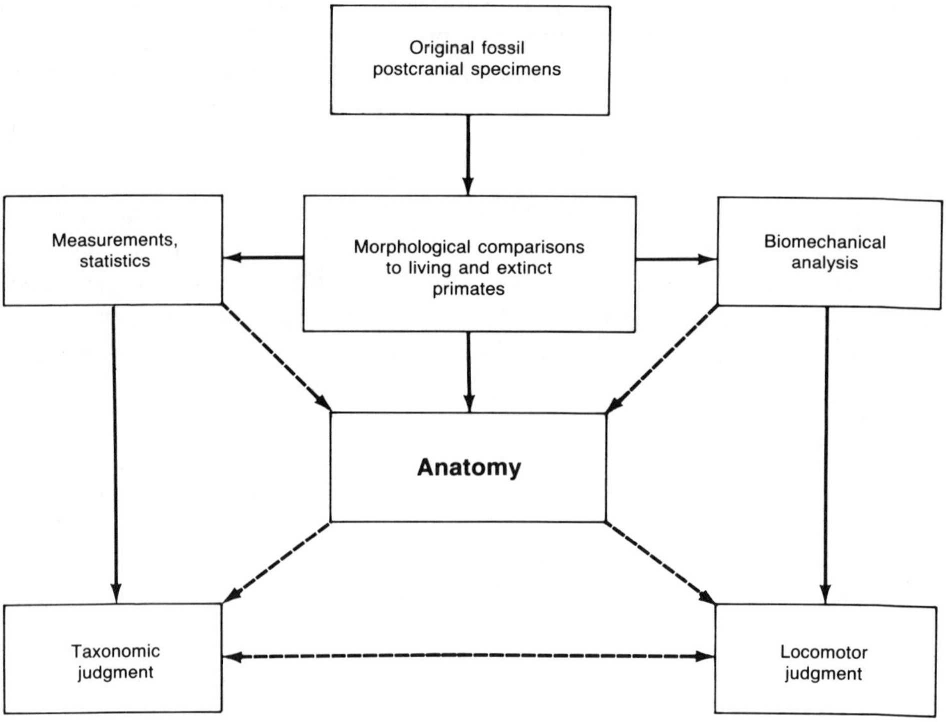

Figure 2. Flow chart modified after Day (1979) illustrating the steps involved in analysis of a postcranial fossil and the methods most commonly used to yield taxonomic and locomotor judgements.

a species tend to channel its habitus characters. Such a view is also adopted here: I believe that a quantifiable, statistically oriented osteometric analysis of the cercopithecoid forelimb skeleton will produce meaningful phylogenetic or heritage determinations, but that a locomotor or habitus overprint will always remain.

Because the forelimb is a complex functional unit with many interrelated elements and cross-cutting functions, it is impossible to isolate single functional parts or to isolate anatomical units without resorting to a statistical methodology. Multivariate analysis is an ideal tool for this situation, since it allows for multiple cross-comparisons between individual measurements among many species. It reduces the complexity of variation of the forelimb and its joints into several more interpretable composite variables or components.

Multivariate analysis is also a very useful technique when dealing with a group of organisms that are all very similar. A. H. Schultz (1970) points out that the Cercopithecoidea is a remarkably uniform group with few divergent specializations when compared with other extant primate groups. Their postcranial skeleton reflects this uniformity; it is very difficult, by classical morphological techniques, to separate species based on observations of the forelimb elements (but see Fleagle, 1977). Yet the few multivariate analyses that have been performed on exclusively cercopithecoid samples (see Epps, 1968; S.N. Booth and Freedman, 1970; Manaster, 1975; Albrecht, 1978; Oxnard, 1984) have produced taxonomically meaningful results.

Univariate, bivariate, and multivariate procedures have been used in this study. Three multivariate morphometric techniques are employed: similarity coefficients analysis (SCA), principal components analysis (PCA), and discriminant functions analysis (DFA). All of these methods are phenetic in nature (Wiley, 1981): they group unknown individuals into known groups based on overall similarity. The relationships that result from these multivariate analyses need not necessarily have a phylogenetic connotation. However, remarkably consistent patterns emerge when these results are interpreted in the context of studies based on phylogenetic analyses of the skulls and teeth.

Principal components analysis yields vector loadings which express the relative contribution of each measurement to the overall dispersion of the groups. These loadings can be interpreted in a cladistic framework through outgroup comparison, providing an independent phylogenetic check of the morphological characters analyzed. Jolicoeur (1959), Pimentel (1979), and Bookstein et al. (1985) have firmly established the usefulness of multivariate analysis in evolutionary studies in a wide array of diverse taxa.

The research presented in this monograph overlaps in many ways with studies undertaken over the past 30 years on the evolution of cercopithecoids and other primate groups. Morbeck's (1972) landmark study on the forelimb of Miocene hominoids emphasized that this "dental ape" group was actually very monkey-like in the postcranial skeleton. Starting from a firm understanding of the differences between the forelimb skeletons of living cercopithecoids and hominoids, she applied this dichotomy to Miocene species, using both comparative anatomical and multivariate methodologies. Conroy's (1976) monograph on the primate postcranial remains of the Oligocene Fayum region of Egypt identifies, describes, functionally interprets, and phyletically sorts all the isolated primate limb remains from two decades of research in the Fayum. Using a variety of techniques with particular emphasis on multivariate analyses, Conroy has made significant contributions to our understanding of the evolution of the primate postcranial skeleton and its locomotor correlates during the Oligocene epoch. Birchette (1982) analyzes the colobine postcranial skeleton and the phyletic affinities and locomotor behavior of the giant extinct colobine *Paracolobus*. Though this study is quite exhaustive, no multivariate analyses were employed. Finally, Krentz (1992), in a doctoral dissertation, addresses the evolutionary significance of interspecific osteological variation in the forelimb of extant and extinct species of *Theropithecus*.

Only a few of the multivariate morphometric studies undertaken over the past three decades deal specifically with the Cercopithecoidea. Epps (1968), in her doctoral dissertation, analyzed the forelimb bones of several East African monkeys, using canonical variates analysis. Manaster (1975) attempted a multivariate approach in her analysis of the postcranial skeletons of *Cercocebus* and *Cercopithecus*. Her results and data on other cercopithecoids are further discussed by Oxnard (1978, 1984). Oxnard's (1984) analysis of the postcranial skeleton of cercopithecoids is in full accordance with new biomolecular-based cladograms which indicate a tripartite arrangement of the Old World monkeys.

The most significant multivariate study to date on the Cercopithecoidea is the monographic work of Albrecht (1978) on the craniofacial morphology of the macaque monkeys of Sulawesi. On the basis of gross morphology and geographic distribution, Fooden (1969) had divided these macaques into 7 allopatric phenotypically distinct species. Albrecht analyzed 24 craniofacial variables on 149 Sulawesi macaques and compared them with 325 non-Sulawesi macaques composing 28 local populations of 11 of 12 extant species. The results of his canonical and principal components analyses confirmed the existence of 7 morphologically distinct species. Furthermore, Albrecht notes that the craniometric variation within Sulawesi macaques is more shape-related than is seen across all 11 non-Sulawesi species. Finally, Albrecht suggests an ancestral stock from which to derive the Sulawesi macaques and a Pleistocene deployment scenario for their arrival. Albrecht's study is a fine demonstration of the applicability of multivariate morphometrics and numerical systematics to taxonomic/phylogenetic problems.

Several significant works on cercopithecoid evolution bear directly on the research reported in this study. Freedman (1957) was the first to recognize the considerable diversity of Plio-Pleistocene Cercopithecoidea, describing 6 genera and 15 species of South African fossil monkeys, though current estimates of this diversity are lower (Szalay and Delson, 1979). Delson (1975), expanding on the views of Freedman (1957), Jolly (1967), and Simons (1970), reviews the entire evolutionary history of the Cercopithecoidea from its origins in the late Oligocene to the development of modern species. He documents this study with cladograms, phylogenies, and numerous photographs. Delson specifically addresses the position of the Eurasian fossil colobines *Mesopithecus* and *Dolichopithecus*, and deals with the placement of Eurasian cercopithecines *Paradolichopithecus* and *Procynocephalus* as well as the African forms *Parapapio* and *Dinopithecus*. In a monographic review, Jolly (1972) discusses the evidence for the natural history and evolution of *Theropithecus (Simopithecus)* in East and South Africa during the Plio-Pleistocene. This giant baboon, sometimes reaching the size of a female gorilla, was highly diverse and is exceedingly prevalent in the fossil record. However, only one species, *T. gelada*, remains today, which is restricted to the arid montane grassland of central Ethiopia (see frontis plate). More current reviews of the diversity and temporal range of the extinct species of *Theropithecus* can be found in Dechow and Singer (1984), Eck and Jablonski (1987), Eck (1987), and in Jablonski, ed. (1993).

Four other recent reviews of African cercopithecoid evolution also deserve mention. M. G. Leakey (1982) describes one new genus, *Rhinocolobus*, and three new species of fossil colobines from the Plio-Pleistocene. Eck and Jablonski (1984) discuss the evidence for the evolution of *Papio* in East Africa and propose that two of its fossil species be transferred to *Theropithecus*, raising the total species for that genus to six and reducing the fossil record of the ubiquitous savanna baboon to a mere trace. Delson (1984) presents a site-by-site biochronology of all African Plio-Pleistocene cercopithecoid localities, which provides the ideal temporal framework necessary for my study. Finally, Strasser and Delson (1987) provide the most detailed cladistic analysis of living and extinct cercopithecoids ever attempted, resulting in the recognition of several new subclades and yielding a new classification of Old World monkeys at the tribal and subtribal level.

Two important symposia on the evolutionary biology of fossil cercopithecoids have also recently been held. At the XIth Congress of the International Primatological Society held in Gottingen, Germany in 1986, a conference entitled 'Dating of the Cercopithecoid Fossil Record' was organized by Martin Pickford (see Pickford, 1987b). This symposium featured contributions on the history and chronology of the fossil Cercopithecoidea of East Africa (Pickford, 1987a), northern Africa (Geraads, 1987), western Europe (Ardito and Mottura, 1987), the Indian subcontinent (Barry, 1987) and China (Pan Yuerong and Jablonski, 1987). Pickford (1987b) provides a post-conference commentary that approaches cercopithecoid evolution from a unique perspective—framing the origin and subsequent spread of monkeys throughout the Old World against the backdrop of global climatic history, zoogeography, paleozoogeography and the comparison of neontological vs. paleontological versions of cercopithecoid phylogeny.

In April of 1990 a symposium sponsored by the Wenner-Gren Foundation entitled '*Theropithecus* as a Case Study in Primate Evolutionary Biology' was convened at the University of Cambridge. Organized by Nina Jablonski (see Jablonski, 1993a), this conference explored all aspects of the biology of *Theropithecus* including its fossil history and evolution, its biogeography, its comparative evolutionary biology and anatomy, and the behavior and socioecology of the living and extinct representatives of the genus. The proceedings of this symposium were just published in the edited volume, *Theropithecus: The Rise and Fall of a Primate Genus* (Jablonski, ed., 1993). Of particular relevance to this monograph are the contributions by Eck (1993) on the first complete description of the cranial remains of *Theropithecus darti* from the Hadar Formation, Ethiopia, by M.G. Leakey (1993) on the evolution of *Theropithecus* in the Turkana Basin, Kenya, by Delson (1993) on the taxonomy of *Theropithecus* fossils from Africa and India, and by Jablonski (1993b) on the phylogeny of *Theropithecus*. Other relevant contributions include Delson and Hoffstetter (1993) on newly described *Theropithecus* fossils from Terefine, Algeria, Krentz (1993) on a species-by-species account of the postcranial anatomy of *Theropithecus*, Dunbar (1993) on the socioecology of the extinct species of *Theropithecus*, Lee and Foley (1993) on the ecological energenics and extinction of *Theropithecus oswaldi*, and Delson and Dean (1993) on the proposed assignment of *Papio baringensis* and *Papio quadratirostris* to the *Theropithecus* clade.

These two symposia contribute to the increasing realization that cercopithecoid monkeys are not just primitive catarrhines, left over from earlier epochs, but are a very specialized radiation of higher primates (*sensu* Andrews, 1982; Pickford, 1987b; Benefit and McCrossin, 1990) with a history of adaptively diverse forms, many of which are quite divergent from any living species (e.g., M.G. Leakey, 1982, 1985). Clearly, the study of Old World monkeys, the most successful primate group to have evolved, is currently undergoing a long overdue renaissance.

2

THE COMPARATIVE SAMPLE

The comparative sample used in this study consists of 13 living species and 7 extinct species, carefully selected to provide a wide spectrum of morphological variation within the Cercopithecoidea as a basis for allocation of the unknown fossil specimens that are the primary focus of the study. To place this comparative sample in perspective, Table 1 gives a current classification of all living and extinct Old World monkeys, and a cladogram consistent with this classification is shown in Fig. 3.

Criteria for the Selection of Extant Species

The 13 species of the extant comparative sample are presented in Table 2. Most are from Africa, since nearly all of the unknown fossil specimens are derived from African Plio-Pleistocene deposits. The cranial remains of these African fossil monkeys reveal that many are closely related to or share direct ancestral-descendant relationships with living African cercopithecoids. Nine of the 13 species in Table 2 are cercopithecines, since these are also the predominant subfamily in the fossil record.

Within the cercopithecines, a wide array of body sizes and locomotor types has been selected, with particular emphasis on the large-bodied terrestrial forms (Papionini), which are the most predominant kind of fossil cercopithecine. Of the large-bodied forms, 5 different *Papio* species (including both savanna and forest-dwelling forms) and the only living species of *Theropithecus* were selected, in order to sample variability within the genus *Papio* and to contrast this group with *Theropithecus*. These species are also used to document the extent and temporal depth of the *Papio-Theropithecus* dichotomy in the African fossil records—a particularly interesting issue in light of a recently proposed phyletic scheme (see Eck and Jablonski, 1984, 1987; and Delson and Dean, 1993). *Cercocebus* and *Cercopithecus* are included because these species may help sort out variation in the less common and little understood small-bodied fossil cercopithecines. In fact, a phylogenetic link between *Cercocebus* and one of these small-bodied extinct forms, *Parapapio*, is a distinct possibility (Szalay and Delson, 1979). *Macaca sylvanus* is included because it is the only African macaque and probably the most primitive species in the genus. Furthermore, the genus *Macaca* has the longest fossil record of any living cercopithecoid, and

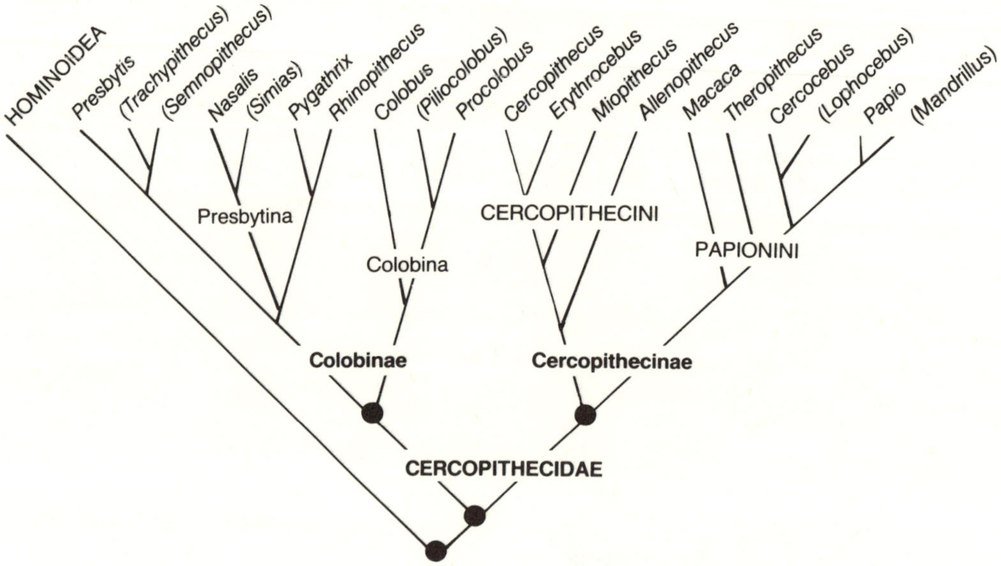

Figure 3. Cladogram depicting relationships of living and extinct cercopithecoids, consistent with the Linnaean classification in Table 1. From Strasser and Delson (1987), redrawn here in modified form.

has recently been dubbed a "living fossil" by Delson and Rosenberger (1984).

Of the 4 living colobines in the comparative sample, 2 are African in distribution and 2 are Asian. Both African species are from the genus *Colobus* and are small-bodied arboreal leapers. Since a number of large, apparently terrestrial, fossil colobines have been described from the African Plio-Pleistocene (e.g., M. G. Leakey, 1982), I included 2 Asian colobines in the sample. Both *Presbytis* and *Nasalis* are large, partially terrestrial colobines that probably have no close evolutionary relationship with the extinct African colobines but can serve as useful functional-locomotor analogues in the osteometric comparisons.

Vital Statistics for Extant Species

The extant comparative sample consists of 252 individuals (see Table 2). I made and recorded all measurements from skeletal materials preserved in museums and universities in Africa, Europe, and the United States. I selected only adult animals lacking obvious pathologies, gunshot wounds, or other bone trauma. Animals were judged to be adult when they exhibited full epiphyseal fusion or were in the final stages of fusion. I tried to include only wild-caught animals in the sample, but this was not possible for some of the rarer species, because I needed a statistically adequate sample size. I made a careful examination of the forelimb elements of animals from zoological parks, and rejected any specimen with the slightest pathology or abnormality. In many

TABLE 1. Classification of the living and extinct Old World monkeys (Cercopithecoidea).*

Order Primates Linnaeus, 1758
  Semiorder Euprimates Goodman, 1975
    Suborder Haplorhini Pocock, 1918
      Hyporder Anthropoidea St. G. Mivart, 1864
        Infraorder Catarrhini E. Geoffroy, 1812
          Superfamily Cercopithecoidea Gray, 1821
            Family Cercopithecidae Gray, 1821
              Subfamily Cercopithecinae Gray, 1821
                Tribe Cercopithecini Gray, 1821
                    Cercopithecus Linnaeus, 1758
                    Miopithecus I. Geoffroy, 1842
                    Erythrocebus Trouessart, 1897
                    Allenopithecus Lang, 1923
                Tribe Papionini Burnett, 1828
                  Subtribe Papionina Burnett, 1828
                    Papio Erxleben, 1777
                    P. (Mandrillus) Ritgen, 1824
                    †*P. (Dinopithecus)* Broom, 1937
                    Cercocebus E. Geoffroy, 1812
                    C. (Lophocebus) Palmer, 1903
                    †*Parapapio* Jones, 1937
                    †*Gorgopithecus* Broom and Robinson, 1949
                  Subtribe Theropithecina Jolly, 1966
                    Theropithecus I. Geoffroy, 1843
                    †*T. (Simopithecus)* Andrews, 1916
                  Subtribe Macacina Owen, 1843
                    Macaca Lacepede, 1799
                    †*Procynocephalus* Schlosser, 1924
                    †*Paradolichopithecus* Necrasov et al., 1961
              Subfamily Colobinae Jerdon, 1867
                  Subtribe Colobina Blyth, 1875
                    Colobus Illiger, 1811
                    Procolobus Rochebrune, 1887
                    P. (Piliocolobus) Rochebrune, 1887
                    †*Libypithecus* Stromer, 1913
                    †*Cercopithecoides* Mollett, 1947
                    †*Paracolobus* R. Leakey, 1969
                    †*Rhinocolobus* M. G. Leakey, 1982
                  Subtribe Presbytina Gray, 1825
                    Presbytis Eschscholtz, 1821
                    P. (Semnopithecus) Desmarest, 1822
                    P. (Trachypithecus) Reichenbach, 1862
                    Pygathrix E. Geoffroy, 1812
                    Rhinopithecus Milne-Edwards, 1872
                    Nasalis E. Geoffroy, 1812
                    N. (Simias) Miller, 1903
              Subfamily Colobinae, *incertae sedis*
                    †*Mesopithecus* Wagner, 1839
                    †*Dolichopithecus* Deperet, 1889
              Subfamily Victoriapithecinae von Koenigswald, 1969
                    †*Prohylobates* Fourtau, 1918
                    †*Victoriapithecus* von Koenigswald, 1969

*Modified from Szalay and Delson (1979), Delson (1984), and Strasser and Delson (1987).
†Extinct taxon.

TABLE 2. The extant cercopithecoid comparative sample.

Species	Sample Size (male/female)	Body Weight (female to male)	Locomotor Mode	Habitat	Geographic Range[8]
Theropithecus gelada	23 (11M/12F)	11.7–19.0 kg[1]	terrestrial quadruped	arid open scrub montaine grassland[6]	northern, central, and southern[26] Ethiopia
Papio cynocephalus	20 (9M/11F)	14.2–22.8 kg[1,2]	terrestrial quadruped	varies from savanna to woodland, usually in hot, lowland areas[7,8]	southern and eastern Africa
Papio anubis	22 (12M/10F)	14.1–25.1 kg[1]	terrestrial quadruped	varies from savanna to woodland, usually in higher, more humid areas[7,8]	central and eastern Africa
Papio ursinus	22 (12M/10F)	14.5–31.2 kg[1,3]	terrestrial quadruped	primarily woodland, but also riparian forest and savanna [8,9,10]	southern Africa
Papio hamadryas	18 (12M/6F)	12.0–21.3 kg[1]	terrestrial quadruped	arid subdesert steppe and alpine grass meadows [8,11]	northeast Africa and southwest Arabian peninsula

Papio sphinx	14 (8M/6F)	11.5–26.9 kg[1]	terrestrial quadruped	rain forest, coastal forest, gallery forest, thick brush and secondary forest[12,13]	equatorial West Africa
Cercocebus albigena	16 (8M/8F)	6.4–9.0 kg[4]	arboreal quadruped	dense primary evergreen forest and swamp forest, always near water[14,15,16]	equatorial Africa from the Atlantic coast to Uganda.
Cercopithecus mitis	20 (8M/12F)	4.5–6.4 kg[3]	arboreal quadruped	rain forest, evergreen or deciduous, riparian forest, montane cloud and dry scrub forest[8,17,18]	central, eastern, and southern Africa
Macaca sylvanus	19 (10M/9F)	11.1 kg[3]	arboreal and terrestrial quadruped	mixed cedar and Holm oak forests, high cedar forests, gorges with scrub vegetation[19]	northern Morocco and Algeria, introduced on Gibraltar
Colobus guereza	22 (11M/11F)	8.0–10.1 kg[2]	arboreal quadruped and leaper	usually riparian or upland forest, also lowland forest, gallery forest, and Acacia woodland and brushland[8, 20, 21]	central Africa from eastern Nigeria to Ethiopia

(continued on next page)

TABLE 2. Continued.

Species	*Sample Size (male/female)*	*Body Weight (female to male)*	*Locomotor Mode*	*Habitat*	*Geographic Range*[8]
Colobus badius	20 (9M/9F/2?)	8.4–8.7 kg[2]	arboreal quadruped and leaper	typically moist lowland or gallery forest, but also dry deciduous forest and savanna woodland[8,20,22]	western, central, and eastern Africa
Presbytis entellus	16 (12M/4F)	10.4–15.2 kg[5]	terrestrial and arboreal quadruped	varies from dry scrub, deciduous forest, to tropical wet forest, to semidesert open park woodland[23,24]	southern Asia from Pakistan in the east to Assam in the west
Nasalis larvatus	20 (9M/11F)	9.8–20.4 kg[4]	arboreal quadruped and leaper, also partly terrestrial	riverine and coastal forests, especially palm-mangrove, swamp, and lowland rain forest[8,25]	island of Borneo and two coastal islands

[1]Dechow (1983)
[2]W.L. Jungers, unpublished data
[3]Rothenfluh (1976)
[4]Jungers (1985)
[5]Clutton-Brock and Harvey (1977)
[6]Crook and Aldrich-Blake (1968)
[7]Maples (1972)
[8]Wolfheim (1983)
[9]Hall (1966)
[10]Pienaar (1967)
[11]Kummer (1968)
[12]Sabater Pi (1972)
[13]Jouventin (1975)
[14]Chalmers (1968)
[15]Jones and Sabater Pi (1968)
[16]Kingdon (1971)
[17]Gartlan and Brain (1968)
[18]C.P. Booth (1962)
[19]Taub (1977)
[20]Struhsaker and Oates (1975)
[21]Oates (1977)
[22]Struhsaker (1975)
[23]Jay (1965)
[24]Vogel (1977)
[25]Kawabe and Mano (1972)
[26]Mori and Belay (1990)

cases, the date of the animal's introduction to the zoo and the date of its death appeared on the museum tag. If the animal had been in the zoo a short period of time (less than six months) and lacked any pathological deformity, it was included in the sample. (See Ciochon, 1986, Tables 51-63, for a complete listing of all specimens included in the extant comparative sample.)

Table 2 summarizes all the vital statistics of the extant comparative sample. A target of 20 individuals for each species, with an equal number of males and females, was attempted, since this sample size proved statistically adequate and was attainable in earlier multivariate morphometric studies (e.g., Corruccini and Ciochon, 1976). In the present study, the target was met or nearly met in 11 of the 13 species. In the remaining 2 species, not enough skeletons could be located in museum collections, so a slightly smaller sample size had to be used. The basic statistics revealed no significant departures from normality, so all 13 species could be used in the multivariate analysis.

Body weight ranges in Table 2 for the extant species were taken from several sources and are not necessarily based on the same individuals as my measured comparative sample. However, they provide a useful indicator of body mass and female and male weight ranges for the various species in the sample. This information is especially important when dealing with the extinct comparative sample and fossil unknowns. Many extinct forms exceed in size, and by analogy in body mass, the largest individuals in the extant sample. Therefore, the greater the amount of information available on the living species, the better estimates one can make on the extinct species in the study.

Locomotor classifications such as those in Table 2 are somewhat arbitrary, since not every individual of a species moves about in its environment in exactly the same way, especially when a species occupies a wide variety of habitats. Furthermore, not every individual of a species uses exactly the same locomotor pattern throughout every day of its existence. A terrestrial quadruped may occasionally climb a tree to escape a predator, or an arboreal quadruped/leaper may occasionally venture to the ground to cross an area devoid of trees. These caveats aside, it is useful to document the *primary* locomotor adaptation of each living species in the comparative sample, since this information is useful when analyzing the fossil unknowns or discussing the hypothetical locomotor adaptations of extinct species in the sample.

A review of the ecological preferences of species in the extant sample revealed that some are quite eclectic in their choice of habitats. Therefore, not every example of a particular habitat utilization pattern could be listed for every species. Knowledge of the primary habitat of each species in the extant sample should be useful when viewed in comparison with paleoecological data derived from Plio-Pleistocene sites yielding the fossil unknowns analyzed in this study.

The current geographic range of each extant species in the sample is taken from Wolfheim (1983). A new range extension for *Theropithecus gelada* is also included (Mori and Belay, 1990).

Criteria for Selection of Extinct Species

The 7 species of the extinct comparative sample are presented in Table 3. These species are derived from late Miocene through Pleistocene localities in Africa and Europe. The single major criterion for selection of the forelimb postcranial material in this sample was its firm taxonomic identification. This was accomplished by including (1) postcranial specimens associated with identifiable cranial/dental remains, usually from the same individual; and (2) postcranial specimens from fossil localities or stratigraphic levels where only one taxon had been identified on the basis of cranial/dental remains.

The first 3 species in Table 3 are all cercopithecines. *Theropithecus brumpti* and *T. oswaldi* are the predominant fossil monkeys of the African Plio-Pleistocene, especially in the Shungura Formation of the Lower Omo Basin in southern Ethiopia (Eck, 1977, 1987; Eck and Jablonski, 1987). Associated cranial and postcranial elements at single localities or stratigraphic levels often consisted almost entirely of one taxon, so it was possible to construct pooled samples of these extinct *Theropithecus* species that closely matched species of the extant sample in overall size. These first 2 extinct species are thus very useful in a statistical sense for analysis of the fossil unknowns as well as for studies of their affinities to the other living and extinct members of the comparative sample. The pooled *Theropithecus* samples should also be helpful in tracing the temporal range of these 2 extinct species through the identification of fossil unknowns from other well dated stratigraphic levels.

The third cercopithecine, *Paradolichopithecus geticus*, is represented only by a single individual, comprising a portion of the elbow joint that is associated with a sample of 8 identified partial crania and mandibles. It was included in this study because of its Eurasian distribution and its affinities with the macaque clade (Szalay and Delson, 1979): it may approximate the ancestral cercopithecine morphotype (see Delson, 1975; Delson and Rosenberger, 1984). Fossil macaques are otherwise not well known in the African Plio-Pleistocene.

The remaining 4 species of the extinct comparative sample are colobines. *Dolichopithecus* and *Mesopithecus* have a Eurasian distribution, while *Paracolobus* and *Rhinocolobus* are known only from East Africa. All except *Mesopithecus* are large-bodied forms and should be particularly useful in identifying the postcranial remains of large extinct African colobines previously known only from cranial remains. The taxonomic relationships and locomotor affinities of these colobine taxa will be discussed in Chapter 6 prior to their use in the analysis of the fossil unknowns presented in Chapter 7.

The utility of including an extinct comparative sample in this study is threefold: (1) by interpolating known extinct species into the analysis of the fossil unknowns, the identification of these unknown forelimb elements should be greatly enhanced, since many of the extinct species are probably more closely related to the unknowns than are any species of the extant comparative sample; (2) by including extinct species of known taxonomic affinity in the analysis, it is possible to further substantiate the

TABLE 3. The extinct cercopithecoid comparative sample.

Species	*Sample Size*	*Estimated Body Weight*	*Comparable Body Size*[2]	*Geographic Range*	*Temporal Range*
Theropithecus brumpti	21	50 kg[1]	very large-bodied	Kenya and Ethiopia[3,4]	middle Pliocene[3]
Theropithecus oswaldi	17	96 kg[1]	very large-bodied	northern and eastern Africa, southern Africa[3,5]	late Pliocene to middle Pleistocene[3,5,10]
Paradolichopithecus geticus	1	23 kg[1]	large-bodied	Spain, France, Romania, and Russian Central Asia[4,6,7]	Pliocene[6,7]
Dolichopithecus rusinensis	6	18 kg[1]	large-bodied	southern and central eastern Europe[4,8]	?latest Miocene and Pliocene[4,8]
Mesopithecus pentelici	6	8 kg[1]	small-bodied	southern and central Europe to Iran[4,8]	late Miocene and Pliocene[4,8]
Paracolobus chemeroni	1	35 kg[1]	very large-bodied	Kenya and Ethiopia[4,9]	Pliocene[4,9]
Rhinocolobus turkanaensis	1	21 kg[1]	large-bodied	Kenya and Ethiopia[9]	Pliocene[9]

[1]Fleagle (1988)
[2]Designations based on comparisons of like parts of the extant comparative sample with known body weights. Very large-bodied species are larger, on average, than any members of the extant comparative sample.
[3]Eck (1977, 1987), Eck and Jablonski (1984)
[4]Szalay and Delson (1979)
[5]Dechow and Singer (1984)
[6]Aguirre and Soto (1978)
[7]Delson and Plopsor (1975)
[8]Delson (1975)
[9]M. G. Leakey (1982, 1987)
[10]Deino and Potts (1990)

relationships of these taxa to the species of the extant sample, testing current phylogenetic hypotheses; and (3) by comparing the functional anatomy of the forelimb elements of the extinct sample and fossil unknowns with that of the species in the extant sample, some judgments can be made about the locomotor behavior of extinct cercopithecoid species. For example, the large terrestrial colobines of the African Plio-Pleistocene have no modern analogue within the living Colobinae and are said to resemble some living cercopithecines postcranially (Szalay and Delson, 1979).

Vital Statistics for Extinct Species

The extinct comparative sample consists of 53 individuals (see Table 3). I made and recorded all measurements from fossil specimens preserved in museums and universities in Africa, Europe, and the United States. Before including a specimen in the extinct comparative sample or fossil unknown group, it was carefully screened for pathologies, bone trauma, or patterns of breakage that might obscure the accurate collection of measurements. If the state of fossilization of the bone was poor, or if the specimen had been rolled or abraded during fossilization, it was rejected. Only adult animals with fully fused epiphyses were included in the study. Those specimens lacking epiphyseal fusion were almost always incomplete, due to separation of the epiphysis and diaphysis during fossilization. A complete listing of all fossil specimens included in the extinct comparative sample is presented in Tables 47-50 in the Appendix.

3

THE MEASUREMENT SET

Anatomy and Joint Structure

The cercopithecoid primate forelimb is a complex series of interconnected joints comprising the wrist, elbow, and shoulder regions. Movements across these joints occur at the ulnar-carpal, radio-carpal, humero-ulnar, humero-radial, and scapulo-humeral articulations. In this numerical systematic study, 87 continuous variables were designed to chart the movement potentials of these joints and to describe their morphology. These variables were carefully chosen to reflect the size and shape of the joint surfaces at the wrist, elbow, and shoulder; the extent of muscle attachments across the joints; and the size, robusticity, and curvature of the long bone shafts connecting them. In past morphometric studies (Ciochon and Corruccini, 1975, 1977b; Corruccini, Ciochon, and McHenry, 1975, 1976; Corruccini and Ciochon, 1976), I have found that analysis at the level of the joint surface itself yielded the most meaningful taxonomic and phylogenetic information.

Since all movement a cercopithecoid primate is capable of making occurs across the joints of its body, it seems most likely that it is these areas of the postcranial skeleton that have been most strongly selected for throughout the course of this group's evolution. More abstract data derived from angular relationships (e.g., Oxnard, 1967; Ashton et al., 1981) or from stress analysis (e.g., Preuschoft, 1971, 1973) have not been used in this study because their application to fragmentary fossil remains is limited. By following the methodology presented in this study, as long as a portion of the joint of a fossil limb bone remains intact, the data from its joint surfaces can be used to analyze and allocate the specimen.

This study began with a series of wet dissections of the cercopithecoid primate forelimb designed to relate functional anatomy with phylogenetic relationships and locomotor adaptations. The structure of joints and the arrangement of muscles and their potential actions were carefully noted. In most cases, fresh cadavers, not yet preserved in formalin, were used to ascertain accurately the full ranges of movement across joints. There have been a number of important studies of the myology of the primate forelimb (e.g., Ashton and Oxnard, 1963; Grand, 1968; Maier, 1971; Ziemer, 1978, M. Schultz, 1986; and Larson and Stern, 1986, 1989), but there has been no attempt to analyze myological data with reference to the fossil primate record, in most cases because of insufficient data on the osteological manifestations of muscle attachments. In my dissections, therefore, I kept careful notes on the bony attachments of muscles so that this information could later be quantified and applied to the analysis of fossil remains. Fig. 4 illustrates the muscle attachments of the forelimb for one member of the extant comparative sample.

Analysis at the level of the joint surface necessitates an understanding of the structural connections between joints and an overall view of their movement capabilities across the entire forelimb. The anatomy of each joint is directly related to the motions between the bony elements it connects. In this sense, the forelimb and, indeed, the whole body can be viewed as an open-chain system of links or lengths of bone that rotate about joint centers (Dempster, 1955, 1961). Movement at these biomechanical joint centers exhibits a wide variety of angular positions. Yet the movement across each joint can be measured through visual analysis of living animals, manipulations of anesthetized animals, and wet dissections of fresh specimens. Within the joints of the forelimb, especially the elbow joint, there is a very close correspondence between the actual hyaline cartilage joint surface and its underlying bony manifestation. Fig. 5A represents a cercopithecoid primate in quadrupedal stance, and 5B is a "Dempster-like" model of potential angular ranges of movement across its joints. Such information was very useful in designing many of the variables described and illustrated below.

Rationale for Selection of Measurement

In selecting variables for an evolutionary study of a component of the body such as the forelimb, there is always a problem of separating those that reflect habitus from those that reflect heritage. Given the assertion in Chapter 1 that every forelimb anatomical feature has both a phylogenetic base and a functional-locomotor overprint, it can be argued that no matter what variable is chosen it should still contribute *some* phylogenetic component to the analysis.

I have chosen to emphasize the phylogenetic import of a particular anatomical feature by measuring it from several different perspectives. Since the landmarks (endpoints) of each measurement will be different, and sometimes even entail the combination of other related anatomical features, there will be a much greater information content for input into the multivariate analysis. It is an axiom of numerical systematics that "the greater the content of information in the taxa of classification and the more characters on which it is based, the better a given classification will be." (Sneath and Sokal, 1973:5). Therefore, my approach of multiple, though not entirely duplicating, measurements of key anatomical features should emphasize uniquely phylogenetic factors while lessening the effects of any functional locomotor componentsonce all measurements have been analyzed together in a multivariate fashion.

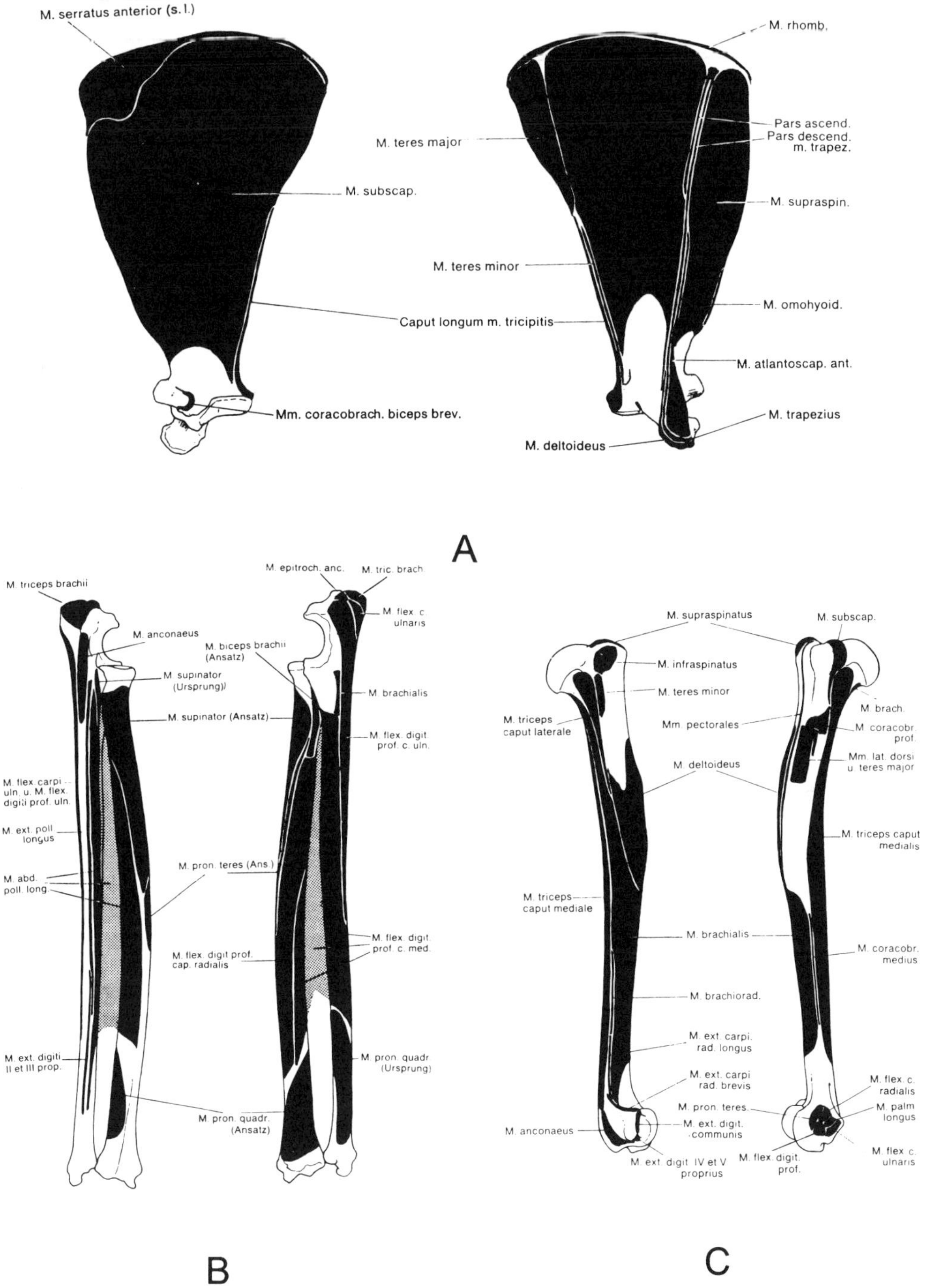

Figure 4. Muscle maps showing origins and insertions of the major forelimb muscles of *Theropithecus gelada*, a terrestrially adapted cercopithecoid primate. A. right scapula; B. right radius and ulna; C. right humerus. These muscle maps were used in the selection of some measurement endpoints. Adapted from Maier (1971). About one-half natural size.

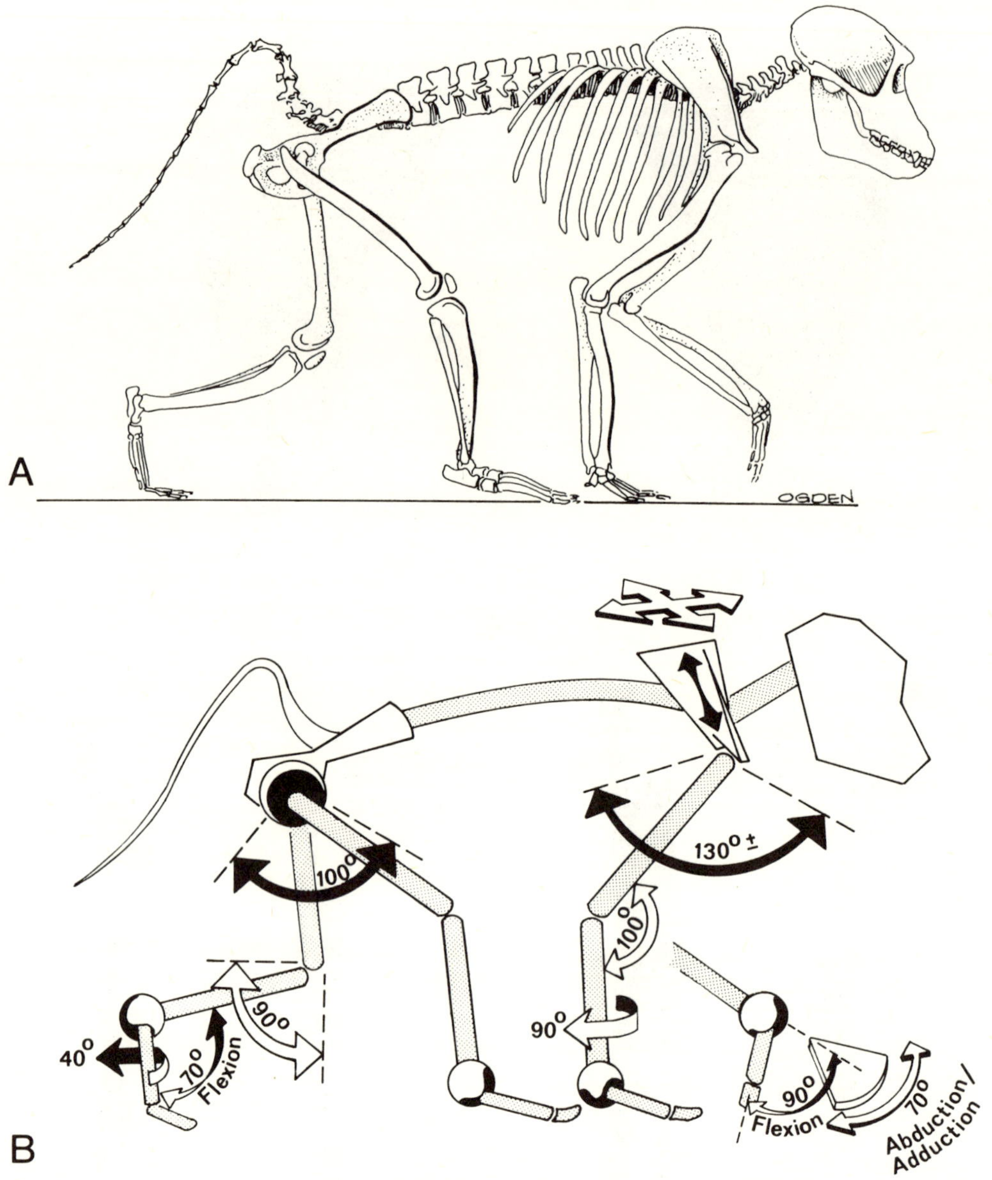

Figure 5. Free body diagram of a terrestrially adapted cercopithecoid, emphasizing potential ranges of joint movement in the extremities. A. Skeleton in quadrupedal stance. B. Schematic view of skeleton depicting actual ranges of movement across joint centers. Free body diagram is derived from wet dissections and from research by S.L. Washburn (personal communication). Diagram concept is based on the work of Dempster (1955, 1961).

Measuring Techniques for Data Collection

All measurements were made with standard dial calipers or with an attachable goniometer. For the smaller measurements, a Helios 16-cm dial caliper was used. The larger measurements were made with a Mitutoyo 30-cm dial caliper. The two angular measurements were taken with a Swiss GPM goniometer mounted on a three-legged horizontal stand. The Helios dial caliper had a special device attached to its base so that depth measurements could be taken with the sharpened rod that projects down as the jaws are widened. All measurements were recorded to one-tenth of a millimeter, or to the nearest degree in the case of the two angular measurements. The data were recorded directly onto standardized 80-column computer data-collection sheets in a format suitable for data entry without intermediate steps.

For each specimen, the left scapula, humerus, radius, and ulna were measured, unless there was damage or obvious pathology, in which case the right elements were substituted. Since each individual of the extant comparative sample was screened before including it in the sample, there were very few instances where data could not be collected because the endpoints were obscured by breakage, pathology, or lack of symphyseal fusion. In the few cases where this occurred (less than 2% of the measurements of all 252 individuals making up the extant sample), the measurement was either estimated or left unrecorded. Missing data were more prevalent in individuals of the 7 extinct comparative samples and in the sample of unknown fossil specimens, because of the greater incidence of damage and breakage during the process of fossilization. If damage to a fossil specimen was such that more than 10% of the potential measurements could not be taken, this specimen was not included in the study.

A test series of individuals of *Papio anubis* and *Colobus guereza* was measured to help standardize the measurement technique and screen for error. Some initial difficulty was encountered regarding the precise location of some measurement endpoints, especially when dealing with cross-subfamily comparisons. Once those difficulties were resolved, the test series was measured twice. Comparison of the two data sets showed that measurement error was very small, considering the variance associated with differences among individual specimens. The error was always less than 1% of the total variance, and thus judged as inconsequential. Measurements were then made on the entire comparative sample and unknown fossil specimen set with the confidence that there would be a high degree of reproducibility for all variables measured.

Measurement Descriptions

Anatomical descriptions of the measurements follow. All measurements were taken with reference to the anatomical positions illustrated in Fig. 6. In order to insure reproducibility during the data collection process, precise orientations of how the measurements were taken accompany each description.

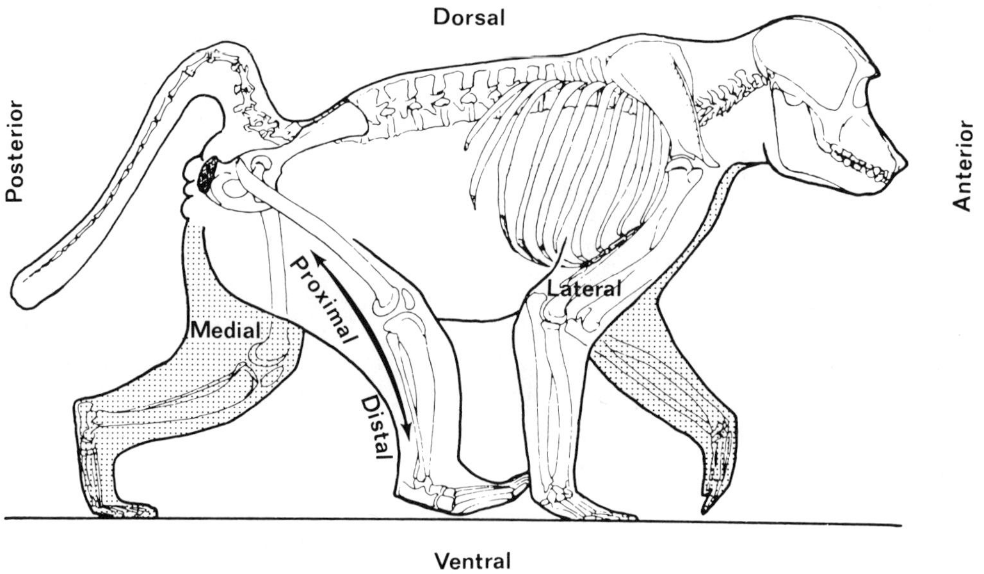

Figure 6. Cercopithecoid primate in quadrupedal stance, illustrating the nomenclature followed in anatomical descriptions of the 87 variables detailed in the text, illustrated in Figs. 7-10, and utilized in all subsequent discussions. Modified after figure in Maier (1972a).

Ulna Measurements

1. Radial Notch Width (Fig. **7D**): Maximum linear width of the radial notch from its most medial extent on the side of the coronoid process (f) to its most lateral extent (h), taken perpendicular to the shaft long axis.

2. Radial Notch Height (Fig. **7D**): Maximum height of the radial notch from its most proximal extent (g) to its most distal extent (i), taken parallel to the shaft long axis.

3. Trochlear Notch Medial Height (Fig. **7A**): Maximum height of the trochlear notch on its medial aspect, taken from the most proximo-medial extent (b) to the most disto-medial extent (h) and measured parallel to the long axis of the trochlear notch.

4. Trochlear Notch Midline Height (Fig. **7A**): Minimum height of the trochlear notch at its midline, taken from the most distal extent of the trochlear articular surface on the coronoid process (g) to the midpoint of the most proximal projection of the trochlear articular surface (c) and measured parallel to the notch long axis.

5. Coronoid/Olecranon Height (Fig. **7B**): Height of the olecranon process, measured from the proximal midpoint of the olecranon (b) to the most distal projection of the coronoid process on its undersurface (g) and taken approximately parallel to the long axis of the olecranon process.

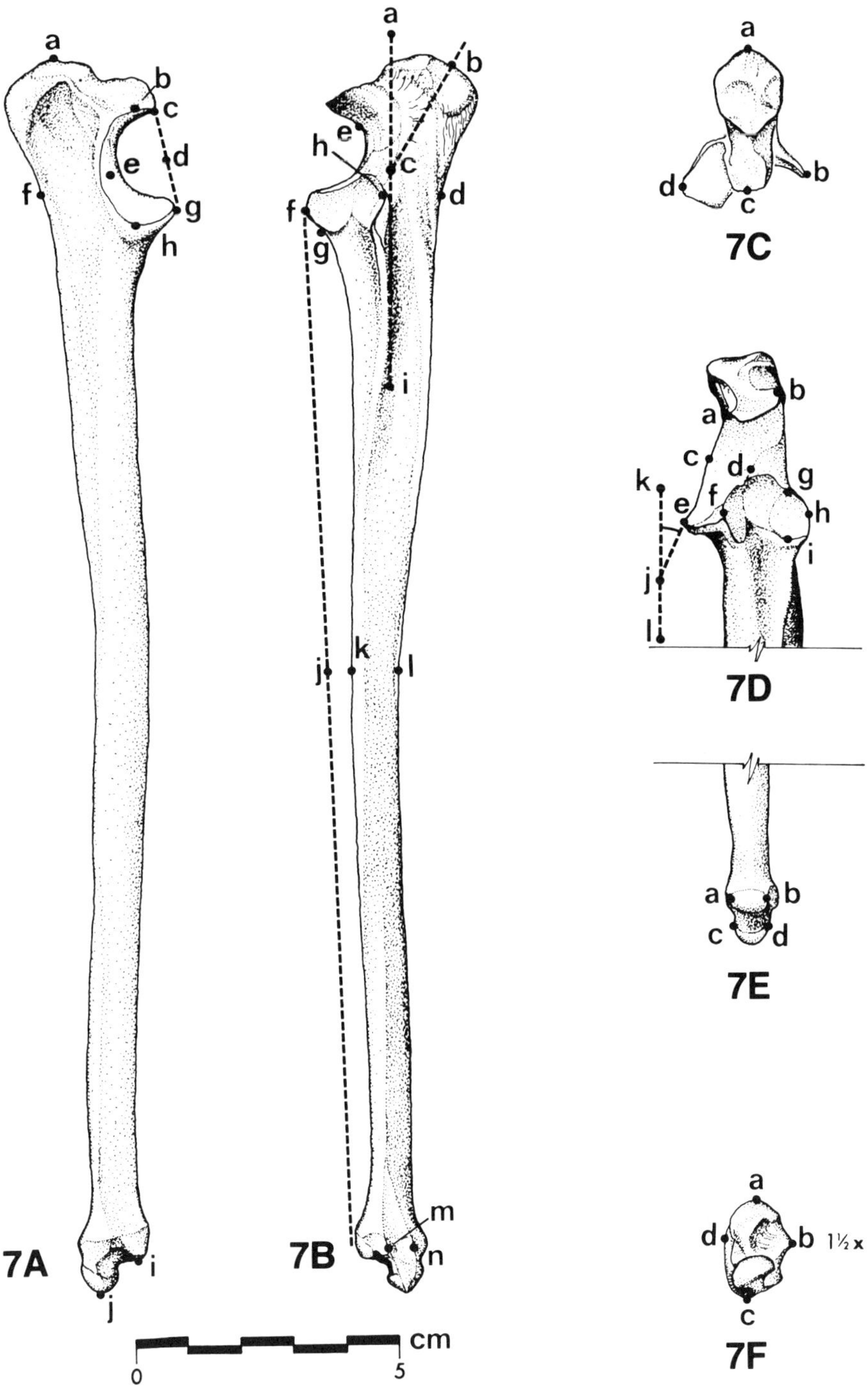

Figure 7. Left cercopithecoid ulna, illustrating the measurement endpoints adopted for this study. **7A**. Medial view; **7B**. lateral view; **7C**. dorsal (superior) view; **7D**. anterior view, proximal end; **7E**. anterior view, distal end; **7F**. ventral (inferior) view. Illustration based on an ulna of *Theropithecus gelada*.

6. Trochlear Notch/Olecranon Length (Fig. **7B**): Length of the olecranon process, measured from the proximal midpoint of the olecranon (b) to the proximo-posterior surface of the trochlear notch (e) and taken as a minimum distance.

7. Trochlear Notch/Olecranon Width (Fig. **7C**): Maximum width of the olecranon process, measured from its most posteriorly directed surface (a) to the most anterior projection of the proximal portion of the trochlear notch (c) and taken approximately perpendicular to the olecranon long axis.

8. Proximal Trochlear Notch Width (Fig. **7D**): Maximum width of the proximal portion of the trochlear notch articular surface, taken from its most proximo-medial extent (a) to its most proximo-lateral extent (b).

9. Middle Trochlear Notch Width (Fig. **7D**): Minimum width of the middle portion of the trochlear notch articular surface from its lateral extent (d) to its medial extent (c), taken perpendicular to the notch long axis.

10. Distal Trochlear Notch Width (Fig. **7D**): Maximum width of the distal portion of the trochlear notch articular surface, taken from its most disto-medial extent (e) to its most disto-lateral extent (f).

11. Coronoid Process Projection (Fig. **7B**): Maximum anterior projection of the coronoid process, taken from its anteriorly directed tip (f) to the posterior surface of the ulnar shaft (d) and measured perpendicular to the shaft long axis.

12. Trochlear Notch Depth (Fig. **7A**): Maximum depth of the trochlear notch, measured from the center point (d) on the line connecting the most antero-proximal and antero-distal projections of the trochlear notch articular surface to a perpendicularly-projected point at the notch center (e).

13. Trochlear/Radial Notch Combined Width (Fig. **7C**): Maximum combined linear width of the trochlear and radial notches from the most medial projection (d) to the most lateral projection (b), taken perpendicular to the shaft long axis.

14. Styloid Process Length (Fig. **7A**): Maximum length (distal projection) of the styloid process, taken from the antero-distal surface of the ulnar head (i) to the most distal projection of the styloid process (j) and measured parallel to the shaft long axis.

15. Styloid Process Width (Fig. **7E**): Maximum transverse width of the styloid process, taken from its most medio-posterior projection (c) to its most latero-anterior projection (d).

16. Ulnar Head Articular Surface Width (Fig. **7E**): Maximum width of the ulnar head articular surface from its most medial extent (a) to its most lateral extent (b), taken approximately perpendicular to the shaft long axis.

17. Extensor Carpi Ulnaris Groove Width (Fig. **7B**): Maximum width of the groove for the extensor carpi ulnaris tendon, taken from the most anterior extent of the groove (m) to its most posterior extent (n).

18. Antero-Posterior Distal Ulnar Width (Fig. **7F**): Maximum antero-posterior width of the distal ulna, measured from the most anterior projection of the ulnar head articular surface (a) to the most posterior projection of the styloid process at its base near the epiphyseal union (c) and perpendicular to the shaft long axis.

19. Medio-Lateral Distal Ulnar Width (Fig. **7F**): Maximum distance from the medial surface of the distal ulna lying between the styloid process and ulnar head articular surface (d) to its lateral surface at the projection of the extensor carpi ulnaris groove (b), taken perpendicular to measurement 18.

20. Olecranon Angulation (Fig. **7B**): Amount of angulation in degrees the olecranon process deviates in an anterior or posterior direction from the shaft long axis, which is defined here by the line connecting endpoints a, c, and i. The measurement is taken by determining the degree of olecranon inflection at angle abc. When the olecranon is inflected anteriorly the degree value is positive, and when it is inflected posteriorly (retroflexed) the degree value is negative. If the line defined by endpoints b and c becomes superimposed on the shaft long axis, there is no olecranon inflection, and angle abc is equal to 0 degrees.

21. Trochlear Notch Angulation (Fig. **7D**): Amount of angulation in degrees that the trochlear notch (a, c, e) deviates in a medial direction from the shaft long axis (k to l), measured by determining the angular projection of the trochlear notch medial surface at angle ejk. Angle ejk is always expressed as a positive value in degrees.

22. Midshaft Antero-Posterior Diameter (Fig. **7B**): Maximum diameter in an antero-posterior plane at the shaft midpoint from its anterior surface (k) to posterior surface (l), taken perpendicular to the shaft long axis.

23. Midshaft Curvature (Fig. **7B**): Maximum curvature of the shaft at its midpoint, measured as the perpendicular distance from the shaft anterior surface (k) to a point (j) on a vertical line connecting the most anterior projections of the coronoid process and the head of the ulna.

24. Radial Notch Posterior Width (Fig. **7B**): Minimum width of the shaft posterior to the radial notch, measured from a central point on the lateral portion of the radial notch articular surface (h) to a point on the posterior surface of the shaft immediately behind the notch (d) and taken approximately perpendicular to the shaft long axis.

25. Trochlear Notch Posterior Width (Fig. **7A**): Minimum width of the shaft posterior to the trochlear notch, measured from the center of the trochlear notch articular surface (e) to a point on the posterior surface of the shaft immediately behind the notch (f) and taken approximately perpendicular to the olecranon process long axis.

26. Ulna Length (Fig. **7A**): Maximum length of the ulna from the most proximal projection of the olecranon process (a) to the most distal projection of the styloid process (j), taken parallel to the shaft long axis.

Radius Measurements

27. Maximum Medio-Lateral Head Diameter (Fig. **8C**): Maximum diameter of the radial head, taken from its most medial extent (d) to its most lateral extent (b).

28. Minimum Antero-Posterior Head Diameter (Fig. **8C**): Minimum diameter of the radial head, taken from its most anterior extent (c) to its most posterior extent (a).

29. Head Maximum Articular Surface Width (Fig. **8D**): Maximum extent of the articular surface of the antero-medial aspect of the radial head, taken from its most distal projection (b) to its most proximal projection (a).

30. Head Minimum Articular Surface Width (Fig. **8B**): Minimum extent of the articular surface on the postero-lateral aspect of the radial head, taken from its most distal projection (b) to its most proximal projection (a).

31. Medio-Lateral Neck Width (Fig. **8A**): Minimum width of the radial neck in the medio-lateral plane, taken from the point on the medial surface of the neck nearest the shaft midline (b) to the point on its lateral surface nearest the shaft midline (c).

32. Antero-Posterior Neck Width (Fig. **8B**): Minimum width of the radial neck in the antero-posterior plane, taken from the point on the anterior surface of the neck nearest the shaft midline (c) to the point on its posterior surface nearest the shaft midline (d).

33. Maximum Neck Length (Fig. **8D**): Greatest length of the radial neck on its anterior aspect, taken from the most proximal extent of the bicipital tuberosity (c) to the center-most proximal point on the anterior lip of the radial head (a) and measured parallel to the shaft long axis.

34. Bicipital Tuberosity Width (Fig. **8D**): Maximum width at the approximate center of the bicipital tuberosity, taken from its most medial projection (d) to its most lateral projection (e) and measured perpendicular to the shaft long axis.

35. Styloid Process Length (Fig. **8A**): Maximum length (distal projection) of the styloid process, taken from the center-most medial extent of the lunate articular surface immediately adjacent to the ulnar notch (k) [also (a) in Fig. **8F**] to the most distal projection of the styloid process (j) and measured parallel to the shaft long axis.

36. Distal Articular Surface Length (Fig. **8F**): Maximum length in the medio-lateral plane of the combined articular surfaces for the scaphoid and lunate, taken from the center-most medial extent of the lunate articular surface (a) to the center-most lateral extent of the scaphoid articular surface (e).

37. Lunate Articular Surface Width (Fig. **8F**): Maximum antero-posterior width of the articular surface for the lunate, taken from its most anterior extent (c) to its most posterior extent (h) and measured approximately perpendicular to measurement 36.

38. Maximum Distal Radial Width (Fig. **8F**): Maximum width of the distal radius, taken from the most posterior projection of the tubercle of Lister (g) to the most

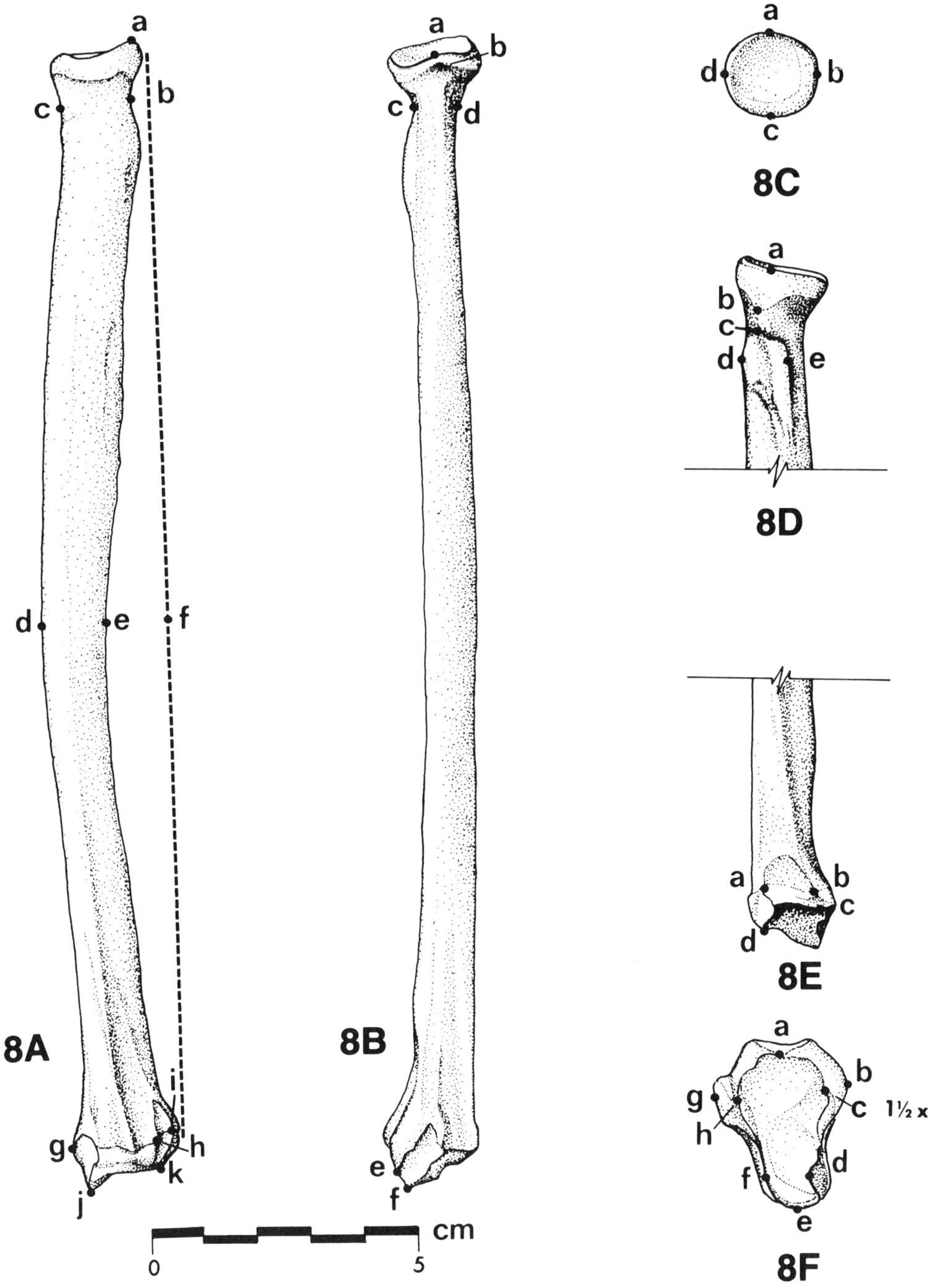

Figure 8. Left cercopithecoid radius, illustrating the measurement endpoints adopted for this study. **8A**. lateral view; **8B**. anterior view; **8C**. dorsal (superior) view; **8D**. medial view, proximal end; **8E**. posterior view, distal end; **8F**. ventral (inferior) view. Illustration based on a radius of *Theropithecus gelada*.

anterior projection of the border surrounding the anterior articular surface of the lunate (b) and measured parallel to measurement 37 and perpendicular to the shaft long axis.

39. Scaphoid Articular Surface Width (Fig. **8F**): Minimum antero-posterior width of the articular surface for the scaphoid, taken from the anterior border of the articular surface just lateral to the anterior tubercle (d) to the nearest point on its posterior border (f) and measured approximately perpendicular to measurement 36.

40. Lister's Tubercle Placement (Fig. **8A**): Relative placement of Lister's tubercle in the medio-lateral plane, determined by measuring from the most lateral projection of the lateral tubercle (g) to the most medial projection of Lister's tubercle on its medial border adjacent to the groove for extensor digitorum (communis) (h) and taken as a minimum distance.

41. Lateral Tubercle Distal Projection (Fig. **8B**): Most distal projection of the radial styloid process (expressed in relation to its position with the lateral tubercle), measured from the distal tip of the radial styloid (f) to the most distal projection of the lateral tubercle (e) and taken as a minimum distance. The greater the value, the more the radial styloid process projects distally.

42. Extensor Digitorum Groove Width (Fig. **8A**): Maximum width of the groove for tendons of extensor digitorum (communis), taken from the medial border of Lister's tubercle at the approximate level of epiphyseal union (h) to an adjacent point at the same level on the posterior border of the ulnar notch articular facet (i).

43. Ulnar Notch Articular Facet Width (Fig. **8E**): Maximum width of the ulnar notch articular surface from its most anterior extent (b) to its most posterior extent (a), taken approximately perpendicular to the shaft long axis.

44. Lunate Posterior Ridge Projection (Fig. **8E**): Maximum distal projection of the posterior border of the lunate articular facet, taken from the most antero-distal extent of the lunate articular facet (c) to its most postero-distal projection (d) and measured parallel to the shaft long axis.

45. Midshaft Medio-Lateral Diameter (Fig. **8A**): Maximum diameter in a medio-lateral plane at the shaft midpoint from the shaft medial surface (e) to its lateral surface (d), taken perpendicular to the shaft long axis.

46. Midshaft Curvature (Fig. **8A**): Maximum curvature of the shaft at its midpoint, measured as the perpendicular distance from the medial surface of the shaft (e) to a point (f) on a vertical line connecting the most medial projections of the radial head and the ulnar notch articular facet.

47. Radius Length (Fig. **8A**): Maximum length of the radius from the most proximal projection of the radial head (a) to the most distal projection of the styloid process (j), taken parallel to the shaft long axis.

Humerus Measurements

48. Antero-Posterior Head Diameter (Fig. **9C**): Maximum diameter of the humeral head in the antero-posterior plane, taken from the antero-center surface of the floor of the bicipital groove (b) to the most posteriorly projecting surface of the humeral head (g) and measured perpendicular to the shaft long axis.

49. Medio-Lateral Head Diameter (Fig. **9D**): Maximum diameter of the humeral head in the medio-lateral plane, taken from the most medial projection of the head articular surface immediately adjacent to the lesser tuberosity (c) to its most lateral projection immediately adjacent to the greater tuberosity (e) and measured perpendicular to measurement 48.

50. Head Maximum Proximal Projection (Fig. **9D**): Greatest proximal (dorsal) projection of the humeral head measured by taking the minimum linear distance from the center-most proximal extent of the head articular surface (b) to its center-most distal extent (d).

51. Head Articular Surface Width (Fig. **9C**): Measure of the degree of anterior extension of the humeral head articular surface, taken from a standardized point defined by a line connecting the most posterior extents of the greater and lesser tuberosities across the proximal (dorsal) aspect of the head. From the center of this line (f) a perpendicular is dropped, extending to or through the bicipital groove to the most antero-distal extent of the head articular cartilage surface (d).

52. Bicipital Groove Width (Fig. **9A**): Minimum width of the bicipital groove at its center, taken from the antero-medial crest of the greater tuberosity (c) to the antero-lateral crest of the lesser tuberosity (b) and measured perpendicular to the shaft long axis.

53. Bicipital Groove Depth (Fig. **9C**): Maximum depth of the bicipital groove at its approximate center, measured by taking the maximum perpendicular distance from a line connecting the most antero-medial (a) and antero-lateral (c) projections of the greater and lesser tuberosities to the most posterior (deepest) point (b) in the bicipital groove.

54. Greater Tuberosity Proximal Projection (Fig. **9D**): Maximum proximal (dorsal) projection of the greater tuberosity above or below the level of the humeral head, taken from the most proximal projection of the greater tuberosity (a) to the most proximal projection of the head articular surface (b) and measured exactly parallel to the shaft long axis. This measurement is recorded as negative if the greater tuberosity lies above the level of the humeral head, and positive if it lies below.

55. Greater Tuberosity Maximum Width (Fig. **9C**): Maximum antero-posterior width of the greater tuberosity, taken from the most antero-medial projection of the tuberosity (a) to its most postero-lateral projection (e) and measured approximately perpendicular to the shaft long axis.

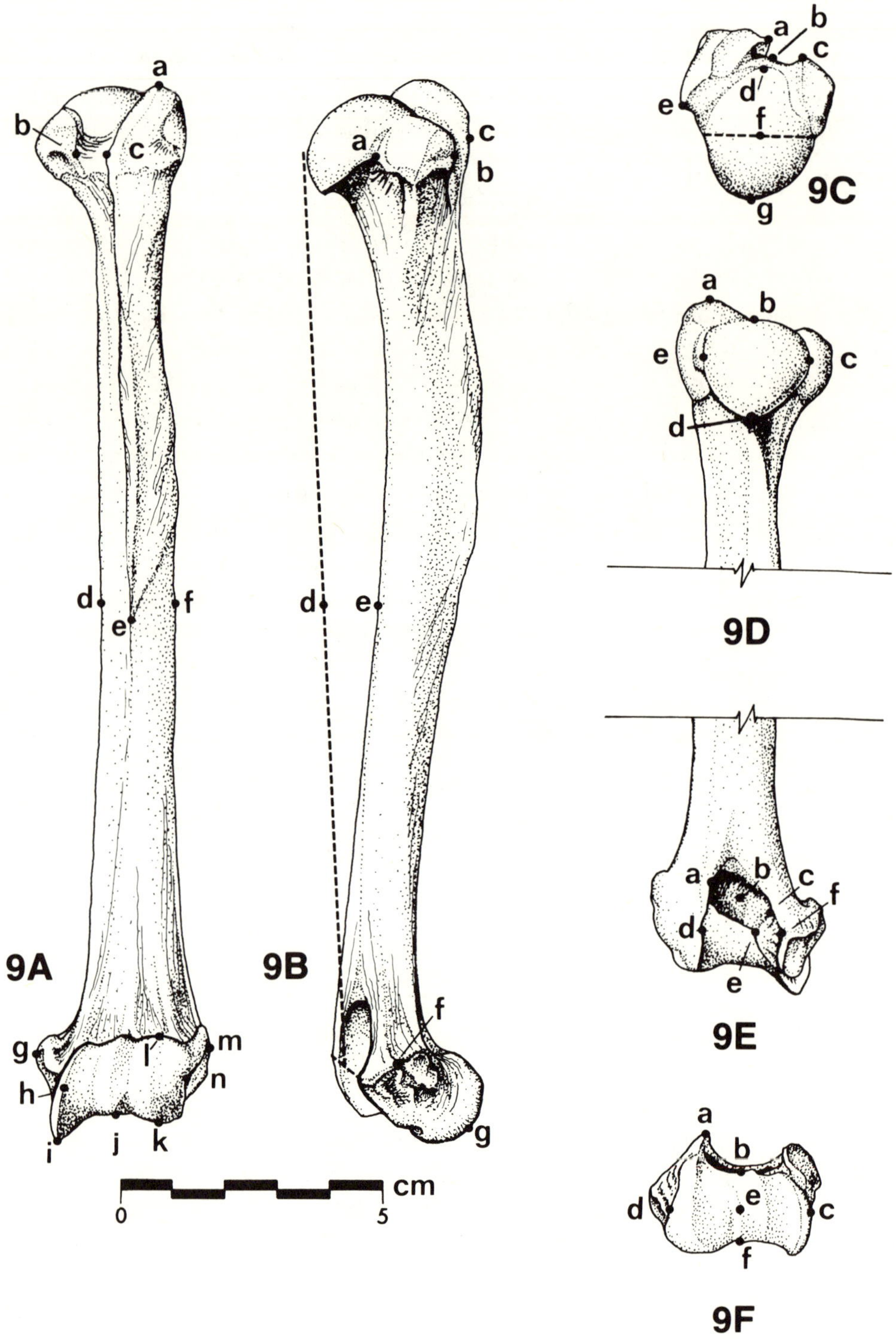

Figure 9. Left cercopithecoid humerus, illustrating the measurement endpoints adopted for this study. **9A**. Anterior view; **9B**. medial view; **9C**. dorsal (superior) view; **9D**. posterior view, proximal end; **9E**. posterior view, distal end; **9F**. ventral (inferior) view. Illustration based on a humerus of *Theropithecus gelada*.

56. Lesser Tuberosity Maximum Width (Fig. **9B**): Maximum antero-posterior width of the lesser tuberosity, taken from the most anterior (and lateral) projection of the tuberosity (b) to its most posterior (and medial) extent (a) and measured perpendicular to the shaft long axis.

57. Lesser Tuberosity Placement (Fig. **9B**): Relative placement of the lesser tuberosity in the antero-posterior plane, measured by determining its position in relation to the greater tuberosity by taking the minimum projected distance from the most anterior extent of the greater tuberosity (c) to the most anterior extent of the lesser tuberosity (b) and executing this measure parallel to measurement 51.

58. Biepicondylar Width (Fig. **9A**): Maximum width across the distal humerus between the epicondyles, taken from the most medial projection of the medial epicondyle (g) to the most lateral projection of the lateral epicondyle (m) and measured perpendicular to the shaft long axis.

59. Medio-Lateral Articular Surface Width (Fig. **9A**): Maximum medio-lateral width of the articular surface across the anterior surface of the distal humerus, taken from the most centro-medial extent of the trochlea (h) to the most centro-lateral extent of the capitulum (n) and measured perpendicular to the shaft long axis.

60. Medial Trochlear Distal Projection (Fig. **9A**): Maximum distal (ventral) projection of the medial side of the trochlea, measured by taking the projected distance from the distal surface of the lateral trochlear ridge (j) to the most distal extent of the medial trochlear ridge (i), always executed parallel to the shaft long axis.

61. Distal Trochlear Width (Fig. **9F**): Greatest width of the trochlea across its distal (ventral) articular surface, taken from the most lateral extent of the trochlea (e) to its most medial extent (c) and measured perpendicular to the shaft long axis.

62. Distal Capitular Width (Fig. **9F**): Greatest width of the capitulum across its distal (ventral) articular surface, taken from the most lateral extent of the capitulum (d) to its most medial extent (e) and measured perpendicular to the shaft long axis and parallel to measurement 61.

63. Antero-Posterior Lateral Trochlear Diameter (Fig. **9F**): Maximum antero-posterior diameter of the lateral side of the trochlea, taken from the most anterior extent of the lateral trochlear ridge (f) to its most posterior extent (b) and measured perpendicular to the shaft long axis.

64. Maximum Capitular Height (Fig. **9A**): Maximum height of the capitulum in the dorso-ventral plane, taken from the most dorsal (proximal) extent (l) to the most ventral (distal) extent (k) and measured parallel to the shaft long axis.

65. Medial Epicondyle/Trochlear Projection (Fig. **9B**): Measure of the maximum size and projection of the medial trochlear ridge, taken from the center of the shelf-like proximo-posterior surface of the medial epicondyle (f) to the most disto-anterior extent of the medial trochlear ridge (g) and measured as a maximum distance.

66. Olecranon Fossa Depth (Fig. **9E**): Greatest depth of the olecranon fossa, measured as a perpendicular distance from a line connecting the proximal border of the olecranon fossa with the posterior trochlear articular surface, to the most anterior (deepest) point in the fossa (b).

67. Olecranon Fossa Lateral Margin Projection (Fig. **9F**): Maximum posterior projection of the lateral margin of the olecranon fossa away from the posterior trochlear articular surface, measured by taking the perpendicular distance from the most posterior extent of the olecranon lateral margin (a) to the most anterior (deepest) point on the posterior trochlear surface (b).

68. Olecranon Fossa Width (Fig. **9E**): Maximum width of the olecranon fossa in the medio-lateral plane, taken from the medial border of the olecranon fossa (c) to its lateral border (a) and measured as a maximum distance not necessarily perpendicular to the shaft long axis.

69. Posterior Trochlear Articular Surface Width (Fig. **9E**): Greatest medio-lateral width of the trochlear articular surface on its posterior aspect, taken from the most medial extent of the trochlear articular surface adjacent to the posterior side of the medial epicondyle (e) to the most lateral extent of the articular surface on the olecranon fossa lateral margin (d) and measured perpendicular to the shaft long axis.

70. Trochlear Gutter Width (Fig. **9E**): Maximum medio-lateral width of the trochlear gutter on the posterior side of the distal humerus, measured by taking the maximum distance between the center of the most posterior extent of the medial epicondyle (f) to the lateral margin of the olecranon fossa (d) and always executed perpendicular to the shaft long axis. This measurement is also an indicator of the degree of medial projection of the medial epicondyle.

71. Midshaft Medio-Lateral Diameter (Fig. **9A**): Maximum diameter in the medio-lateral plane at the shaft midpoint from the shaft medial surface (d) to the shaft lateral surface (f), taken perpendicular to the shaft long axis.

72. Delto-Pectoral Crest Length (Fig. **9A**): Maximum length of the delto-pectoral crest on the anterior aspect of the humeral shaft, taken from the most distal extent of the surface for the attachment of the deltoid muscle (e) to the most proximal extent of the greater tuberosity (a) and measured parallel to the shaft long axis.

73. Midshaft Curvature (Fig. **9B**): Maximum curvature of the shaft at its midpoint, measured as the perpendicular distance from the shaft posterior surface (e) to a point (d) on a vertical line connecting the most posterior extent of the humeral head and the posterior projections of the medial epicondyle and olecranon fossa lateral margin.

74. Humerus Length (Fig. **9A**): Maximum length of the humerus from the most proximal projection of the greater tuberosity (a) to the most distal projection of the medial trochlear ridge (i), taken parallel to the shaft long axis.

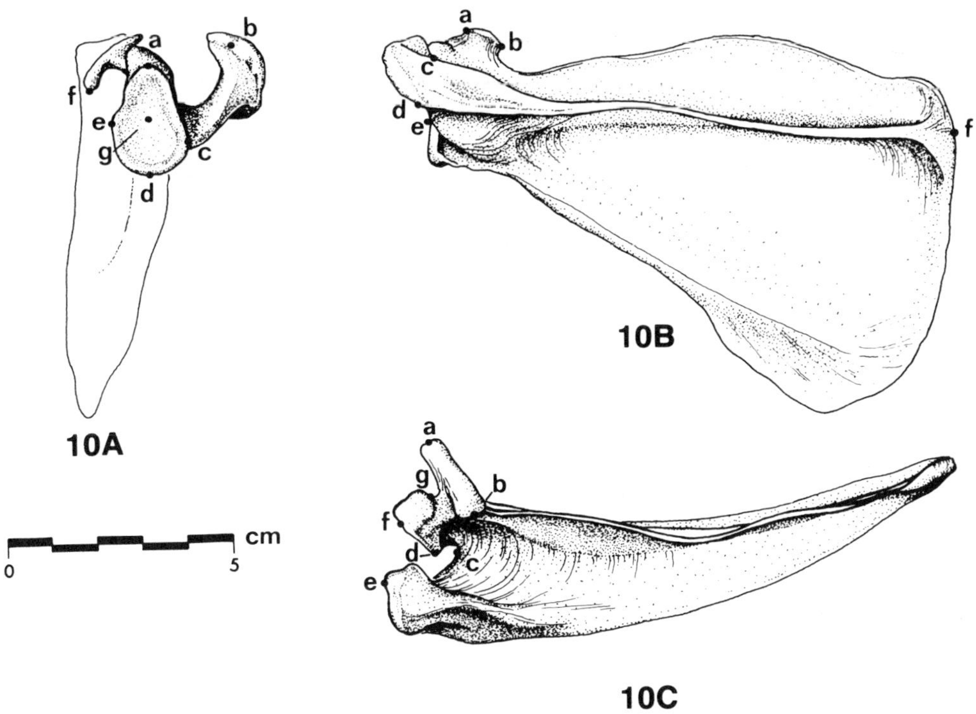

Figure 10. Left cercopithecoid scapula, illustrating the measurement endpoints adopted for this study. **10A**. ventral (inferior) view; **10B**. lateral view; **10C**. anterior view. Illustration based on a scapula of *Theropithecus gelada*.

Scapula Measurements

75. Coracoid Process Length (Fig. **10C**): Maximum length of the coracoid process along its axis, taken from its most dorso-lateral projection (b) to its most ventro-medial projection (a).

76. Coracoid Process Maximum Width (Fig. **10B**): Maximum width of the coracoid process at its lateral (proximal) projection, taken from its dorsal surface (b) to its ventral surface (a) and measured perpendicular to the coracoid long axis.

77. Acromion Width (Fig. **10B**): Greatest width of the acromion dorsal to the clavicular articulation, taken from its antero-medial surface (c) to its postero-lateral surface (d) and measured approximately perpendicular to the acromion long axis.

78. Acromion Length (Fig. **10C**): Greatest length of the acromion in a ventrally projecting direction, taken from the most dorsal (deepest) point in the great scapular notch (c) to the most ventral projection of the acromion (e) and measured parallel to the scapular spine.

79. Great Scapular Notch Depth (Fig. **10C**): Maximum depth of the great scapular notch, taken from its most dorsal extent at the confluence of the supraspinous and infraspinous fossae (c) to a point at the approximate center of the lateral border of the glenoid cavity (d) and measured parallel to the scapular spine.

80. Coraco-Acromial Projection (Fig. **10A**): Maximum ventral projection of the coracoid process and acromion past the glenoid cavity, measured by taking the perpendicular distance from the approximate center of the line connecting the most ventral points on the coracoid process (f) and acromion (b) to the most ventral projection of the supraglenoid tubercle (a). In some cases, recorded as a negative value when the supraglenoid tubercle projects further ventrally than either the coracoid process or the acromion. (For additional illustration and discussion of this measurement, see Ciochon and Corruccini, 1977a.)

81. Supraglenoid Tubercle Length (Fig. **10C**): Maximum length of the supraglenoid tubercle along its axis, taken from its most dorso-medial projection (g) to its most ventro-lateral projection (f).

82. Glenoid Cavity Width (Fig. **10A**): Greatest width of the glenoid cavity articular surface, taken from its most medial projection (e) to its most lateral projection (c) and measured perpendicular to the glenoid cavity midline (defined by a, g, d).

83. Glenoid Cavity Length (Fig. **10A**): Greatest length of the glenoid cavity articular surface from its most anterior extent at the supraglenoid tubercle (a) to its most posterior extent immediately adjacent to the infraglenoid tubercle (d), taken at the glenoid cavity midline.

84. Infraglenoid/Coracoid Process Distance (Fig. **10A**): Relationship of the coracoid process to the infraglenoid tubercle, depicted by measuring the minimum distance from a point at the most posterior extent of the glenoid cavity articular surface immediately adjacent to the infraglenoid tubercle (d) to the most ventral projection of the coracoid process (f).

85. Infraglenoid/Acromion Distance (Fig. **10A**): Relationship of the acromion to the infraglenoid tubercle, depicted by measuring the minimum distance from a point at the most posterior extent of the glenoid cavity articular surface immediately adjacent to the infraglenoid tubercle (d) to the most ventral projection of the acromion (b).

86. Glenoid Cavity Depth (Fig. **10A**): Greatest depth of the glenoid cavity at its approximate center, measured by taking the perpendicular distance from a line connecting the most ventral projections of the glenoid cavity articular surface (a and d) at its midline to the most dorsal (deepest) point at the approximate center of the glenoid cavity (g).

87. Scapula Morphological Length (Fig. **10B**): Maximum dorso-ventral length of the scapula (defined as the scapula morphological length), taken from the approxi-

mate center of the glenoid cavity (e) to a central point on the vertebral border at its intersection with the scapular spine (f).

Measurements at Second Level of Abstraction

Mention can now be made of the functional categories initially ascribed to each variable. These categories are obviously not wholly discrete. Just as some measurements reflect more than one anatomical feature, others also cross-cut the functional categories I have selected. In any case, I feel it is still a useful exercise to view the functional categories as follows:

1. Wrist Joint (Radio-Ulnar Component)
    - A. Measurements describing joint surfaces and their locations:
        - 14. Ulnar Styloid Process Length
        - 15. Ulnar Styloid Process Width
        - 16. Ulnar Head Articular Surface Width
        - 18. Antero-Posterior Distal Ulnar Width
        - 35. Radial Styloid Process Length
        - 36. Radial Distal Articular Surface Length
        - 37. Lunate Articular Surface Width
        - 38. Maximum Distal Radial Width
        - 39. Scaphoid Articular Surface Width
        - 43. Ulnar Notch Articular Facet Width
        - 44. Lunate Posterior Ridge Projection
    - B. Measurements reflecting muscle attachments and structures acting on or across joints:
        - 17. Extensor Carpi Ulnaris Groove Width
        - 19. Medio-Lateral Distal Ulnar Width
        - 40. Lister's Tubercle Placement
        - 41. Lateral Tubercle Distal Projection
        - 42. Extensor Digitorium Groove Width

2. Elbow Joint
    - A. Measurements describing joint surfaces and their locations:
        - 1. Radial Notch Width
        - 2. Radial Notch Height
        - 3. Trochlear Notch Medial Height
        - 4. Trochlear Notch Midline Height
        - 8. Proximal Trochlear Notch Width
        - 9. Middle Trochlear Notch Width
        - 10. Distal Trochlear Notch Width
        - 11. Coronoid Process Projection

12. Trochlear Notch Depth
13. Trochlear/Radial Notch Combined Width
21. Trochlear Notch Angulation
24. Radial Notch Posterior Width
27. Maximum Medio-Lateral Radial Head Diameter
28. Minimum Antero-Posterior Radial Head Diameter
29. Head Maximum Articular Surface Extent
30. Head Minimum Articular Surface Extent
31. Medio-Lateral Radial Neck Width
32. Antero-Posterior Radial Neck Width
59. Humerus Medio-Lateral Articular Surface Width
60. Medial Trochlear Distal Projection
61. Distal Trochlear Width
62. Distal Capitular Width
63. Antero-Posterior Lateral Trochlear Diameter
64. Maximum Capitular Height
65. Medial Epicondyle/Trochlear Projection
66. Olecranon Fossa Depth
67. Olecranon Fossa Lateral Margin Projection
68. Olecranon Fossa Width
69. Posterior Trochlear Articular Surface Width
70. Trochlear Gutter Width

B. Measurements reflecting muscle attachments and structures acting on or across joints:

7. Ulnar Trochlear Notch/Olecranon Width
19. Medio-Lateral Distal Ulnar Width
20. Olecranon Angulation
33. Radial Maximum Neck Length
34. Bicipital Tuberosity Width
58. Biepicondylar Width

3. Shoulder Joint

A. Measurements describing joint surfaces and their locations:

48. Humerus Antero-Posterior Head Diameter
49. Medio-Lateral Head Diameter
50. Head Maximum Proximal Projection
51. Head Articular Surface Extent
79. Great Scapular Notch Depth
80. Coraco-Acromial Projection
82. Glenoid Cavity Width
83. Glenoid Cavity Length
84. Infraglenoid/Coracoid Process Distance

85. Infraglenoid/Acromion Distance
86. Glenoid Cavity Depth

B. Mcasurements reflecting muscle attachments and structures acting on or across joints:

52. Bicipital Groove Width
53. Bicipital Groove Depth
54. Greater Tuberosity Proximal Projection
55. Greater Tuberosity Maximum Width
56. Lesser Tuberosity Maximum Width
57. Lesser Tuberosity Placement
72. Delto-Pectoral Crest Length
75. Coracoid Process Length
76. Coracoid Process Maximum Width
77. Acromion Width
78. Acromion Length
81. Supraglenoid Tubercle Length

4. Ulnar Shaft

A. Robusticity, curvature, and size features:

22. Midshaft Antero-Posterior Diameter
23. Midshaft Curvature
26. Ulna Length

5. Radial Shaft

A. Robusticity, curvature, and size features:

45. Midshaft Medio-Lateral Diameter
46. Midshaft Curvature
47. Radius Length

6. Humeral Shaft

A. Robusticity, curvature, and size features:

71. Midshaft Medio-Lateral Diameter
73. Midshaft Curvature
74. Humerus Length

7. Scapula

A. Size feature:

87. Scapula Morphological Length

4

METHODS OF ANALYSIS

Introduction to the Multivariate Techniques

A great diversity of multivariate analytical methods (for a summary, see Rao, 1952; Anderson, 1958; Rohlf and Sokal, 1965; Blackith and Reyment, 1971; Sneath and Sokal, 1973; Oxnard, 1973, 1984; Pimentel, 1979; Corruccini, 1975, 1978a, 1983, 1987; Manly, 1986; Sundberg, 1989; and references cited therein) has been developed over the past 40 years by adherents of numerical systematics and other morphometric techniques. This is a clear indication that no *single* technique is yet completely satisfactory for exploring a complex data set. The outcome of multivariate analysis is the result of an interplay between the technique and the data set. Each technique makes assumptions about the data, and each data set can only approximately meet these restrictions.

Multivariate analysis can be thought of as a type of artificial intelligence—the mathematical modeling of the human process of pattern classification. The diversity of techniques reflects the many aspects of the human approach to classification. Therefore, just as subjective taxonomic classification is the result of time-tested observation and contemplation by many individuals with different viewpoints, credible results from numerical classifications must also come from approximate agreement of results from a number of techniques.

This study employs three multivariate approaches in the analysis of the cercopithecoid forelimb and the identification of unknown fossils:

1. Similarity coefficient analysis (SCA)
2. Principal component analysis (PCA)
3. Discriminant function analysis (DFA)

SCA is the simplest and mathematically most straightforward of the three approaches. It involves calculating the pairwise similarity of a number of items (fossil or extant forelimb elements) by reducing all the variables describing the differences between items to a single number. PCA, unlike DFA, makes no assumptions about specimen groupings; therefore, its results are independent of the user's assumptions and previous classifications of the data set. This allows a more objective view of the number and degree of overlap of taxonomic groupings and their relationships to fossil unknowns. DFA assumes a priori knowledge of the taxonomic groupings of the items entered into the analysis. This knowledge is used to maximally separate the groups. The program used here also had a provision for classification of unknown specimens. A brief explanation of each technique follows.

Similarity Coefficient Analysis (SCA)

In general terms, a similarity coefficient is a single number that indicates either similarity or dissimilarity of two items, based on some number of attributes (variables). A great many similarity coefficients have been developed (see Sneath and Sokal, 1973). The coefficient used in this study (Sokal's "d") is the average taxonomic distance between points in the original variable space (ibid.). Fig. 11 gives a geometric portrayal of this coefficient, as well as the formula employed. Obviously, if two points are close together in variable space, their coefficient will be small. As these points are drawn farther apart, the distance between them becomes larger. Thus the average taxonomic distance can vary from ZERO (total similarity) to near INFINITY (total dissimilarity).

Note the inclusion of the divisor N in the formula (Fig. 11). This allows the comparison of two items that do not have the same number of variables, i.e., there are missing data. For comparing items that have the same number of variables, N is a constant and is irrelevant to the formula.

The average taxonomic distance (Sokal's "d") between a fossil unknown and a fossil or extant group is determined as the distance between the unknown and the centroid of the nearest group (see Fig. 12). There is a price to be paid for the conceptual and mathematical simplicity of similarity coefficients analysis. Item similarity is in the original variable space. Furthermore, some variables are very likely to be correlated, and this inclusion of redundant information can introduce systematic errors. Other techniques, such as PCA and DFA, have ways of treating this problem.

Because similarity coefficients only measure similarity between pairs of single taxa in variable space, they give no information regarding the actual degree of membership of a single fossil unknown to an extant or fossil group. The situation could arise where a fossil unknown lies equidistant from two groups, yet in reality would be a member of only one of the groups. Membership in a group can only be arrived at if the dispersion of the group can in some way be determined. Therefore, in comparing a single specimen to a centroid in SCA, membership can only be *suggested.* The techniques that take group dispersion into account, such as PCA membership and DFA classification (see below), give a more accurate picture.

Information loss is inherent in the use of similarity coefficients. By definition, a similarity coefficient reduces the differences between two items (both with a number of variable attributes) to a single number. This introduces the important and often

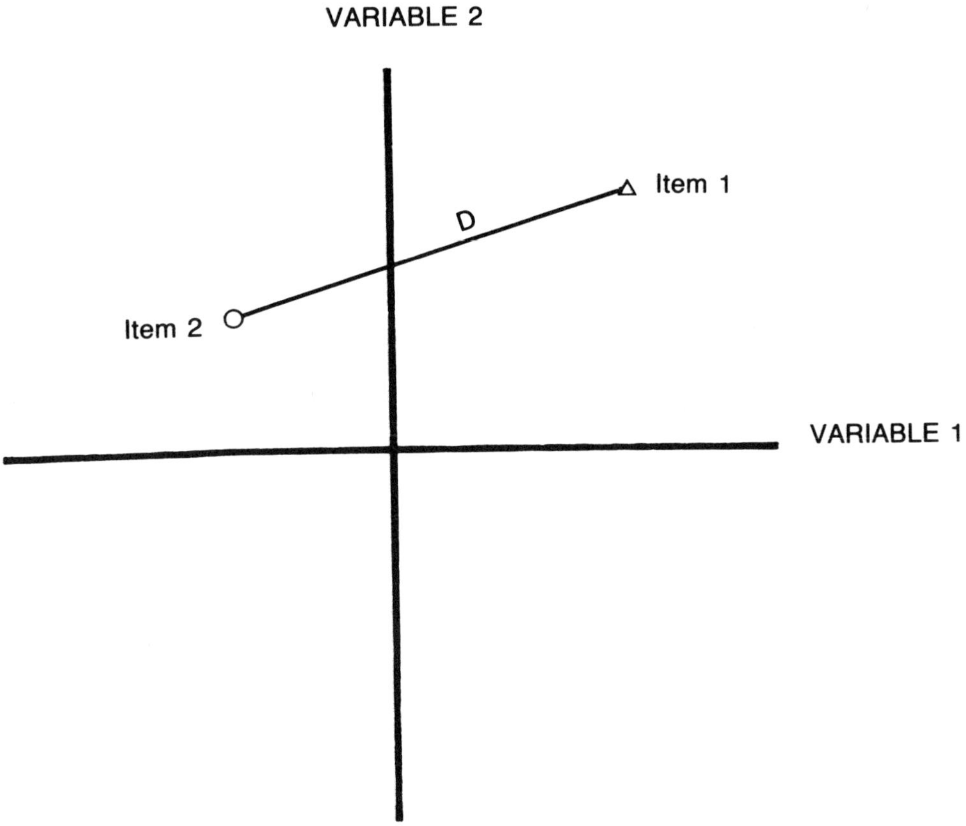

Figure 11. The formula for the average taxonomic distance coefficient is:

$$S_{ij} = (([\sum_{n=1}^{N} X_{in} - X_{jn})^2]) / N)^{½}$$

In the above case:

$$D = S_{i2} = \sqrt{((X_{item\ 1} - X_{item\ 2})^2 + (Y_{item\ 1} - Y_{item\ 2})^2)/2}$$

The above formula can be extended to any number of variables. The average taxonomic distance formula will simply contain more terms under the square root. For example,

$$D = \sqrt{((X_1 - X_2)^2 + (Y_1 - Y_2)^2 + (Z_1 - Z_2)^2)/3}$$

would be the form for 3 variables (dimensions). In the above formula, note that S_{ij} must equal S_{ji}. The quantities under the square root are *squared*. Therefore $(X_1 - X_2)^2 = (X_2 - X_1)^2$. The distance from point 1 to point 2 must equal the distance from point 2 to point 1.

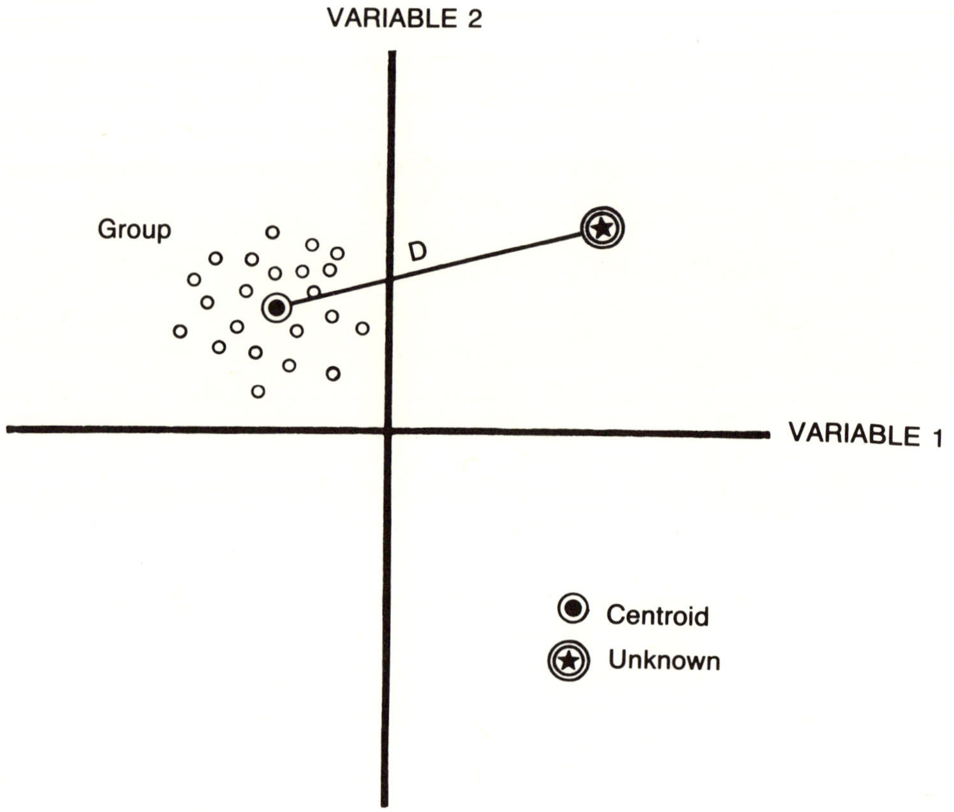

Figure 12. This illustrates the average taxonomic distance (Sokal's "d") between a fossil unknown and a known extant or fossil group. In the above figure, D is the average taxonomic distance between the group centroid and the unknown. The centroid is the average of all points in the group determined by the following equations:

$$C_x = \frac{\sum_{i=1}^{N} X_i}{N} \qquad C_y = \frac{\sum_{i=1}^{N} Y_i}{N}$$

where C_x is the average X coordinate (variable 1) of the group's members and C_y is the average Y coordinate (variable 2) of the group's members.

undesirable property that single large differences in particular variables will come to dominate several small differences. Thus the arithmetic combination of a number of variables often "swamps" the individual significance of each variable. Because of the simplicity and possible flaws in this approach, SCA often does not produce definitive results. It is, though, an excellent first step in examining relationships among items, and should roughly agree with analysis of the same data employing more sophisticated multivariate methods.

Principal Components Analysis (PCA)

PCA is one of the most widely used multivariate techniques in physical anthropology (Andrews and Williams, 1973; Corruccini, 1978a, 1983, 1987). The technique is straightforward: given a series of items, each with the same number of variable attributes, a new data space is created that eliminates redundant information from the data. Inevitably, in any data set, variables are correlated to some degree. The new space created by PCA is in essence a new "variable" space where none of the variables are correlated, i.e., there is no longer any redundancy of information. The overall result of this transformation is that a smaller set of latent, uncorrelated variables is traded for a larger set of meaningful variables. These nameless variables are called *components*. They are ranked, by percentage of variance explained, with regard to the total amount of variation in the data set for which they account. Usually, the first few components describe the entire set of original variables. This greatly facilitates interpretations of items in any multivariate study. The relationship between the original variables and the principal components is expressed in quantities known as "component loadings" or "vector loadings." These values are converted to correlation coefficients between each variable and the component, expressing how much each of the original variables contributes to the distribution of items (taxa) along each component.

Fig. 13 shows a set of data representing a number of items, each described by two variables. The component axes are oriented in such a direction that they express the greatest amount of variance of the data set. In this case, the first component encompasses most of the variance in the data, and the components are made up of a combination of the original variables. Thus, the result of the PCA of the data in Fig. 13 is the reduction of variables from two raw attributes to two transformed, nameless variables.

If distinct groupings of items (i.e., species samples) are entered into a PCA, these groupings will be evident in the plots of each component against the others. Usually, because of the decreasing significance of the components, it is only necessary to examine the first two or three components in order to visualize the intergroup pattern (Corruccini, 1973, 1975). In principal component space, each group is a multidimensional "cloud" of points. How can single points of unknown group affinity be related to these biologically and taxonomically defined clouds? Two approaches have been used in this study: PCA Euclidean Distance and PCA Membership.

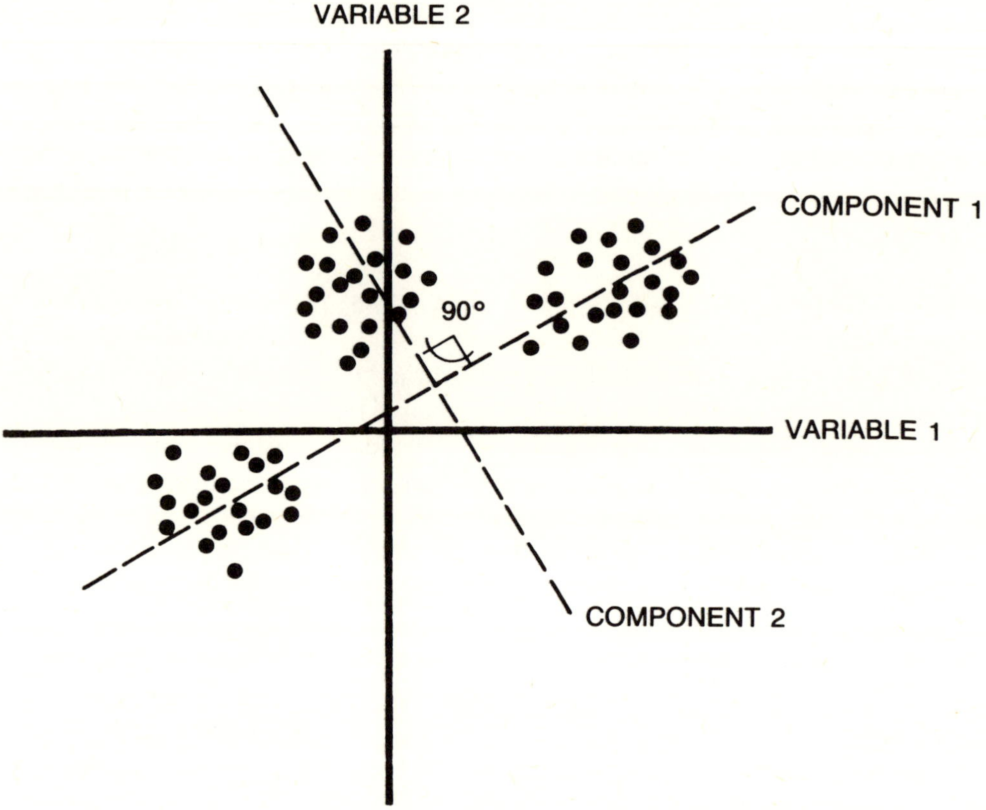

Figure 13. In this two-dimensional case, the separation of points into groups could be made without the use of PCA. The technique becomes more useful when there are many variables describing each item. In such a case, reduction to two or three components is essential for visualizing the relationships among points. Note that in the above diagram the points in the original variable space are correlated; in component space they are not. The result is that the first component explains much more of the variation in the data than either of the original variables alone. Also note that in PCA, component 2 is always projected perpendicular to component 1.

PCA Euclidean Distance

Fig. 14 illustrates a hypothetical example of an unknown point and two group centroids in a PC two-space. If the unknown is closer to one centroid than to the other, it can be judged morphologically more similar to that group, i.e., proximity in principal components space is a measure of similarity. This proximity is measured as a Euclidean distance. As in SCA, a PCA Euclidean distance of 0 implies identity. As the distance between unknowns and group centroids increases, this distance also increases in an unbounded fashion. Considering only the distance shown in Fig. 14, one can conclude that it is more likely that the unknown is affiliated with group 1.

Note that the formula in Fig. 14 is nearly identical to the formula used to compute

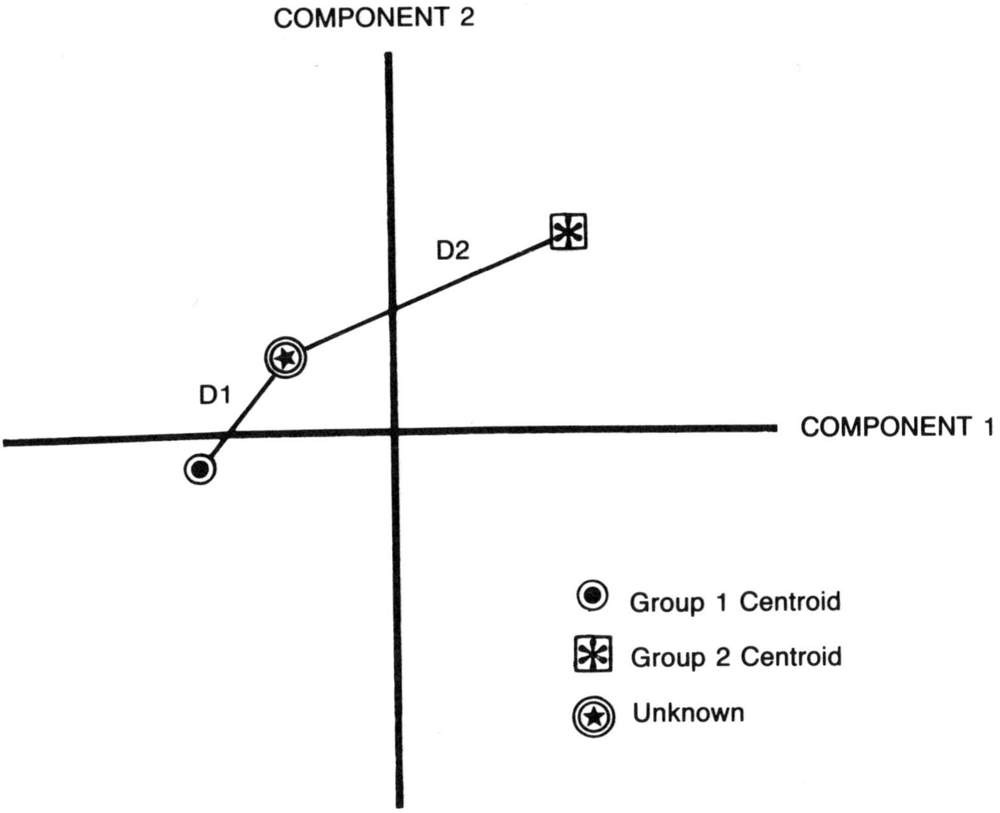

Figure 14. The Euclidean distances ($D_1 < D_2$) from the unknown to the group centroids in component space in the above graph suggest that the unknown is more closely related to group 1. The formula for computing these distances is:

$$\sqrt{D = (C_{1\ \text{unknown}} - C_{1\ \text{group}})^2 + (C_{2\ \text{unknown}} - C_{2\ \text{group}})^2}$$

average taxonomic distance. It is for all practical purposes identical; both are measures of Euclidean distance. But the use of the distance formula in component space overcomes two of the shortcomings of SCA. First, because of the nature of PCA, the redundancy of information in the raw variables has been reduced. Second, the preprocessing of data in PCA, i.e., the scaling to standard deviation units, enters all variables into the analysis equally. One or two of the variables may still dominate the analysis; biologically, these are often variables primarily reflecting size. (A correction to eliminate the adverse effects of this factor will be discussed below.) However, even when size factors appear to dominate, the information reflective of shape is not lost, but is often simply carried into the higher components.

Even though PCA Euclidean distance overcomes several of the shortcomings of SCA, it still does not take group dispersion into account. As explained in the section on SCA, distances between a point and a group can only suggest membership, but never certify it. A more thorough determination must include some measures of group dispersion.

PCA Membership Quotient

The PCA membership quotient takes the idea of PCA Euclidean distance one step further. This technique, which was developed specifically for this study, has all the advantages that PCA Euclidean distance had over SCA, *with* the added feature of accounting for group dispersion. Fig. 15 illustrates two group centroids and an unknown in principal component space, in a relationship similar to that presented in Fig. 14. In this case, however, individual group members have also been plotted. Now, rather than belonging to group 1 as indicated by PCA Euclidean distance, the unknown is seen to be a member of group 2; the greater dispersion of group 2 makes this the most reasonable choice.

A simple statistic was devised to help determine the group to which an unknown most likely belonged. It takes into account both distances separating an unknown from group centroids and dispersions around each centroid. This statistic is based upon the definition of an arbitrary group boundary. If it can be assumed that, for any group, the distribution of points from its centroid is normal (Gaussian), one can define a distance from the centroid of 1 standard deviation greater than the mean distance. This distance, geometrically represented by a circle in two-dimensional space, must, by definition, contain approximately 84% of a group's points within its boundaries. For graphic justification of this statement, see Fig. 16 A-D.

The final step in the construction of this statistic involves normalization. The PCA membership quotient (Q) for a particular unknown and a particular group is the distance from the unknown to the group centroid, divided by the distance of that centroid to its group "boundary" (see Fig. 17). This quotient can be interpreted as follows:

$Q = 1$ implies the unknown is on the group boundary.
$Q > 1$ implies the unknown is outside of the group.
$Q < 1$ implies the unknown is within the group.

Thus, this quotient ranges, as does any Euclidean distance, from ZERO (the unknown is at the group centroid) to INFINITY (unknown is totally unrelated to the group). Note that the PCA membership statistic makes the following assumptions of the groups in component space:

1. The distribution of distances of points from their group centroid is close to a normal distribution.

2. The shape of the group cloud in component space is circular (spherical in three-dimensional space, and hyperspherical in spaces greater than three).

One meaningful test of the above conditions is the agreement between the results of this technique and the others employed in this study. Another is the application of

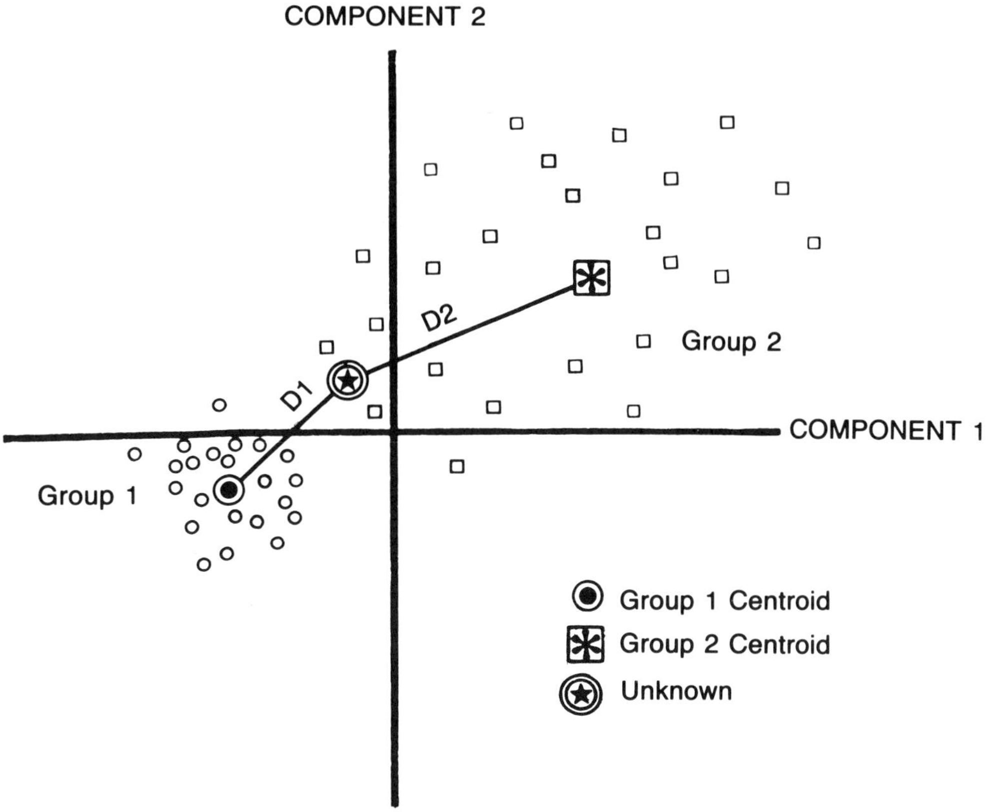

Figure 15. This graph illustrates the applicability of PCA membership. Though $D_1 < D_2$, which places the unknown closer to the centroid of group 1 than to that of group 2, the dispersion of group 2 is much greater and it "encompasses" the unknown. Therefore, the unknown is as likely (or more likely) to be a member of group 2, rather than of group 1 as was the case in Fig. 14.

the PCA membership quotient to points of known group affinity, resulting in a finding of concordance. Both tests confirm the validity of the technique and the satisfaction of its basic assumptions. Additionally, results from statistical tests performed on raw data before applying the PCA membership statistic indicated near normality in all samples.

Both PCA Euclidean distance and PCA membership quotient can be calculated using any number of components. The distance formula used in each technique is easily generalized to any number of dimensions. The decision as to how many components are included in these distance analyses depended on the rate of decrease of their significance as indicated by their eigenvalue percentage contribution of the total variation (percent variance). In preliminary analyses, the first 6 components provided nearly all (95%) of the total variation, and therefore these 6 components were used in all subsequent PCA Euclidean distance and PCA membership runs.

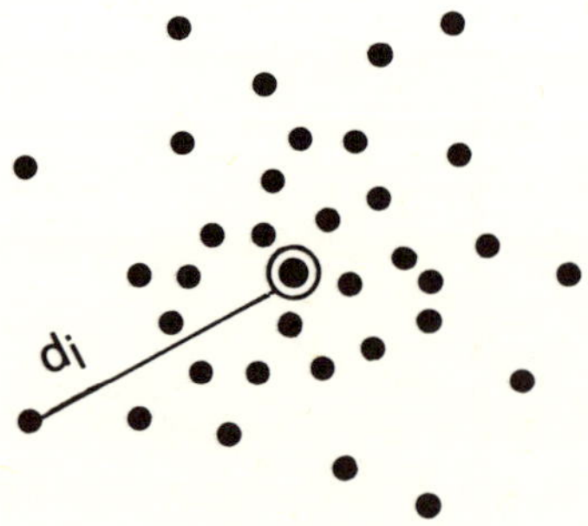

Figure 16A. A graphic illustration of the PCA membership technique. Above is a group dispersion in two-dimensional principal component space where ◉ is the centroid and d_i is the distance (Euclidian) of the *i*th point to the centroid. An average distance to the centroid can be computed as follows:

$$d = (\sum_{i=1}^{N} d_i) / N$$

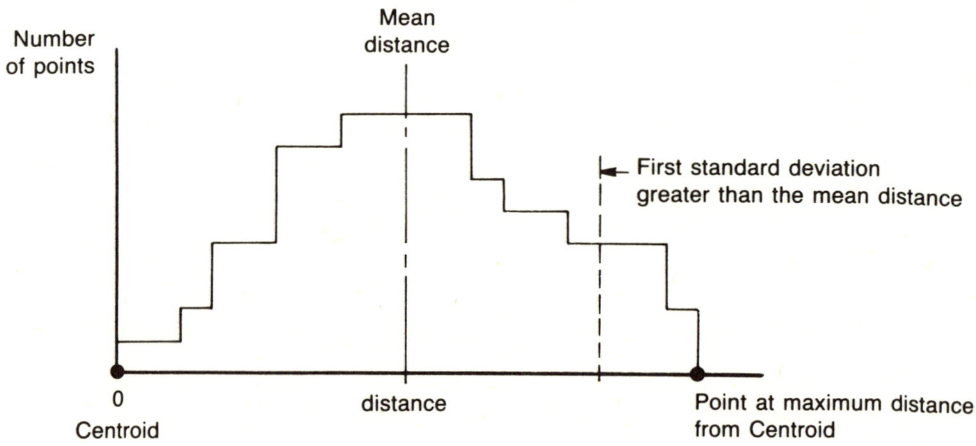

Figure 16B. If the distribution of distances of points to their centroid is normal, a standard deviation of the mean distance can be calculated, and the distribution of points would appear as presented here.

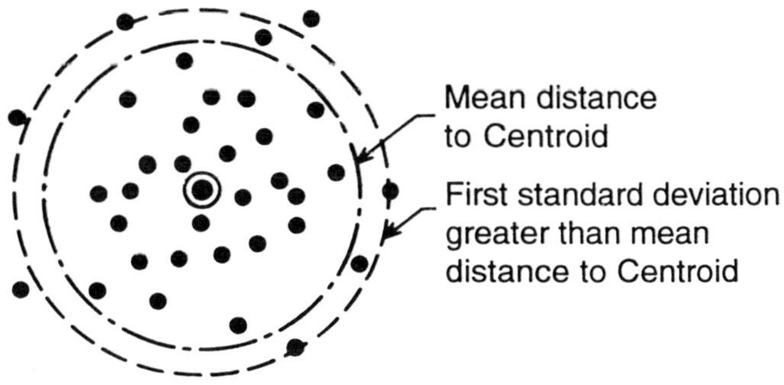

Figure 16C. In two-dimensional component space, the mean and standard deviation would be represented by circles. By definition, the number of points within a standard deviation above or below the mean of a normal distribution is approximately 67%. Therefore, the proportion of points in the dispersion from the centroid to the first standard deviation is 67% + 16.5% = 83.5%.

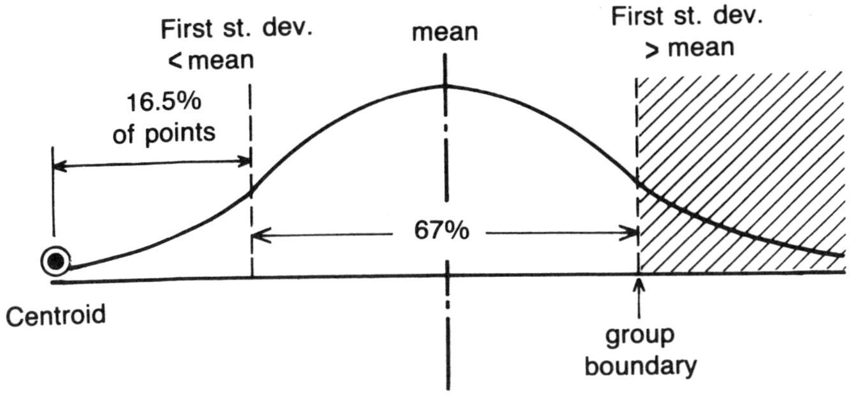

Figure 16D. Thus, for purposes of identifying unknowns, I have arbitrarily established the group's size as that volume containing 83.5% of its points. Naturally, this will not be true for each and every group in component space. On the average, however, if normality assumptions are met (see text for discussion), this will be the case.

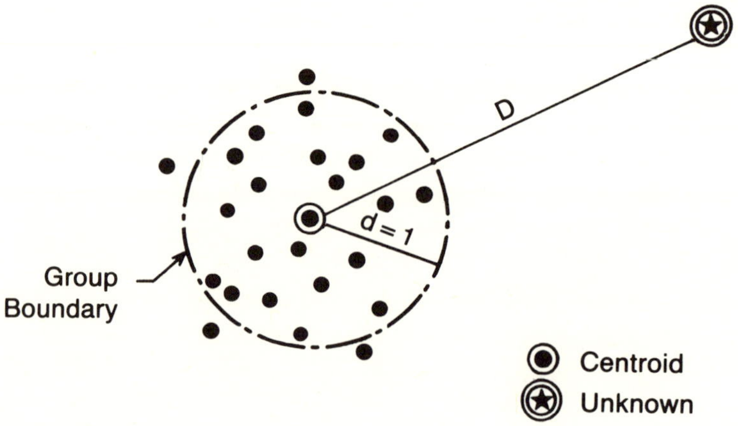

Figure 17. The normalizing technique of PCA membership quotient is depicted here. The Euclidean distance of the unknown from the group boundary (which can also be referred to as the membership boundary) and the centroid can be calculated and expressed as follows:

$$\text{PCA Membership Quotient (Q)} = \frac{D}{d}$$

where:

D = distance from unknown to centroid
d = distance of centroid to group boundary

if value of Q < 1— unknown is within membership boundary.
if value of Q > 1— unknown is outside membership boundary.
if value of Q = 1— unknown is on membership boundary.

Discriminant Function Analysis (DFA)

DFA is a multivariate technique for statistically separating two or more groups of items, each being evaluated on the basis of a number of variables (e.g., Fisher, 1936; Howells, 1973; Albrecht, 1980; Bookstein et al., 1985). Among its several purposes is an estimation of the separation and overlap of user-defined populations, i.e., how distinct are the populations in reality? Furthermore, the sets of functions derived from the discriminating procedure can be used to identify points of unknown group affinity (provided they actually belong to one of the defined populations).

The most important difference between DFA and other classification procedures employed in this study is that this mathematical procedure used to discriminate among groups makes formal use of a priori knowledge of each group's identity and the dispersion of its members. Whether or not this knowledge actually reflects biological (taxonomic) reality usually becomes apparent during the process of analysis.

Fig. 18 is a simple graphic display of DFA. As in PCA, the data points are examined and a new space is defined. However, in DFA the new space is created to maximize discrimination between groups relative to pooled within-group variation (in PCA the new space is determined through iterative variance maximization over all points). Note that the lines shown in Fig. 18, representing the first two discriminant axes, tend to maximally separate the groups. As with PCA, the first function (axis) creates a greater separation than the second, which in turn is greater than the third, etc. The major difference between DFA and PCA can be seen by comparing Fig. 18 with Fig. 13. Fig. 18 contains the same data as Fig. 13, except that each raw data point has been entered into the analysis already assigned to one of the three groups. An a priori knowledge of point groupings allows a maximization of the differences in group means, rather than the equal weighting of individual points as in PCA. Thus the discriminant functions "force" the groups apart as much as possible. Note that successive discriminant functions are not necessarily perpendicular to one another as in PCA.

The "stepwise" method of DFA used in this study is a refinement of the technique described above. Instead of entering all variables into the analysis at once (as in PCA), one can enter variables in a stepwise fashion on the basis of their discriminating power. In many instances the full set of variables will contain redundant information about group differences, or some variables may not be useful in separating groups. By sequentially selecting the "next best" discriminating variable at each step, a reduced set of variables will be found which is almost as good as, and sometimes better than, the full set.

As new variables are selected for inclusion, some of those previously selected may lose their discriminating power. Thus, at the beginning of each step, each of the previously selected variables is tested to determine if it still makes a significant contribution to overall group discrimination. This process is somewhat analogous to the elimination of higher components in PCA because of their insignificant contribution to the explanation of interpoint variance.

The stepwise procedure requires some criterion for determining when entry of a variable increases or decreases the overall discrimination of the groups. In this study the Mahalanobis distance statistic (D^2) (Mahalanobis, 1930, 1936) was chosen as such a criterion, because it forces maximization of the distance between the two *closest* groups in the analysis. This is the natural choice for a study involved in assignment of unknown specimens to known groups, because it favors discrimination among closely related groups. Therefore a more accurate assignment of unknowns to groups that are morphologically very similar becomes possible.

The D^2 (generalized distance) statistic is much-used in multivariate analysis and in physical anthropology (for example, see Rao, 1948; Bronowski and Long, 1952; Hulzinga, 1962; Rightmire, 1969; Constandse-Westermann, 1972; and Sjovold, 1975). It can be used to measure the distance between two groups or a point and a

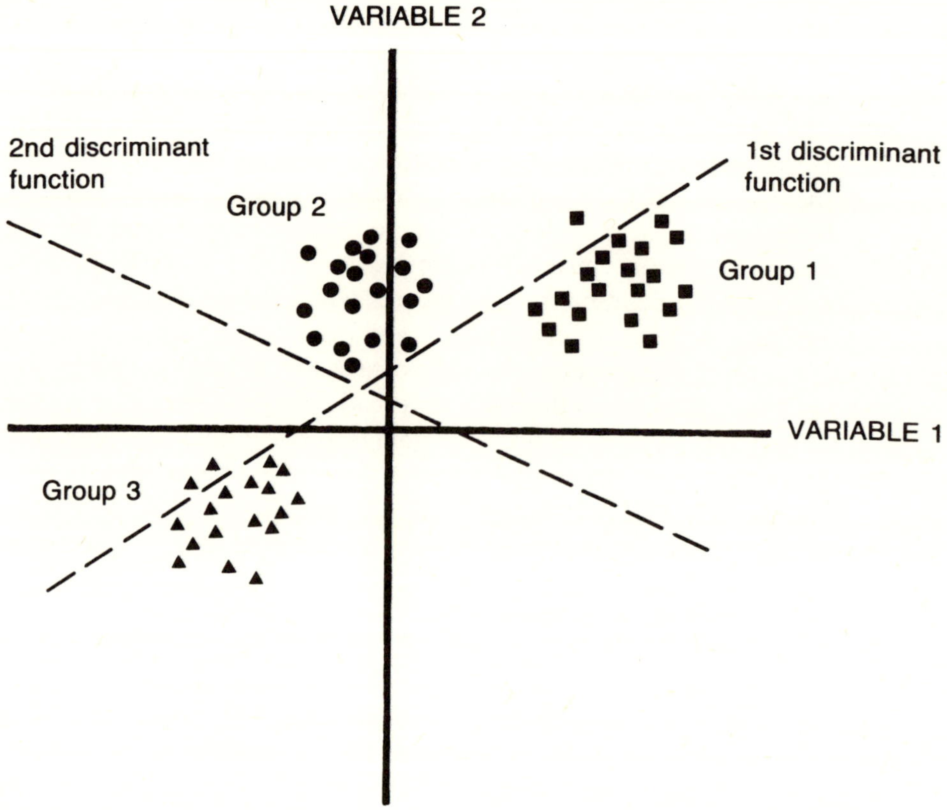

Figure 18. Graphic representation of the mathematical basis of discriminant function analysis (DFA). The first discriminant function serves to separate groups 1 and 3 from group 2. The second separates group 3 from groups 2 and 1. As explained in Fig. 13, a two-dimensional analysis can be performed without DFA, but when many groups and variables are involved, DFA becomes a necessity. Note that discriminant functions 1 and 2 need not be at right angles to one another, as in PCA. Instead, their angular interrelationship is determined by the distribution of the groups in n-dimensional hyperspace.

group of points represented by a centroid. It is perhaps best viewed as a type of Euclidean distance that accounts for within- and between-group dispersion. The D^2 technique is more refined and mathematically justifiable than the PCA membership techniques, though the latter are more readily understandable. The use of both procedures in tandem provides a more credible result than is afforded by either in isolation.

The DFA stepwise package used in this study (Dixon, 1977) produces posterior probabilities that are mathematically related to the D^2 statistic. These are expressed as percentage values ranging from .000 (no chance of membership) to 1.000 (certain membership). The posterior probabilities reflect the likelihood of membership of an unknown to a known group, given the distance between the unknown point and the centroid of the known group, and the group's dispersion.

One can compute discriminant axis loadings in DFA and measure the overall contribution of each variable to group discrimination. But in essence this is performed by the DFA stepwise procedure. Those variables that were chosen earlier in the analysis are more likely to be responsible for group discrimination than those chosen later or those eliminated. A check was made comparing the stepwise input order of variables in DFA with the component loadings in PCA, and the ranks were found to be nearly identical. Therefore, no DFA axis loadings were calculated, since information derived from them is considered redundant with component loadings derived from PCA (see Chapter 6).

When comparing DFA with SCA and PCA, it becomes evident that the main drawback of DFA is also its most important virtue. The degree of discrimination among groups in a DFA is only as good as the amount of overlap of the groups in reality. It is a well-known axiom in biology that just because two closely-related groups (taxa) have been given different Linnaean designations is no assurance that the groups will be 100% separated in all features. A good deal of overlap in some aspect of morphology is likely to exist. Using DFA, it becomes possible to study this between-group similarity as well as the differences which separate the groups.

Associated with DFA is the problem of assigning unknowns from extinct populations to groups of extant populations. In reality, most fossil unknowns belong to none of the modern populations used in DFA. It is possible that both functional convergence, as well as close genetic affinity, will be revealed by applying a set of discriminant functions derived from extant populations to an individual from an extinct population. Obviously, this is a problem of degree. It would be considered ridiculous to use a set of functions derived from modern hippopotamus humeri to determine the relative affinity of a fossil cercopithecoid humerus to any of the hippopotamus groups. However, DFA will assign the fossil cercopithecoid humerus to one of the hippopotamus groups (but with a low probability of membership). This is not the fault of the technique, but that of the user. Caution must be exercised in the choice of the known grouped taxa and in the initial identification of unknown fossil specimens.

Application of the Multivariate Techniques

As mentioned at the beginning of this chapter, no single multivariate technique is totally appropriate for analysis of a complex data set. The approach followed in this study of the cercopithecoid forelimb relies heavily on concordance among all four of the multivariate techniques employed (SCA, PCA Dist, PCA Mem, DFA) to reach accurate taxonomic assessments of the specimens of unknown affinity. Situations do arise where this four-part concordance is not attained. When this occurs, it is important to know which technique is the more "robust" and can be counted on to mathematically provide a more accurate taxonomic assignment of unknown specimens. At the top of the list of the four techniques is discriminant function analysis (DFA). The stepwise program used in this study (BMD07M) is specifically designed to assign unknown individuals to previously designated groups, in part through application of the D^2 statistic. This technique is also mathematically the most sophisticated of the four (Oxnard, 1973, 1984; Campbell, 1978; Albrecht, 1978; Manly, 1986). However, the inherent mathematical complexity of DFA has often resulted in the technique's being grossly misused (see discussions by Corruccini, 1975, and Corruccini and Ciochon, 1976, 1978).

The inherent statistical complexity and problems of misapplication of the DFA technique prompted me, with the aid of computer programmer Michael B. Schwartz, to develop an alternative multivariate method of assigning unknown specimens to known groups. This alternative employs the mathematically simple, relatively nonparametric technique of PCA. PCA membership, as I termed this method, is the second-most robust technique used in this study. It does parallel the DFA technique by taking into account each group's dispersion before assigning an unknown to one of the groups. However, it is essentially a distance statistic, and does not provide an estimate of the probability of assignment of an unknown specimen to the known group of its closest affinity as does DFA. For this reason, and other factors concerning the accurate appraisal of multivariate normality in hyperspace, the PCA membership technique is considered slightly less reliable than DFA.

The technique which is ranked third in accuracy and reliability is the PCA Euclidean distance method. It is solely a measure of straight-line distance between individual points in multivariate hyperspace. It is employed in this study in part to act as a cross-check on the DFA and PCA membership techniques. When the PCA Euclidean distance is used, only the group centroids represented by single points are entered into the analysis, along with unknown specimens. This allows group centroids to be treated as single known individuals. Therefore, PCA Euclidean distance provides the capability to assess the relationship of the unknowns to single known individuals. These single known individuals are all fossil specimens of known taxonomic affinity (usually identified by their association with cranial and dental remains). In most cases, these known fossil taxa are only represented by one or perhaps two individuals. Since they do not constitute a statistically significant biological sample like the other known groups of 18-25 individuals, they cannot be entered into the DFA stepwise procedure or PCA membership. Consequently, the only techniques available for assessing taxonomic affinities of the unknowns vis-à-vis these often important single known fossil specimens are PCA Euclidean distance, and to a much lesser extent, SCA.

The SCA technique is the least accurate and reliable of the four outlined above because it is only a simple measure of similarity between individual points in original variable space. If several of the variables of a particular data set are highly correlated, the distance between individual points (representing unknowns, single known specimens, or group centroids) can be distorted, resulting in an erroneous depiction of their relationships. In this situation, the results from SCA could differ considerably from those of the other three techniques. However, in the absence of highly correlated variables, the results from SCA can be used to corroborate the other three techniques and can also provide an additional way of ascertaining the relationship of unknown specimens to known fossil taxa represented by only a single individual.

Steps in the Analytic Procedure

Several steps were taken to screen raw data for possible measurement and recording errors and to check for departures from normality. To reduce the overriding influence of size so the results would be more reflective of the subtle nuances of shape, I used a procedure similar to that followed in my previous multivariate morphometric studies (e.g., see Ciochon and Corruccini, 1975, 1976, 1977a, 1977b; Corruccini and Ciochon, 1976).

Univariate Statistics and Normality Tests

After screening the tabulated data for obvious errors, the next analytical step involved the application of univariate statistics and normality tests. The basic statistics program employed in this study summarizes the data for each measurement of each species in the comparative sample as follows: number (N), mean ($\bar{X}$), standard error (S_x), standard deviation (S), coefficient of variation (V), variance (S^2), skewness (S_k), kurtosis (K_s), maximum (Max), and minimum (Min). The program also outputs the means for each measurement across all species which would be used later in the PCA Euclidean distance and SCA runs. The basic statistical summary of all the data by measurement and species grouping can be found in Ciochon (1986).

For each measurement in the basic statistics tables, the maximum and minimum values were checked in relation to the mean in order to screen and correct for outliers possibly representing recording or data entry errors. Next, the coefficients of skewness (representing asymmetry of the distribution) and kurtosis (indicating peakedness or flatness of distributions) were examined. These coefficients are considered to be accurate estimates of normality when the population sample is appropriately large (Simpson, Roe, and Lewontin, 1960). The univariate statistical package used in this study sets the parameters for a normal curve equal to a skewness of 0 ($S_k = .00$) and a kurtosis of 3 ($K_s = 3.00$). Each measurement in the basic statistics tables was then screened for significant departures from normality as defined by these parameters.

Very few of the measurements departed significantly from the .00 normality reference for skewness, while a significant number were less than the 3.00 normality reference for kurtosis. This is the expected pattern when dealing with sample sizes in the 20-25 range. Croxton (1953) has shown that a minimum sample size of 100 is needed for kurtosis to be an effective normality test. With sample sizes below 100, the distribution will appear flatter than normal, or platykurtic. However, a sample size of 20-30 should be of significant size to use skewness as a normality test (ibid.).

The skewness column of the basic statistics tables was reviewed, and each apparent significant departure from normality was noted (usually greater than +.50 and less than –.50). The individual data points for the measurement in question were checked against the original data record. If no recording errors were found, and the value or values causing the significant departure from normality exceeded 3.0 standard deviations of the mean, that value and sometimes the entire skeletal element it represented was rejected. The basis for deciding whether or not to reject the element depended on a number of factors such as the relative size of the group sample, the possibility of a non-detectable measurement error, and aberrancy of the individual due to pathology, captivity, or previously unrecognized juvenile status. Since 1% of the individual measurements should exceed 3.0 standard deviations, very few deletions for the comparative sample were actually made. In total, less than 4% of the original sample was either deleted or corrected. The small size of this percentage is almost certainly due to the careful preliminary selection of wild-caught populations exhibiting no apparent pathologies.

The departures from normality indicated by significantly high or low coefficients of skewness for the data (summarized in Tables 71-138 of Ciochon, 1986) are now only slightly greater than would be expected by chance alone. A summary of the average normality for the entire sample can be characterized as follows: The average group sample is slightly skewed to the right (indicated by a predominance of positive signs), with a curve that in reality is probably only slightly platykurtic (indicated by the current predominance of values less than 3.00). In other words, the normal distribution of a linear measurement for the individuals of each sample group should, on the average, be slightly skewed to the right and slightly platykurtic. This average departure from normality is so slight that it should have no effect on the performance of the multivariate statistical techniques employed in this study. Indeed, there is no way to determine whether univariate normality is a condition for multivariate normality, since no useful test has yet been devised for comparing real data with a multivariate normal distribution (Cooley and Lohnes, 1971; Pimentel, 1979; Reyment et al., 1984; Manly, 1986). It is logical, however, that the probability statements (posterior probabilities) generated by the stepwise discriminant functions analysis do in some large way depend on a multivariate normal distribution of the data.

Analysis of Univariate Data

Univariate analysis was undertaken to ascertain what patterns of variation were present when only one or two variables were considered. Initially, when these variables were plotted using a univariate mean/range plot routine developed by Andreasen (1976), it was found that the effects of size were the single overriding pattern. Prior to the multivariate analysis phase of the data set, the influences of size would be factored out using a size/shape correction specially designed for this purpose (see following section); however, in order to isolate patterns of variation in the raw univariate data, a series of ratios was calculated for various pairs of measurements. The measurements analyzed in this fashion were selected from different areas of the forelimb whose functional significance and taxonomic relevance were already documented. The following ratios were calculated:

Ulna Ratios

$$\textbf{Olecranon Process Height Index} = \frac{\textbf{\# 6 Trochlear Notch / Olecranon Length X 100}}{\textbf{\# 4 Trochlear Notch Midline Height}}$$

$$\textbf{Styloid Process Projection Index} = \frac{\textbf{\# 14 Styloid Process Length X 100}}{\textbf{\# 18 Antero-Posterior Distal Ulnar Width}}$$

$$\textbf{Trochlea Notch Shape Index} = \frac{\textbf{\# 10 Distal Trochlear Notch Width X 100}}{\textbf{\# 8 Proximal Trochlear Notch Width}}$$

Radius Ratios

$$\textbf{Radius Head Shape Index} = \frac{\textbf{\# 28 Minimum Antero-Posterior Head Diam. X 100}}{\textbf{\# 27 Maximum Medio-Lateral Head Diameter}}$$

$$\textbf{Radius Styloid Projection Index} = \frac{\textbf{\# 41 Lateral Tubercle Distal Projection X 100}}{\textbf{\# 38 Maximum Distal Radial Width}}$$

$$\textbf{Radius Robusticity Index} = \frac{\textbf{\# 45 Midshaft Medio-Lateral Diameter X 100}}{\textbf{\# 47 Radius Length}}$$

Humerus Ratios

$$\textbf{Trochlea Projection Index} = \frac{\textbf{\# 60 Medial Trochlear Distal Projection X 100}}{\textbf{\#58 Biepicondylar Width}}$$

$$\textbf{Delto-Pectoral Crest Length Index} = \frac{\textbf{\# 72 Delto-Pectoral Crest Length X 100}}{\textbf{\# 74 Humerus Length}}$$

$$\textbf{Olecranon Fossa Shape Index} = \frac{\textbf{\# 67 Olecranon Fossa Lat. Margin Proj. X 100}}{\textbf{\# 70 Trochlear Gutter Width}}$$

$$\textbf{Brachial Index} = \frac{\textbf{\# 47 Radius Length X 100}}{\textbf{\# 74 Humerus Length}}$$

Scapula Ratios

$$\textbf{Glenoid Cavity Shape Index} = \frac{\textbf{\#86 Glenoid Cavity Depth X 100}}{\textbf{\#83 Glenoid Cavity Length}}$$

$$\textbf{Coraco-Acromial Projection Index} = \frac{\textbf{\#80 Coraco-Acromial Projection X 100}}{\textbf{\#83 Glenoid Cavity Length}}$$

Shaft Curvature Indices

$$\textbf{Ulna Curvature Index} = \frac{\textbf{\#23 Ulna Midshaft Curvature X 100}}{\textbf{\#26 Ulna Length}}$$

$$\textbf{Radius Curvature Index} = \frac{\textbf{\#46 Radius Midshaft Curvature X 100}}{\textbf{\#47 Radius Length}}$$

$$\textbf{Humerus Curvature Index} = \frac{\textbf{\#73 Humerus Midshaft Curvature X 100}}{\textbf{\#74 Humerus Length}}$$

The mean/range plots for these ratios appear in Chapter 5. As will be discussed further, these univariate ratios provide a preview of the functional and systematic variation present in the comparative sample and set the stage for the multivariate analyses.

Size/Shape Correction of the Data Set

When attempting to quantify the morphological differences in the postcranial skeleton of a group as inherently uniform as the Cercopithecoidea (A. H. Schultz, 1970), the influence of size tends to overshadow and sometimes distort taxonomically more significant differences in shape. While size as a character in its own right is relevant in the study of closely related populations, alone it is an incomplete criterion of taxonomic relationships and is certainly less meaningful than shape as an indicator of affinity (Steyskal, 1968; Oxnard, 1968, 1969; Gerven, 1972; Corruccini, 1973, 1978a, 1978b, 1987; Bookstein, 1989). That some sort of size correction is mandatory before attempting multivariate analysis is now considered a truism among many biologists (for example, see recent exchange by Somers, 1986, 1989; Rohlf and Bookstein, 1987; Sundberg, 1989; and Bookstein, 1989). Corruccini (1987) provides a comprehensive review and literature survey of this complex issue.

In the past, many researchers tried to avoid the adverse effects of a size-bias in multivariate studies by analyzing indices or ratios of measurements instead of the individual raw measurements. In some cases this method does provide an adjustment for the effects of size, but in others it can produce dramatically distorted results. In a linear regression equation for two variables x and y, $y = a + bx$, simple ratios of y and x will not be linear if the y-intercept, a, differs significantly from zero (Corruccini 1975).

There is also a problem of nonlinear size-related growth distortion such as allometry (Smith, 1980; Sweet, 1980; Jungers, 1985; Corruccini, 1983, 1987; Somers, 1989). Allometry can be defined as "the study of size and its consequences" (Gould, 1966), "differential relative growth or size" (Hopkins, 1966), or "size-dependent morphological change" (Corruccini and Ciochon, 1979). For the purposes of this study, the major allometric problem is one of allomorphosis (Huxley, Needham, and Lerner, 1941), which is relative change in size among variables in a group of individuals *all* at the *same* approximate stage of growth. Corruccini (1972, 1973, 1978b) has developed a size/shape correction based in part on Mosimann's (1970) concept of the "shape variable." This correction has been successfully applied in a number of previous morphometric studies dealing with the postcranial skeleton of the Primates (for a partial review, see Ciochon and Corruccini, 1975, 1977b; Corruccini and Ciochon, 1976, 1978; and McHenry and Corruccini, 1975, 1976, 1983).

The operation of the size/shape correction in the current study can be explained by the following series of 7 steps:

1. For each individual in the study, a standard size reference variable (X) is calculated (Mosimann, 1970), which is the average magnitude of all range-standardized measurements for that individual:

$$X_i = \frac{\sum \frac{x_{ip}}{\max - \min}}{p}$$

where p = number of variables and x_{ip} = measure of variable p on individual i.

2. The effects of allometric size distortion are adjusted first by taking each variable for every known individual of the groups chosen for pooled within-group correction (in this case, the Colobinae and Cercopithecinae) and regressing that against the standard size variable for that individual by reduced major axis method (log-log).

3. By determining the exact relationship between each measurement and the standard size variable throughout all the individuals of the two previously designated groups, a correction factor (regression exponent: parameter $^1/_a$ from the $y = bx^a$ formula) can be derived. If the relationship between the measurement and the standard size variable is isometric, then the correction exponent is one, engendering no change in corrected variables.

4. The correction factors for each individual measurement within each of the two groups are then summed and averaged, resulting in an average correction factor for each measurement. This step is performed because the overall average correction factor has to be applied not only to individuals of known taxonomic affinity (the comparative sample) but also to the unknowns which could be either colobines or cercopithecines (the unknowns were not included in the step involving the actual derivation of the correction factors).

5. Each measurement (from step 2) for each known and unknown individual in the study is at this stage exponentiated to the power determined from the regression formula (Corruccini, 1978b) and then size-adjusted.

6. Each standardized measurement of each individual can then be divided by its standard size variable, resulting in size-adjusted measurements partially corrected for linear size:

$$\bar{x}_{ip} = \frac{x_i}{X_i}$$

where $\bar{x}_{ip}$ = constituents of the shape vector for individual i.

7. The final step in the size/shape correction is to normalize and equally weight all the adjusted shape variables by converting them to standard normal deviates (Z scores). This involves calculation of the overall mean and standard deviation for each shape variable and then subtracting that variable mean from each individual observation and dividing by the standard deviation. The resulting equally weighted, allometrically-adjusted shape variables are now ready for input into the multivariate programs.

This size/shape correction as outlined above was individually applied to each skeletal element or joint unit before attempting the multivariate analyses described in Chapters 6 and 7. As previously reported, this data adjustment technique has been adequately justified in the literature and its use in this study is also addressed elsewhere (Ciochon, 1977, 1986, 1993).

5

RESULTS: UNIVARIATE ANALYSIS OF THE COMPARATIVE SAMPLE

The results of the univariate analysis for the entire comparative sample are presented in this chapter. (The basic statistics on all 87 variables for each of these 20 living and extinct species can be found in Tables 71-145 in Ciochon, 1986. These tables are organized by forelimb element within each species and provide the sample size [N], mean, standard error, standard deviation, coefficient of variation, variance, skewness, kurtosis, and maximum and minimum values for all measurements.)

Results of the Univariate Analysis

Each of the 87 variables in this study was analyzed separately and plotted in an effort to locate patterns of variation reflecting phyletic and functional trends. The univariate plot routine used here was specially adapted for this study by Michael B. Schwartz from PROGRAM FISHPLOT (Andreasen, 1976). These analyses revealed that the size differences separating species in the comparative sample usually overshadowed any shape patterns. Therefore a series of ratios was devised (see Chapter 4) which partially canceled out the effects of size and allowed some variables to be analyzed in a bivariate fashion. Before we proceed to the range/mean plots of these ratios, it is worthwhile to discuss several raw variables that are plotted in their original states without the use of ratios.

Figs. 19-22 represent the comparative length measurements of the ulna, radius, humerus, and scapula, respectively, for all extant species, and for extinct species where these elements were available. All four figures exhibit a similar pattern. The typically large-bodied Papionini are separated from the smaller-bodied Cercopithecini and Colobinae. The only exceptions are in the positions of the papionins *Cercocebus* and *Macaca*, which fall with the smaller-bodied groups, based on limb length. These genera are indeed the smallest members of the Papionini. Among the colobines, the limb size ranges of *Presbytis* and *Nasalis* substantially overlap those of the Cercopithecinae, including the large-bodied papionins. Note, however, an exception in Fig. 22. Though *Nasalis* has nearly twice the body mass of *Colobus*, it is indistinguishable in scapular morphological length from the *Colobus* mean, and its range does not overlap with that of any of the large papionins. This reflects the fact that *Nasalis* has a very medio-laterally compressed scapula.

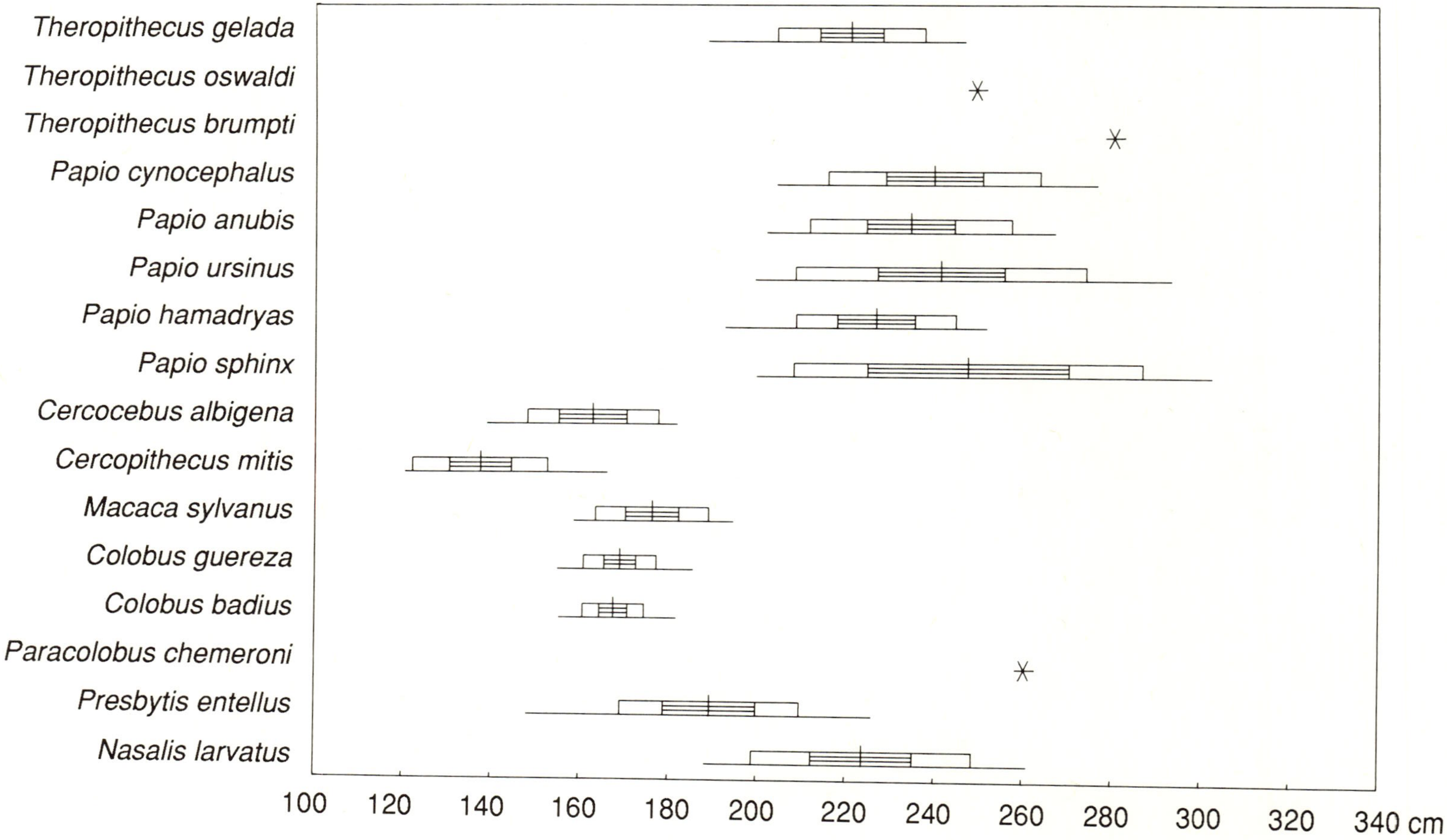

Figure 19. Graphic representation of the mean, two standard errors of the mean, one standard deviation, and total range for ulna length of species in the comparative sample.

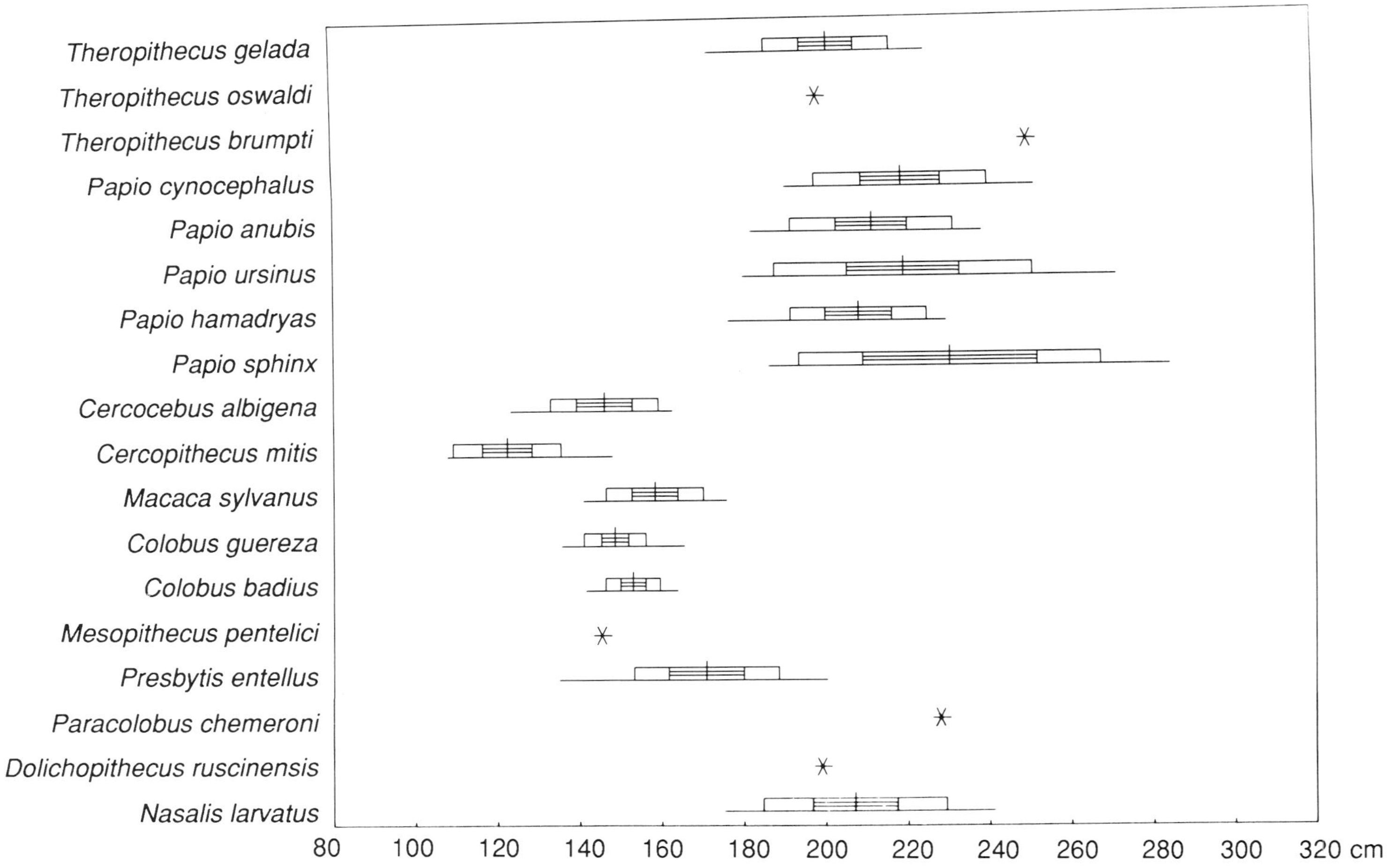

Figure 20. Graphic representation of the mean, two standard errors of the mean, one standard deviation, and total range for radius length of species in the comparative sample.

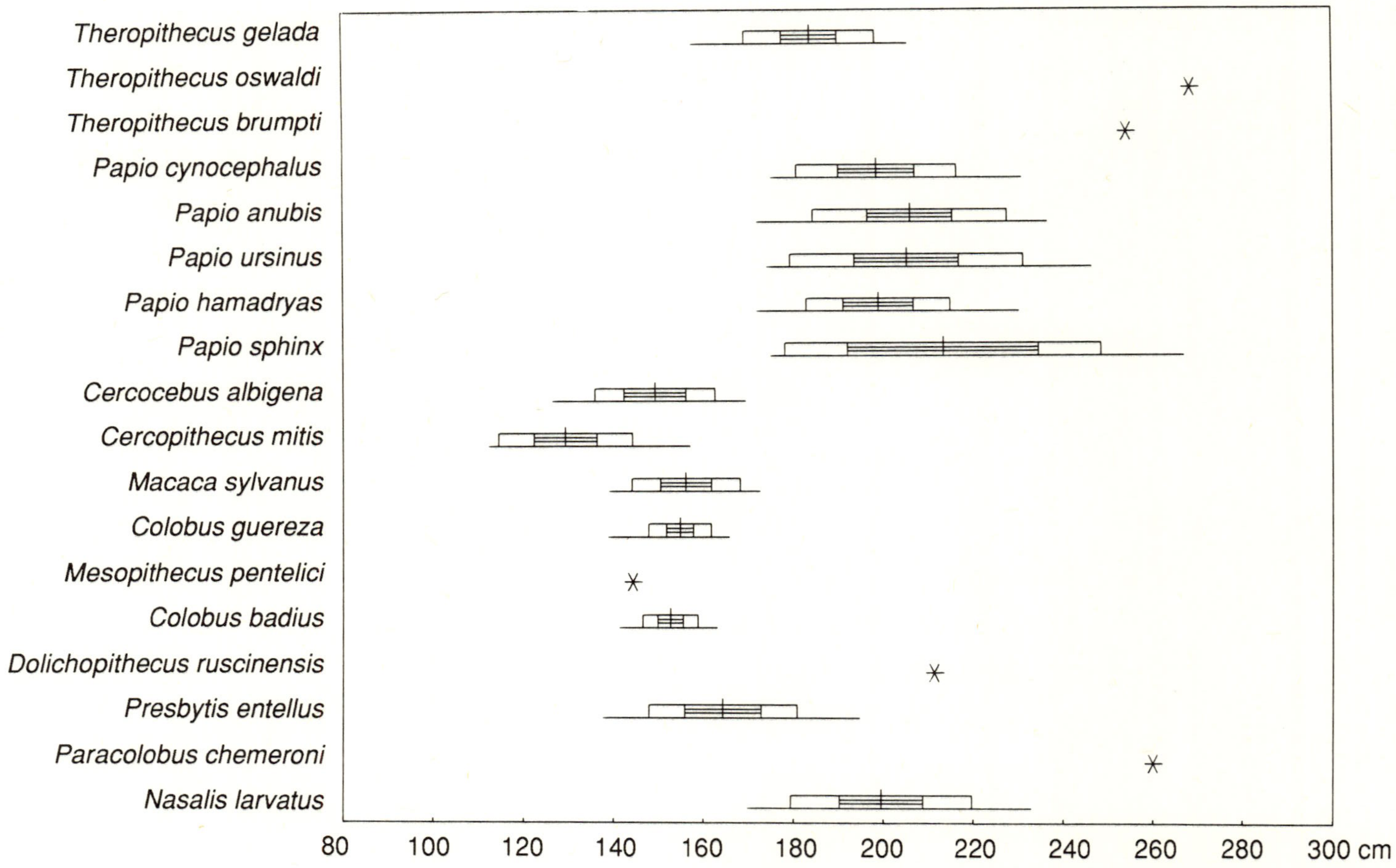

Figure 21. Graphic representation of the mean, two standard errors of the mean, one standard deviation, and total range for humerus length of species in the comparative sample.

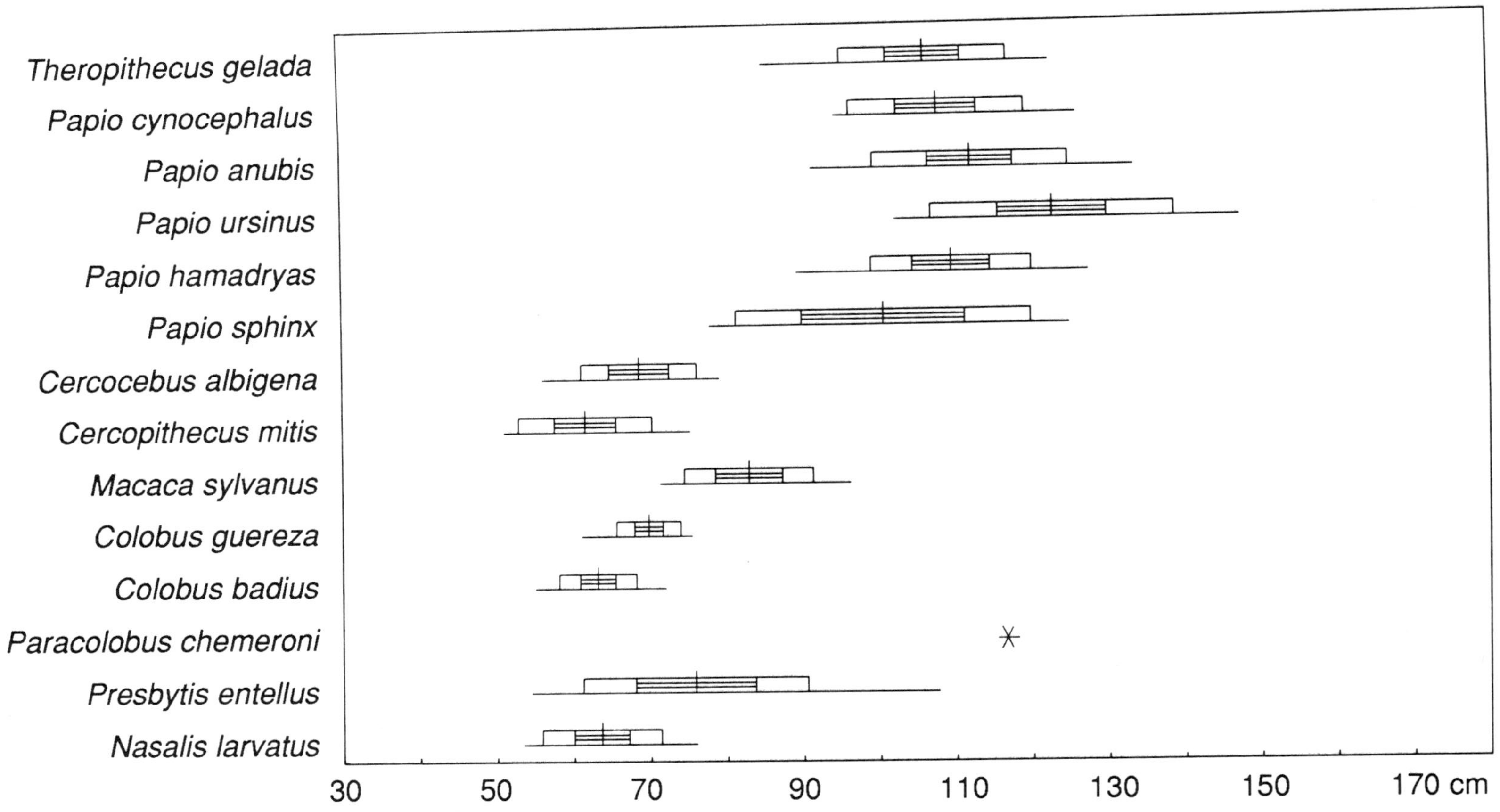

Figure 22. Graphic representation of the mean, two standard errors of the mean, one standard deviation, and total range for scapula morphological length of species in the comparative sample.

Extinct cercopithecoid species are represented in the figures by asterisks. Each asterisk usually depicts a single individual or, at best, the mean of two or three individuals. In no case were there enough complete forelimb elements of an extinct cercopithecoid species to generate a complete range/mean plot for limb length. In ulnar length (Fig. 19), both *Theropithecus oswaldi* and *T. brumpti* lie within the upper range of modern papionin species, but exceed the size range for the only living species of *Theropithecus, T. gelada*. On the other hand, *Paracolobus chemeroni* lies at the edge of the total range of living colobines and is at the upper range for both living and extinct papionins.

Radius length (Fig. 20) reveals a similar pattern. *T. oswaldi* and *T. brumpti* fall with the papionins, though *T. oswaldi* is at the lower range. *Paracolobus* once again is at the upper range for extant colobines. The Eurasian Miocene colobines *Mesopithecus* and *Dolichopithecus* also appear in this plot. *Mesopithecus* falls with the smaller arboreal colobines. *Dolichopithecus* overlaps both of the larger-bodied arboreal/terrestrial colobines, *Presbytis entellus* and *Nasalis*.

T. oswaldi and *Paracolobus* almost exceed the maximum humerus length range for all living cercopithecoids (Fig. 21). Thus, these two fossil species have absolutely (and in the case of *Paracolobus,* relatively) longer humeri than any of the living cercopithecoids in this study. On the other hand, *Mesopithecus* and *Dolichopithecus* fall well within the ranges of *Colobus* and *Nasalis*, respectively.

The only extinct species for which a scapula is complete enough to measure its morphological length is *Paracolobus* (Fig. 22). As measured here, this exceeds those of all living colobines in the study sample. However, Birchette (1982) suggests that past orientations of the *Paracolobus* scapula have been incorrect. He estimates the morphological length (his "length spine base") to be about 94 cm, which would place it at the upper range of *Presbytis entellus*.

The size relationships among species of the comparative sample, based on limb lengths, correlated closely with overall body size and body weight (compare body weights of species in Table 2 of Chapter 2 with Figs. 19-22). Aiello (1981) and Jungers (1985), among others, have argued that there is a close correspondence between body mass and limb length across the order Primates, and indeed across all orders of mammals. Steudel (1982) found that the allometric relationship of skeletal weight and limb length was near isometry in catarrhine primates, especially in closely related species groups exhibiting similar locomotor modes. Since the current study deals with only one family (Cercopithecidae), which contains species characterized by a grossly similar locomotor mode, the prediction of approximate body weight from the limb lengths of fossil species (see Chapter 8) could prove an interesting application of the data presented in Figs. 19-22. (For further discussion of the estimation of body size in fossil catarrhines, also see Steudel, 1985.)

In comparing the limb length differences represented in Figs. 19-22, the separation (both in means and overall ranges) between large- and small-bodied cercopithecoids is more pronounced for the radius and ulna than for the humerus. It can be concluded from this observation that forearm (radius and ulna) length has increased in both a relative and absolute sense in the large-bodied cercopithecoids.

Similar conclusions were reached by Jungers (1984) in a study of the allometry of the catarrhine forelimb skeleton. Since all of the large-bodied papionins occupy terrestrial habitats (see Table 2 in Chapter 2), the lengthening of their radius and ulna can be viewed as a terrestrial adaptation emphasizing a more cursorial locomotor mode. This same allometric trend is also seen among the colobines. The two largest species, *Nasalis* and *Presbytis*, have relatively longer forearms, and both are partially terrestrial, though neither could be considered cursorially adapted.

The brachial index (Fig. 23) is a measure of the relative size of the forearm to the arm (A. B. Howell, 1944). It has long been used in primate studies (see Napier and Napier, 1967), though Jungers (1985) is among the first to provide a useful across-species view of its functional meaning. The interpretation of Fig. 23 is fairly straightforward. Species having a high brachial index, such as the papionins, have relatively long forearms. The large colobine species, such as *Presbytis* and *Nasalis*, also show a trend toward lengthening of the forearm. The smallest species in the comparative sample, *Cercopithecus*, has the lowest brachial index, which suggests a negatively allometric trend toward forearm lengthening.

The really notable exception to the trends seen in the extant sample is in the position of *Paracolobus*. Its brachial index value lies at the bottom of the range for the smallest living species, *Cercopithecus*, yet its humerus length (Fig. 21) indicates that *Paracolobus* was larger than any living colobine, and indeed was probably as large as the largest living cercopithecine. Since a correlation between large body size and a high brachial index in cercopithecoids appears firmly established, the existence of such a low index in *Paracolobus* is particularly unusual. However, Fleagle and Simons (1982) have pointed out that both large and small arboreal quadrupeds *are* characterized by low brachial indices. Perhaps the brachial index of *Paracolobus* is indicative of an arboreal paleohabitat preference? Birchette (1982) concludes that *Paracolobus* was predominantly an arboreal quadruped, but argues that the brachial index is of low diagnostic value for making such judgments.

Limb curvature is another potentially interesting feature very likely correlated with limb length. Figs. 24-26 present limb curvature indices for the ulna, radius, and humerus, respectively. For all three, the higher the index the greater the degree of shaft curvature (anterior concavity). The arboreal colobines and small cercopithecines exhibit greater ulna curvature than the large papionins (Fig. 24), yet the pattern is not wholly consistent. The two large colobines, *Presbytis entellus* and *Nasalis*, overlap considerably with both subfamily groups. The ulna of *Theropithecus oswaldi* is rather more curved than that seen in *T. brumpti* or *T. gelada.*

Fig. 25 shows very little differentiation between large and small cercopithecoids, or between cercopithecines and colobines, in the curvature (lateral concavity) of the radius. There is substantial overlap among all species except for the high degree of curvature seen in *Dolichopithecus* and, to a lesser extent, in *P. entellus. T. oswaldi* and *T. brumpti* are nearly identical, unlike their divergent relationship in the ulna curvature index (Fig. 24). The lack of consistent differences in the radius curvature index of this sample of cercopithecoids accords well with the conclusion reached by Jolly (1965) that there are no significant differences between arboreal and terrestrial

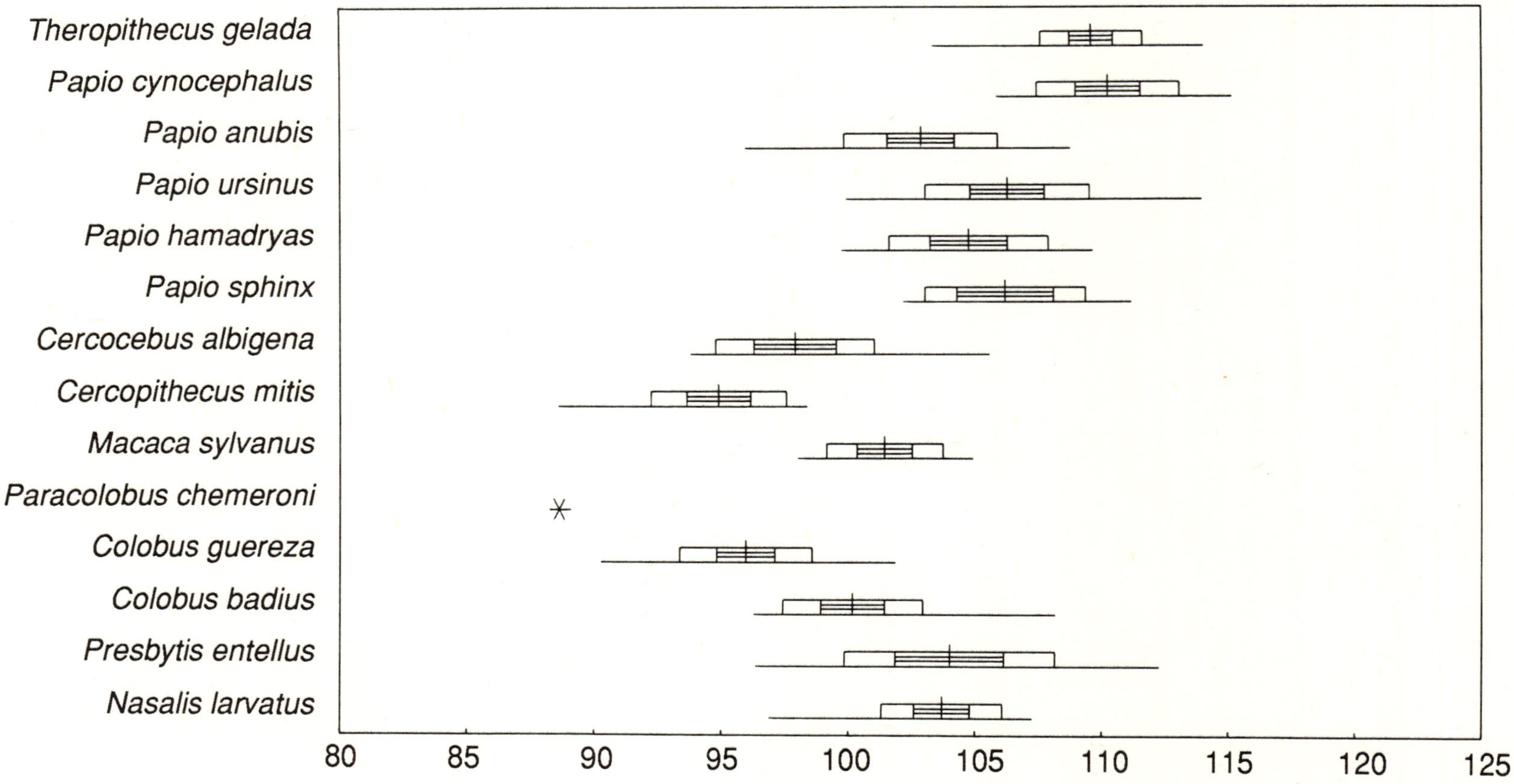

Figure 23. Graphic representation of the mean, two standard errors of the mean, one standard deviation, and total range for brachial index of species in the comparative sample. The brachial index is calculated as follows: Radius Length X 100/Humerus Length.

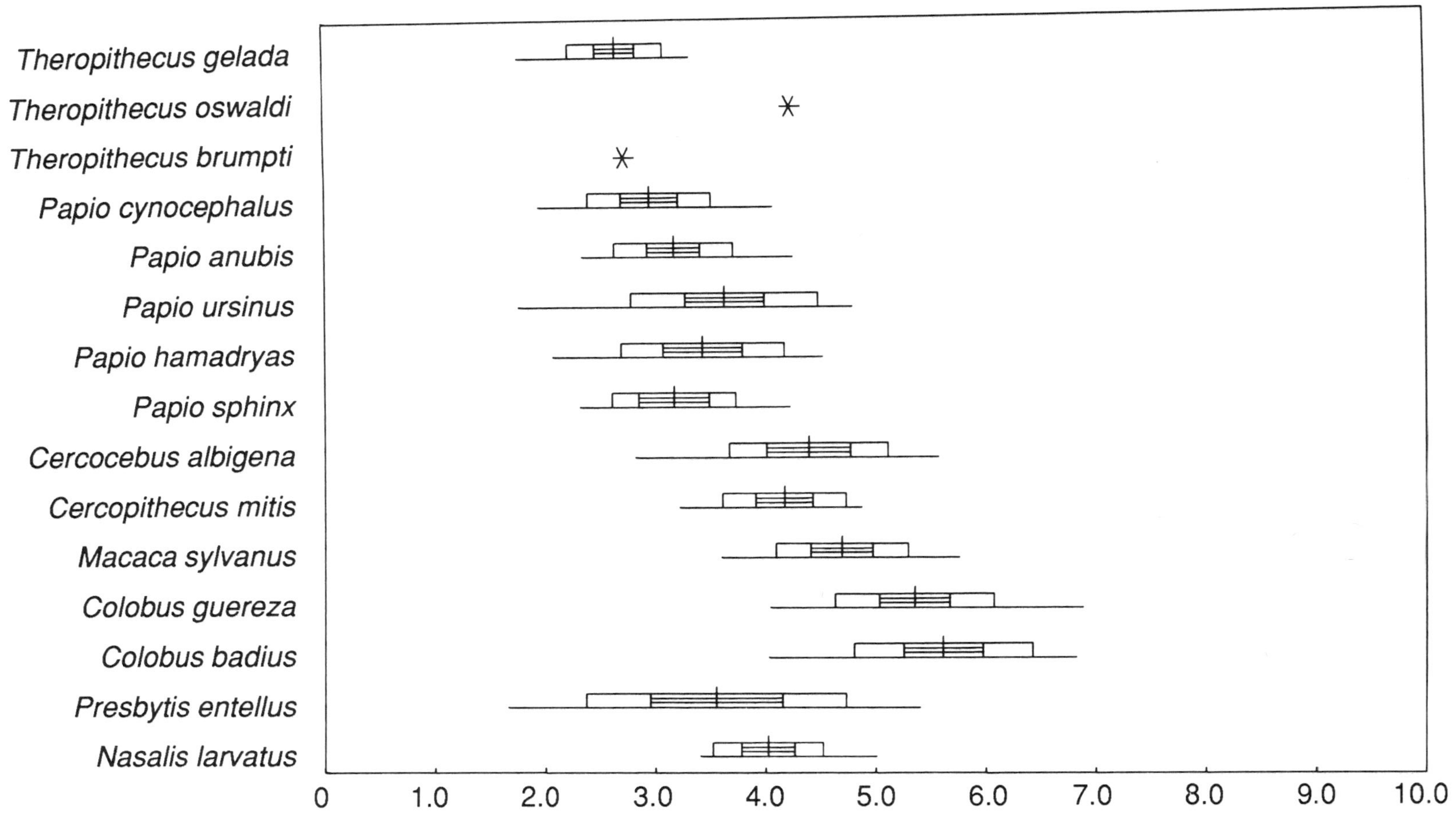

Figure 24. Graphic representation of the mean, two standard errors, one standard deviation, and total range for the ulnar curvature index of species in the comparative sample. The ulnar curvature index is calculated as follows: Ulna Midshaft Curvature X 100/Ulna Length.

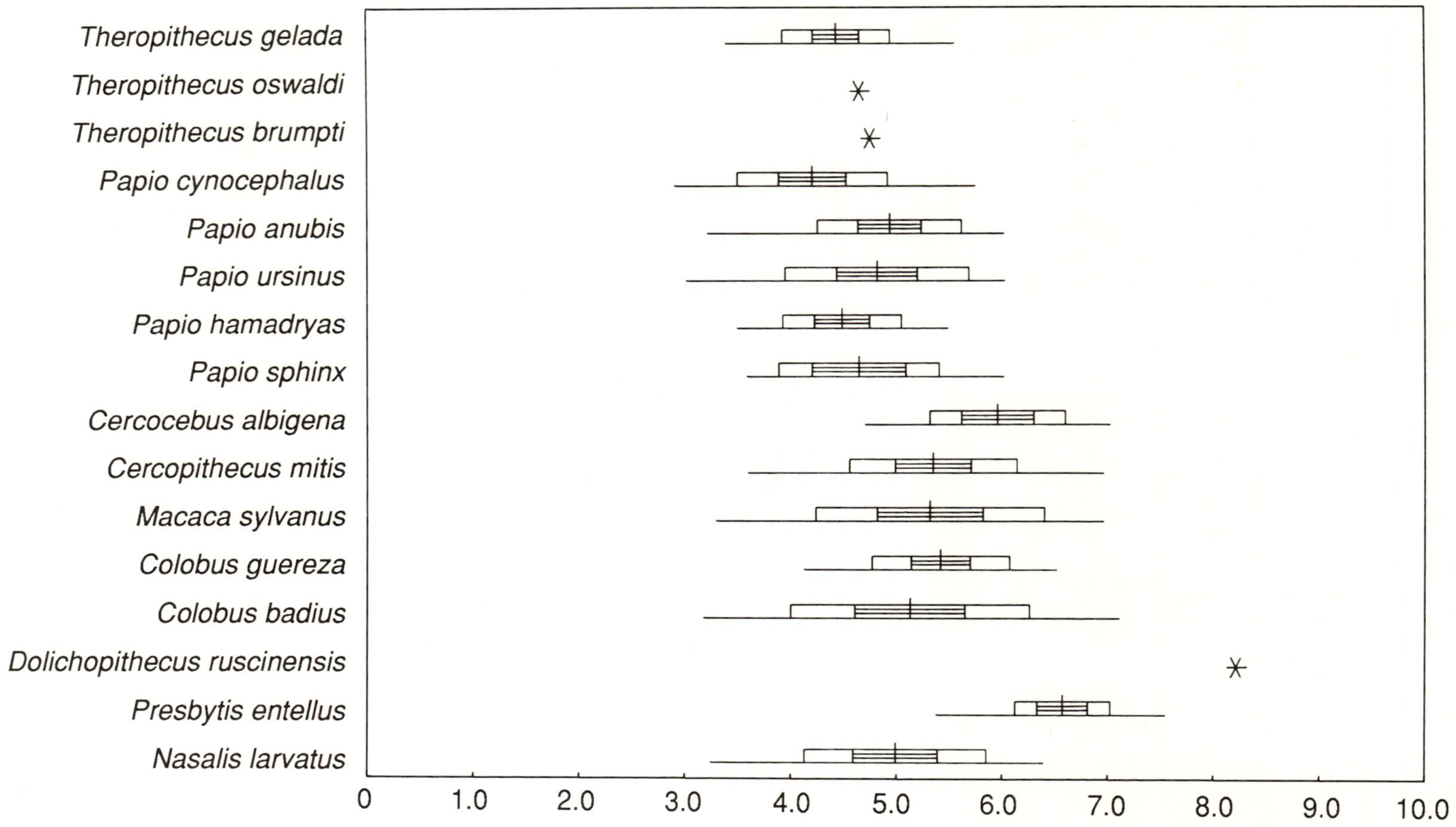

Figure 25. Graphic representation of the mean, two standard errors of the mean, one standard deviation, and total range for the radius curvature index of species in the comparative sample. The radius curvature index is calculated as follows: Radius Midshaft Curvature X 100/Radius Length.

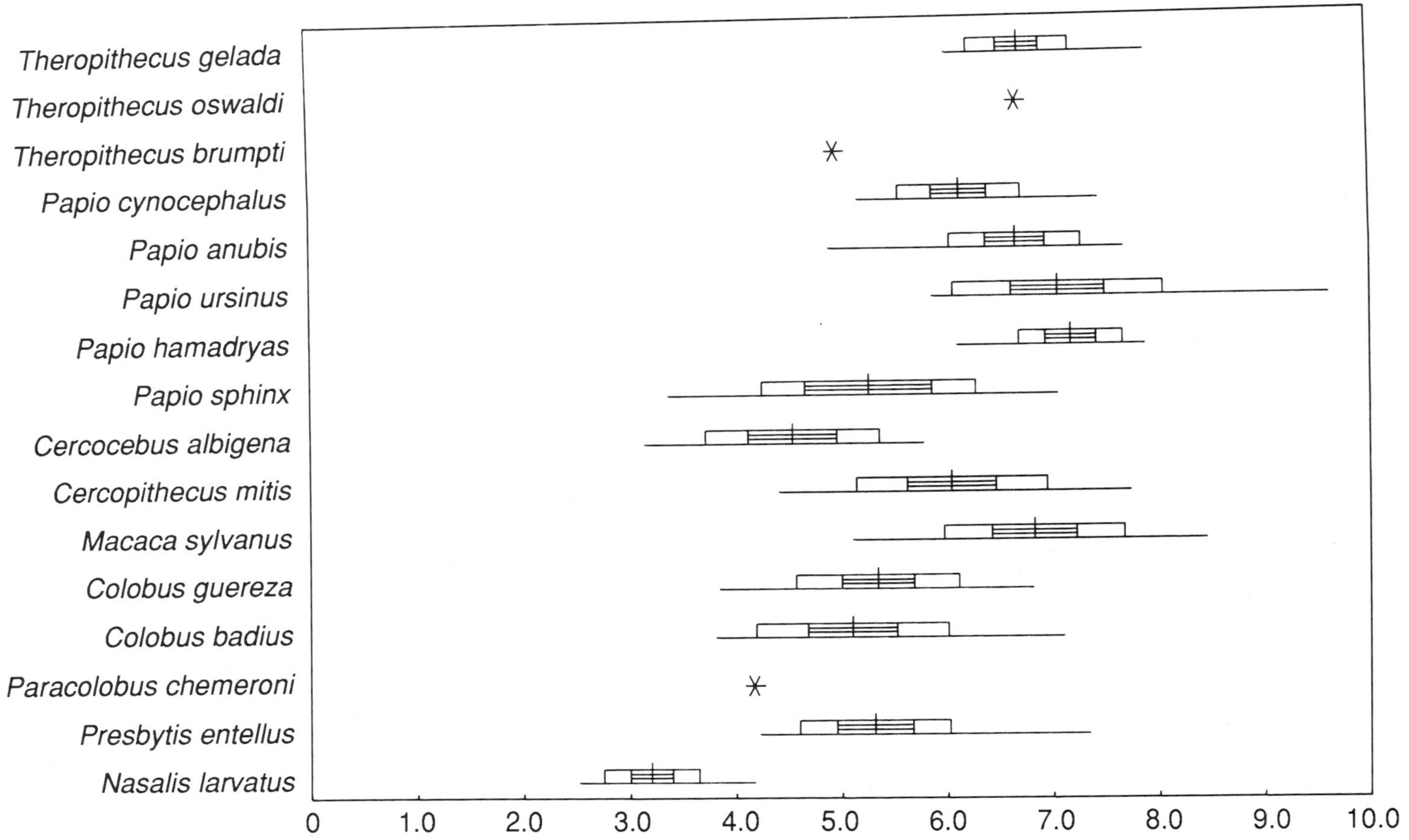

Figure 26. Graphic representation of the mean, two standard errors of the mean, one standard deviation, and total range for the humerus curvature index of species in the comparative sample. The humerus curvature index is calculated as follows: Humerus Midshaft Curvature X 100/ Humerus Length.

cercopithecines with respect to lateral radial shaft curvature.

The papionins and all the other cercopithecines except *Papio sphinx* and *Cercocebus albigena* exhibit a greater degree of humerus curvature (posterior concavity) than do all colobines, including the large extinct African form *Paracolobus* (Fig. 26). The greatest degree of humerus curvature is seen in *Papio ursinus*, and the least in *Nasalis larvatus*. *T. oswaldi* falls at the mean of *T. gelada*, and *T. brumpti* is divergent (less curved). The functional significance of humerus curvature might be related to a change in angulation of the humeral shaft to give the deltoid and pectoralis muscles a better mechanical advantage at their insertion. This trend is most readily seen in the large-bodied papionins, which have bowed humeri and a massive area for muscle attachment along the delto-pectoral crest.

Of the 87 measurements analyzed in this study of the cercopithecoid forelimb, two of the variables—no. 20, Olecranon Angulation and no. 21, Trochlear Notch Angulation—could not be included in the multivariate phase of the study because they were measured in degrees and not in centimeters. Combining different units in a multivariate analysis does not meet statistical requirements for the procedure (Blackith and Reyment, 1971) and could lead to erroneous results. The implications of these variables with regard to the evolution of the cercopithecoid forelimb are discussed in more detail elsewhere (Ciochon, 1986; Ciochon and Nisbett, 1993).

One way to interpret the ulnar olecranon process in a functional sense is by describing its projection (height or length in a dorso/posterior direction) beyond the elbow joint proper. The olecranon process height index (Fig. 27) measures the amount the olecranon projects behind or above the pivotal trochlear articular surface (signoid notch) of the proximal ulna. A high index value indicates a relatively long olecranon process, and a low value, a short process. There is a significant separation between cercopithecines and colobines in Fig. 27. All of the cercopithecines are closely linked here: the Cercopithecini and small-bodied papionins are undifferentiated from the large-bodied papionins. *Presbytis entellus* once again substantially overlaps both cercopithecines and colobines. All of the fossil colobines fall near the mean values of living colobines. There is almost complete overlap between *T. gelada* and *T. oswaldi*, yet *T. brumpti* is quite divergent, with the largest mean olecranon height index of the entire sample.

The functional significance of olecranon height is closely linked with the degree of olecranon angulation (see Ciochon and Nisbett, 1993). As ulnar retroflexion increases the mechanical advantage of triceps at full extension, so does a relatively long olecranon *when* it occurs in species with a posteriorly inclined olecranon. However, what is the functional significance of a relatively long olecranon in those species without a posteriorly inclined, olecranon, such as the small-bodied Papionini or Cercopithecini? Fleagle et al. (1975) and Conroy (1976) argue that a long olecranon process *per se* can increase the leverage of triceps when the elbow is *partially* flexed. Thus, depending on whether the elbow joint is operated at fully extended or partially flexed position, the mechanical advantage of triceps can be increased by either inclining the olecranon posteriorly and/or lengthening the olecranon. The fundamental point is that, in cercopithecines, the farther the olecranon is distant from its pivotal

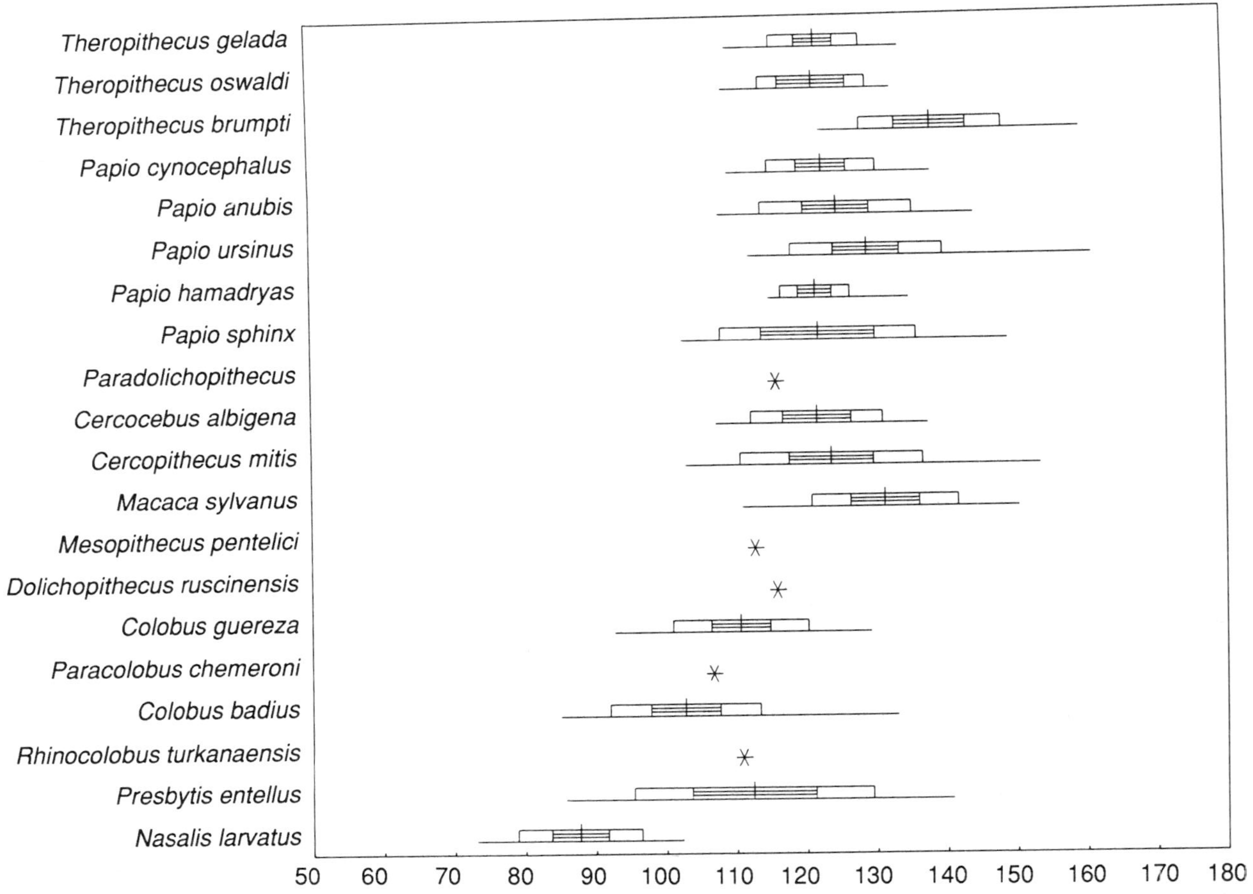

Figure 27. Graphic representation of the mean, two standard errors of the mean, one standard deviation, and total range for the olecranon process height index of all species in the comparative sample. The olecranon process height index is calculated as follows: Trochlear Notch-Olecranon Length X 100/Trochlear Notch Midline Height.

arc about the point of humeral-ulnar articulation, the more significant the degree of leverage the triceps brachii can impart to the forearm.

Colobines have relatively short olecranon processes which are often anteriorly inclined. This morphology requires a different functional explanation. Since most colobines are arboreal leapers, they require an elbow joint that can easily be fully extended to facilitate a safe landing. However, a powerful thrust to full extension, as can be accomplished by the large terrestrial papionins, is not necessary in colobines, since it is their hindlimb (not their forelimb) that provides the major propulsive force during locomotion. Colobines, therefore, have evolved the ability to extend or hyperextend the forearm by reducing the olecranon process and widening the humeral olecranon fossa, resulting in a less stable but fully extendable elbow joint.

The trochlear notch shape index (Fig. 28) charts the relative width of the articular surface of the trochlear notch. The large-bodied papionins, except for *Papio sphinx*, form a very tight group. The remaining cercopithecines, together with *Paradolichopithecus*, substantially overlap both the large-bodied papionins and living and extinct colobines, and occupy an intermediate position with respect to both groups. *Dolichopithecus, Paracolobus,* and *Rhinocolobus* all have very similar indices, yet all lie near the outside range limit of extant colobines. *Mesopithecus*, on the other hand, falls right at the mean of *Colobus guereza* and outside the low range for all cercopithecines.

The styloid process projection index (Fig. 29) charts the relative degree of projection of the styloid process, with high index values indicating a relatively long styloid and lower values a shorter styloid. Except for *Papio sphinx* and *Macaca sylvanus,* all cercopithecines have a relatively longer styloid process than do colobines. *T. oswaldi* and *T. brumpti* fall within the range of *T. gelada*. Within the colobines, *Nasalis* has a rather skewed sample that overlaps both colobines and cercopithecines, with the mean falling near that of the large-bodied papionins; all other colobines have significantly shorter styloid processes. *Rhinocolobus* had a very short styloid process falling at the mean of *Colobus badius*, which is the most habitually arboreal species in the extant sample.

The ulnar styloid of cercopithecoids articulates by way of a convex facet on its distal, anterior, and posterior surfaces, with the articular cup formed by the pisiform and triquetral bones. The functional implications and evolutionary significance of this articulation for the divergence of cercopithecoids and hominoids have been the subject of much debate (e.g., Lewis, 1971, 1972; O'Connor, 1975; Corruccini, Ciochon, and McHenry, 1975, 1976). Most authors agree that the function of a long styloid process in direct contact with the pisiform and triquetral bones is to provide stability to the wrist joint while limiting adduction (ulnar deviation). In the context of cercopithecines vs. colobines, it therefore seems reasonable to argue that the long and robust styloid process of cercopithecines limits the capacity for adduction while providing stability to the wrist joint—important for terrestrial quadrupedal locomotion. The shorter, more gracile styloid process of colobines is more advantageous in arboreal leaping, where a more mobile wrist joint with greater movement capabilities is needed. The shorter styloid also allows a greater range of adduction in colobines, a useful adaptation for reaching upward when climbing.

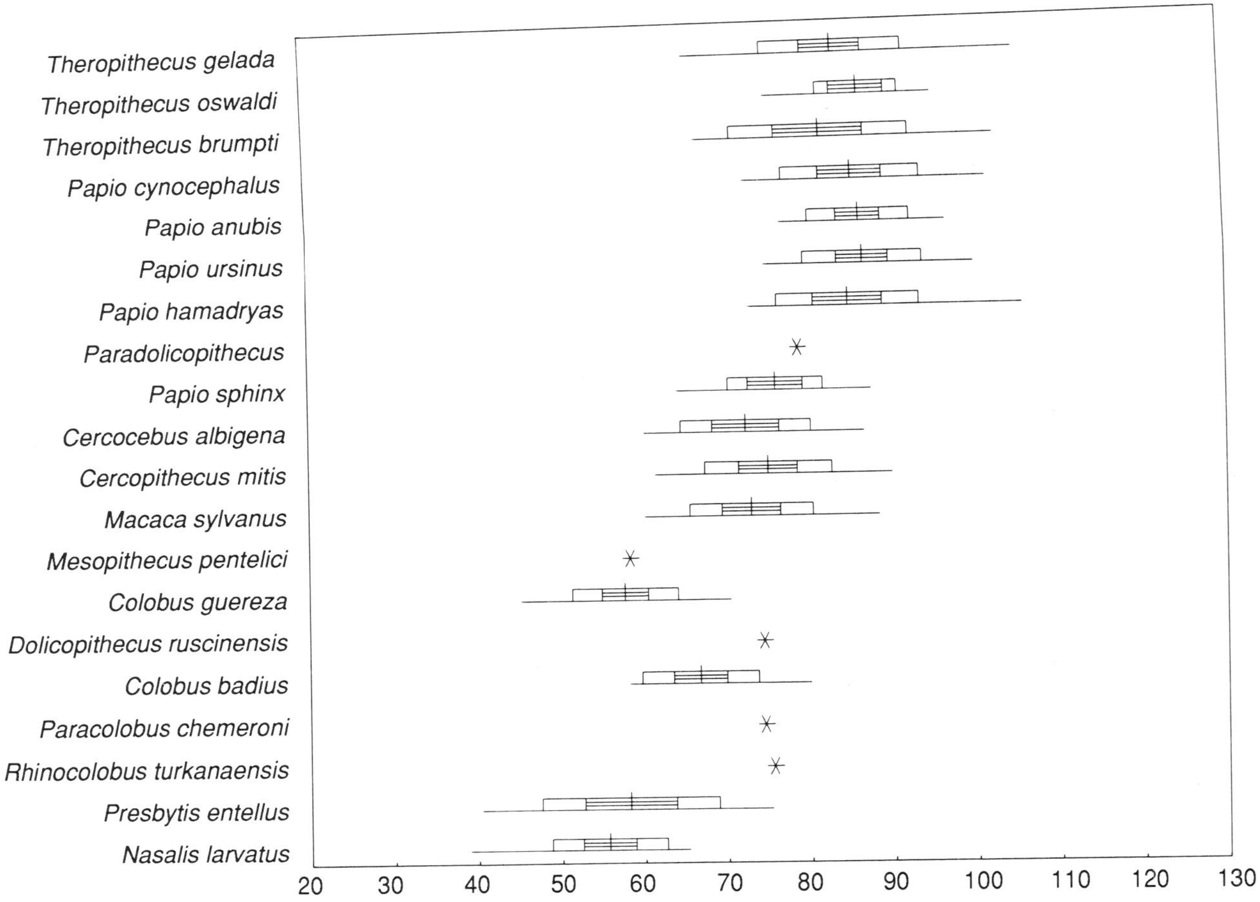

Figure 28. Graphic representation of the mean, two standard errors of the mean, one standard deviation, and total range for the trochlear notch shape index of species in the comparative sample. The trochlear notch width index is calculated as follows: Distal Trochlear Notch Width X 100/Proximal Trochlear Notch Width.

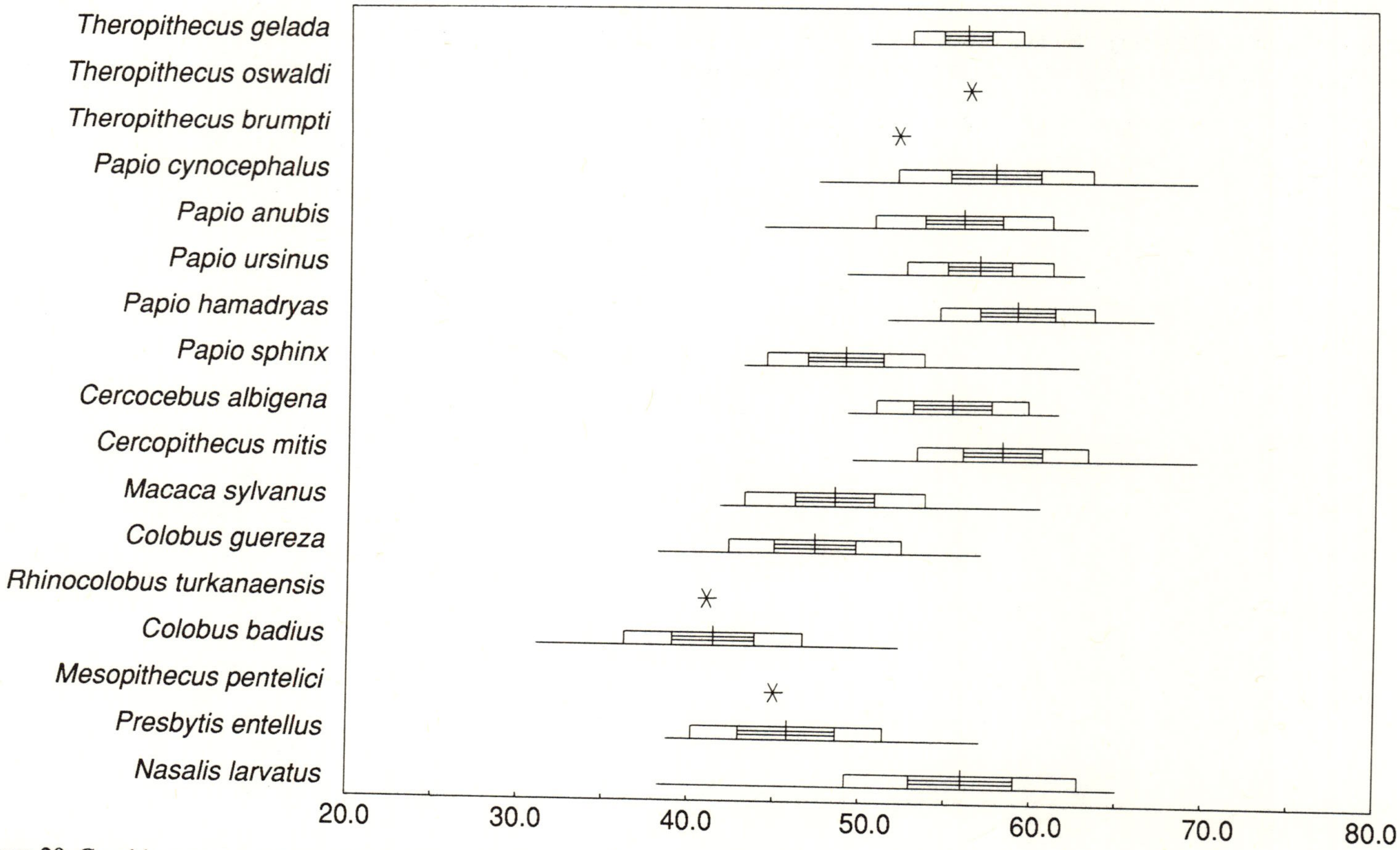

Figure 29. Graphic representation of the mean, two standard errors of the mean, one standard deviation, and total range for the styloid process projection index of species in the comparative sample. The styloid process projection index is calculated as follows: Styloid Process Length X 100/Antero-Posterior Distal Ulnar Width.

The projection of the styloid process of the radius is functionally related to the projection of the ulnar styloid. The beak of the radial styloid articulates with the scaphoid, providing a brace and stability to the anterior (or lateral) side of the wrist joint. A distally projecting styloid also limits abduction (radial deviation) at the wrist. The radius styloid projection index (Fig. 30) depicts the relative projection of the radial styloid process in a distal direction from the lateral tubercle. A pattern very similar to that seen in Fig. 29 emerges. However, here (Fig. 30) all cercopithecines form a group, characterized by relatively longer radial styloid processes. Colobines, on the other hand, have relatively shorter styloids, the one exception being *Presbytis entellus*, which substantially overlaps the cercopithecines. *Mesopithecus* and *Dolichopithecus*, the only two fossil colobines analyzed in this plot, both fall at the middle to upper range of *P. entellus*, an Asian colobine with a terrestrial/arboreal locomotor adaptation. Delson (1973) points out a similar relationship with regard to *Mesopithecus*, but suggests that *Dolichopithecus* was very baboon-like in its adaptation. Note that in Fig. 30 *Dolichopithecus* also falls within the range of several baboon groups, especially *Papio anubis*.

In the radial head shape index (Fig. 31), smaller values indicate a more circular head shape and larger values a more elliptical (or square) shape. All 13 cercopithecines form a tight cluster in Fig. 31, united by their common possession of a relatively elliptical radial head. All of the colobine species have relatively rounder radial heads, but there is overlap with the cercopithecines, especially in *Colobus badius* and *Nasalis*. All of the fossil colobines fall within the range of living colobines, with *Mesopithecus*, unique in this feature, exhibiting the most nearly round radial head shape.

Conroy (1976) and Birchette (1982) also calculated an index of radial head shape in their respective analyses of the primate skeleton. The results presented here agree closely with those of Conroy, but not with those of Birchette, who apparently calculated the index in an alternate manner and possibly took the measurements in a different fashion. Birchette detected considerable interspecific variability in radial head shape within his colobine comparative sample, much greater than that found here. However, the sample size used in the present study is considerably larger. Birchette also found that *Paracolobus* fell near the mean index value for *Colobus badius* and that *Mesopithecus* was nearest the mean of *C. guereza*, both of which are in close agreement with the results shown in Fig. 31. However, Birchette (1982:208) supports Jolly's (1965) contention that "no clear morphological distinction of the radius exists between arboreal and terrestrial cercopithecoids." My results (Fig. 31) point strongly to the contrary, and the functional arguments for this viewpoint are hereby presented.

The functional interpretation of radial head shape is straightforward. An overview of mammalian outgroups such as the Artiodactyla and Perissodactyla reveals that these habitually pronograde species have elliptical or "squared" radial heads, while groups with greater forelimb mobility, associated with a pronounced capacity for supination, have nearly circular or "round" radial heads (A. B. Howell, 1944). An

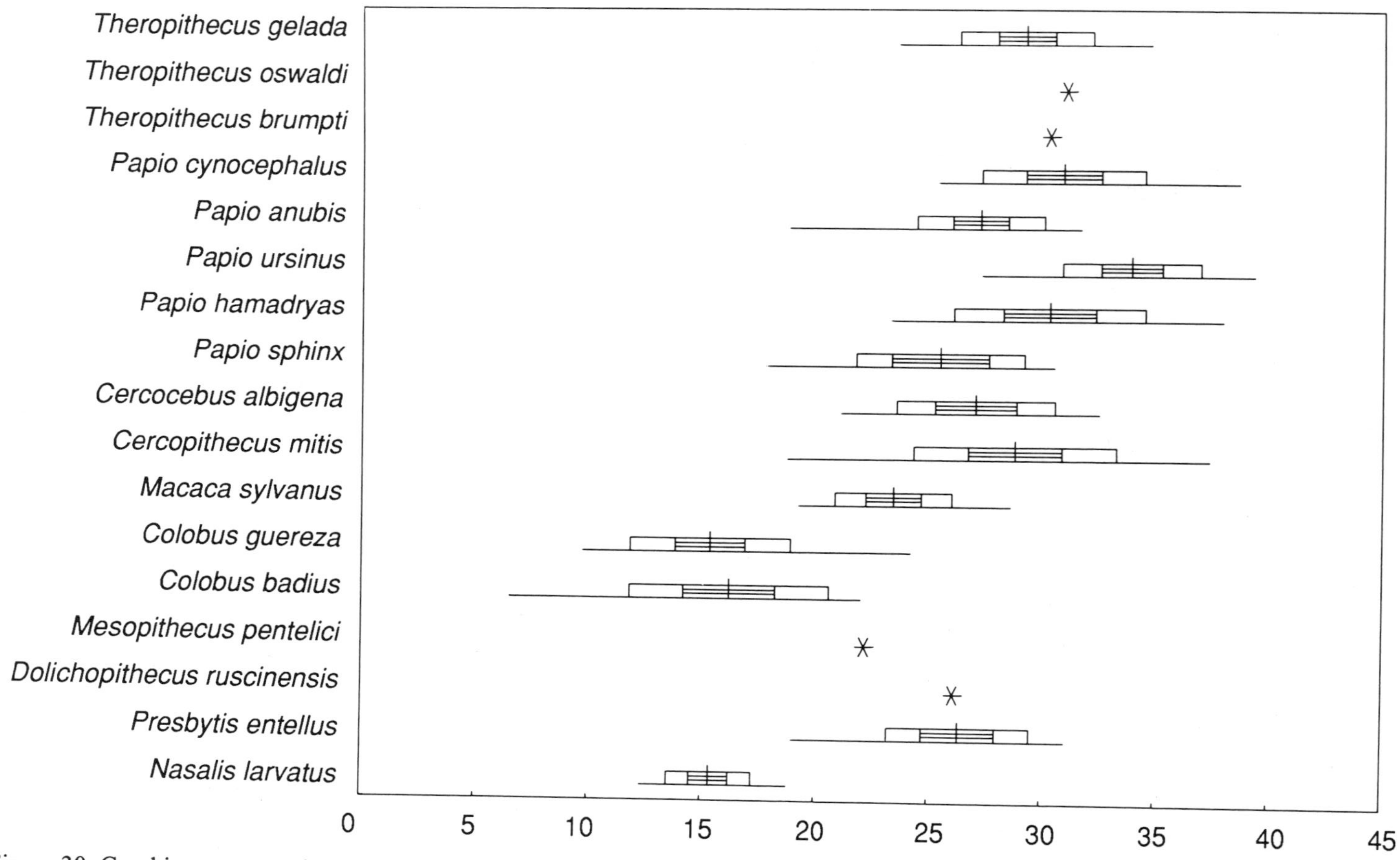

Figure 30. Graphic representation of the mean, two standard errors of the mean, one standard deviation, and total range for the radius styloid projection index of species in the comparative sample. The radius styloid projection index is calculated as follows: Lateral Tubercle Distal Projection X 100/Maximum Distal Radial Width.

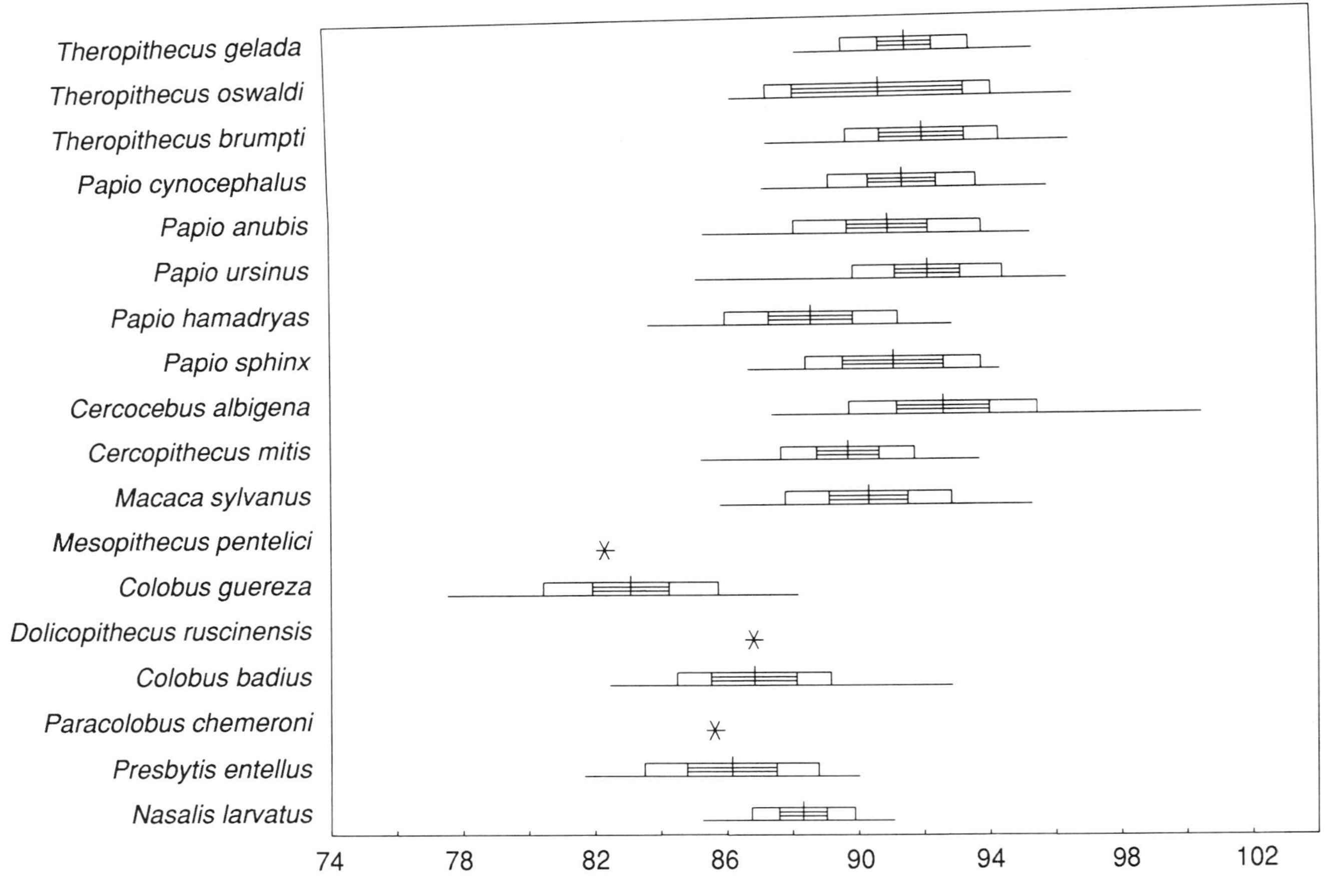

Figure 31. Graphic representation of the mean, two standard errors of the mean, one standard deviation, and total range for the radius head shape index of species in the comparative sample. The radius head shape index is calculated as follows: Minimum Antero-Posterior Head Diameter X 100/Maximum Medio-Lateral Head Diameter.

elliptical radial head articulates with a widened and relatively flat humeral capitulum; it thus can function more easily as a weight-bearing strut with a limited range of pronation/supination. On the other hand, a circular radial head articulates with a more "ball-shaped" humeral capitulum, emphasizing that a wide range of supination/pronation is associated with reaching, climbing, hoisting, and hand-manipulation abilities. Cercopithecines, especially the papionins, locomote on terrestrial and other substrates with their forearms in a habitually pronated position, and are characterized by wide and flat humeral capitulae and elliptical radial heads. In contrast, colobines, which are more arboreal and emphasize climbing and leaping in their locomotor repertoire, have narrower and more bulbous capitulae and near-circular radial heads. This interpretation is based on functional/locomotor differences that separate cercopithecines and colobines; yet, as will be demonstrated, these differences have an underlying phylogenetic basis that is at the root of the evolutionary separation of these two groups.

The radius robusticity index (Fig. 32) is the only index of robusticity calculated for the forelimb elements in this study; in most cases, it reflects the overall trend of robusticity seen in the forelimb. This index depicts the relative stoutness of the radius among species of the comparative sample—the higher the index value, the more stout or robust the element. Living colobines generally have a more slender radius than do cercopithecines; however, the separation or distinctiveness of the two groups is not compelling. The two fossil cercopithecines, *T. oswaldi* and *T. brumpti*, are very far removed from the range of *T. gelada*, and are nearly outside the high range for all living cercopithecoid monkeys. The fossil species of *Theropithecus* had very robust forelimb bones. A look at the colobine group reveals a similar pattern; the index value for the extinct species *Paracolobus* falls outside the high range for living colobines and overlaps that of several cercopithecine groups. The radii of these three extinct species are very robust, but this is due as much to their relatively short length (see Fig. 20) as to any other factor.

Birchette came to a similar conclusion on the radius of *Paracolobus* when compared with that of known colobine species. He reflects (1982:214), "Increased stoutness—relative to general body size—does imply an overall increase in the amount of compressive stress to be withstood by the bone, which in turn may imply more massive musculature associated with particular locomotor activities." With respect to *Paracolobus, T. oswaldi,* and *T. brumpti,* this probably implies that there are no wholly adequate living analogues for these extinct species.

Two indices for the distal humerus, the trochlea projection index (Fig. 33) and the olecranon fossa shape index (Fig. 34), chart different aspects of the same functional complex in the elbow joint. Fig. 33 shows the degree of medial trochlear ridge distal projection expressed as a function of the humeral biepicondylar width. The higher the index value, the greater the distance the medial trochlear ridge projects distally. Uniformly, cercopithecines have more-projecting trochlear ridges than do colobines. Cercopithecines form a cohesive group, though the small papionins *Cercocebus albigena* and *Macaca sylvanus* depart somewhat from the group mean.

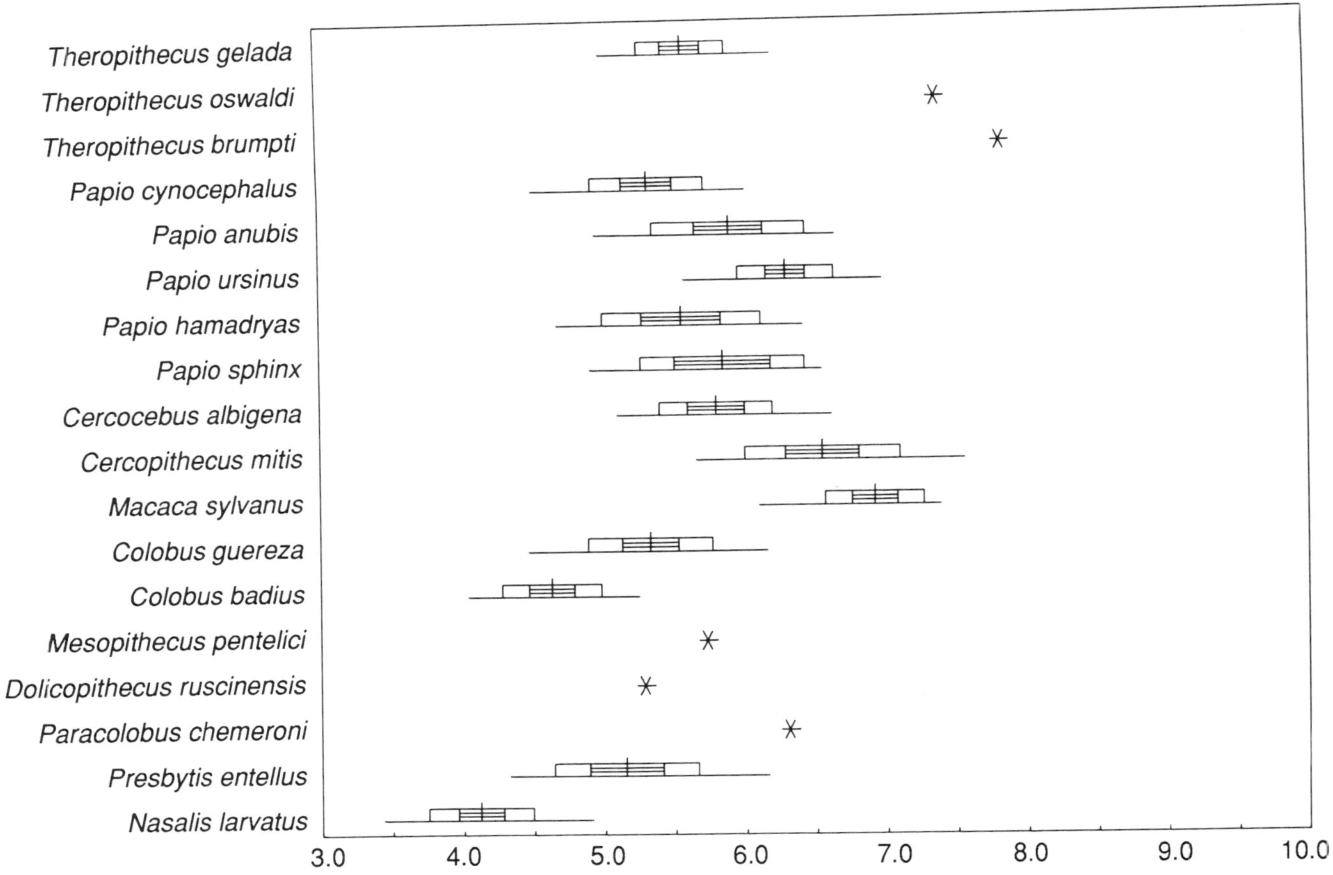

Figure 32. Graphic representation of the mean, two standard errors of the mean, one standard deviation, and total range for the radius robusticity index of species in the comparative sample. The radius robusticity index is calculated as follows: Midshaft Medio-Lateral Diameter X 100/ Radius Length.

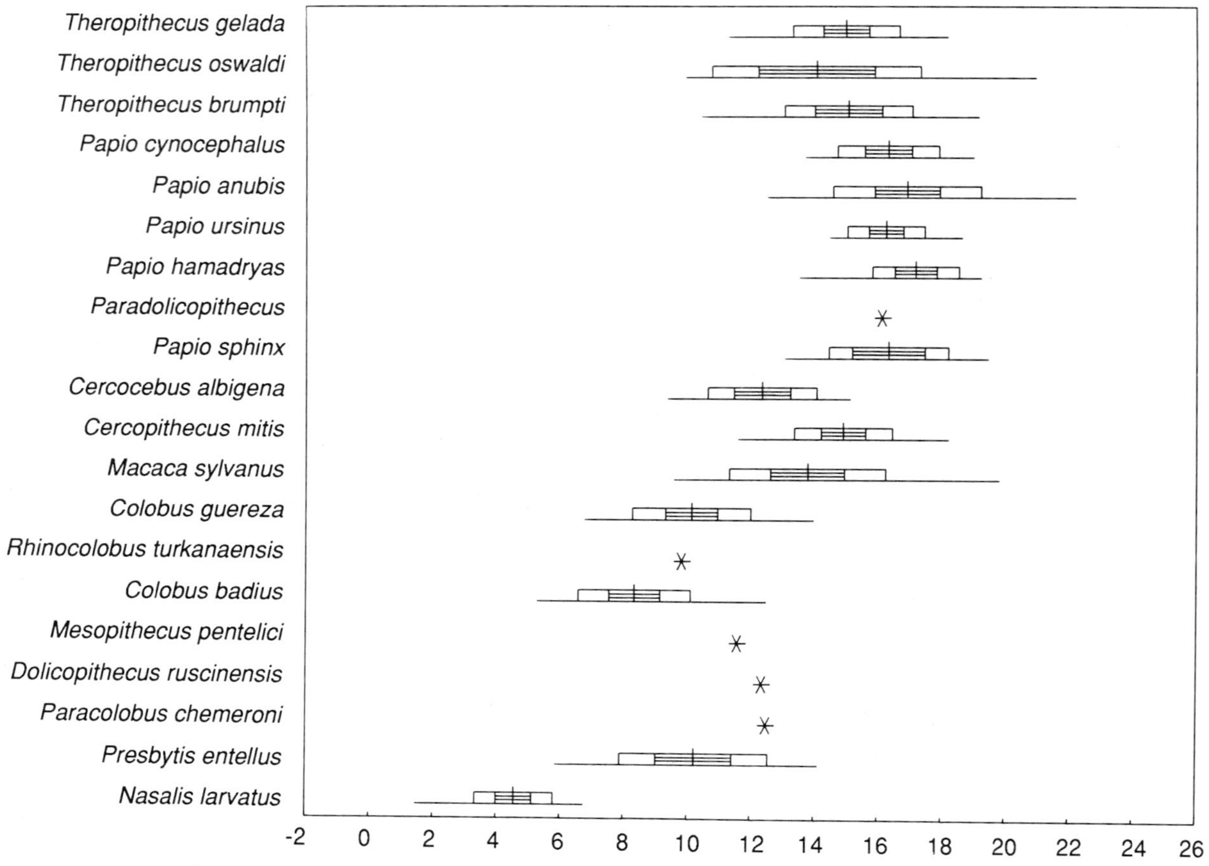

Figure 33. Graphic representation of the mean, two standard errors of the mean, one standard deviation, and total range for the humerus trochlea projection index of species in the comparative sample. The humerus trochlea projection index is calculated as follows: Medial Trochlear Distal Projection X 100/Biepicondylar Width.

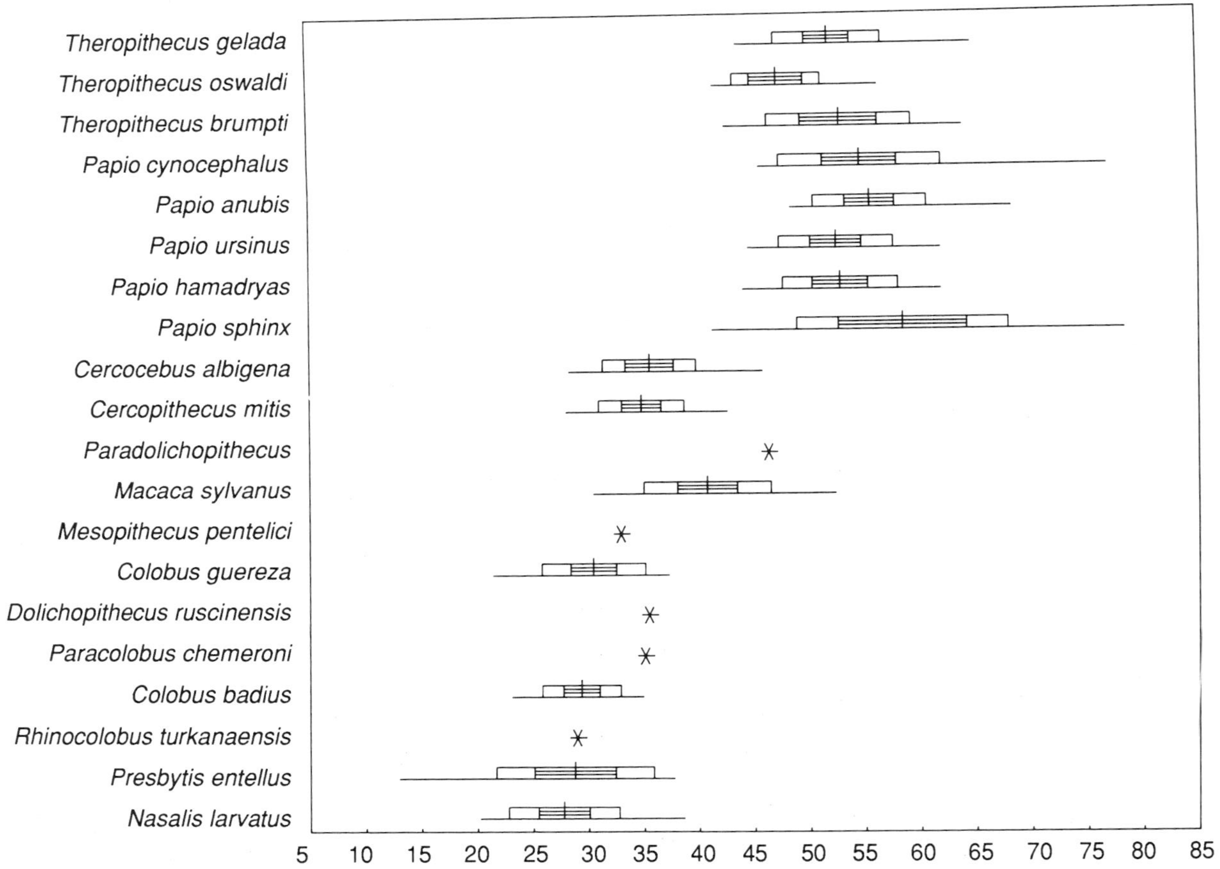

Figure 34. Graphic representation of the mean, two standard errors of the mean, one standard deviation, and total range for the olecranon fossa shape index of species in the comparative sample. The olecranon fossa shape index is calculated as follows: Olecranon Fossa Lateral Margin Projection X 100/Trochlear Gutter Width.

Paradolichopithecus and the extinct *Theropithecus* species are indistinguishable from the other large-bodied papionins. Ranges for three of the four colobine species overlap those of cercopithecines, but *Nasalis* is unique in having exceedingly small trochlear ridge projection values. Not only is the mean value for *Nasalis* very low, but there is little overlap between its range and that for the rest of the comparative sample. Most of the fossil colobines lie at the upper range of living colobines and therefore overlap the cercopithecines. The one exception is *Rhinocolobus*, which falls at the mean of *Colobus guereza*. Because of the high incidence of medial trochlear ridge projection (= "projection of the trochlear flange") in *Paracolobus*, Birchette (1982:167) concludes that this species "may have engaged in periodic terrestrial locomotion." Such functional/locomotor interpretations will be discussed further in Chapter 8.

The olecranon fossa shape index (Fig. 34) graphs the extent of articular surface posterior projection onto the lateral margin of the olecranon fossa as a function of trochlear gutter width. The higher the index value, the more the lateral margin of the olecranon fossa projects in a posterior direction. This index provides a 100% separation of all the large-bodied papionins from living and extinct colobines, the former having a significantly greater incidence of olecranon fossa lateral margin projection. The small-bodied papionins and *Cercopithecus* occupy an intermediate position. All the extant colobine species have nearly identical group means. As in Fig. 33, three of the extinct colobines, *Mesopithecus, Dolichopithecus,* and *Paracolobus*, fall at the upper range for living colobines and thus also within the range of some cercopithecines. *Rhinocolobus*, however, falls right at the mean for living colobines.

The dispersion pattern for Figs. 33 and 34 is quite similar, because both of these indices are measuring the same functional complex. The projecting medial trochlear ridge of the humerus, the rearward displacement (retroflexion) of the medial epicondyle, and the projecting lateral margin of the olecranon fossa have evolved in cercopithecines, and particularly in large-bodied papionins, to resist the adducting torque generated by the carpal and digital flexors during simultaneous extension of the elbow joint and flexion of the wrist. Such movements occur in the forelimb of cercopithecines during running and jumping activities usually associated with terrestrial quadrupedalism. Without this stabilizing mechanism in the elbow joint, the upper forearm and the manus would be continually displaced in a lateral direction, severely limiting the efficiency of cursorial activities. Jenkins (1973:292-294) provides an excellent functional characterization of this complex:

> Flexion of the wrist and digits increases volar pressure to provide frictional resistance, and at the end of a propulsive limb movement actually provides forward thrust. During these movements, those flexor muscles that arise from the medial epicondyle exert a medial torque at the elbow joint in proportion to their distance from the radio-ulnar axis. Although such torque may be counterbalanced in part by extensor muscles arising from the lateral epicondyle, the extensors have neither the size nor mechanical advantage of the flexors. Instead, the tendency to twist the extending ulna on the humerus is resisted by the configuration of the humero-ulnar

articulation. The asymmetrical expansion at the dorsolateral and ventromedial aspects coincides with the areas where torquing forces generated by the carpal and digital flexors are brought to bear.

To summarize these features in cercopithecines: (1) on the distal aspect of the elbow joint, the projecting medial trochlear ridge of the humerus abuts the overhanging articular surface of the ulnar coranoid process; and (2) on the posterior aspect of the elbow joint, the expanded proximo-lateral articular surface of the ulnar trochlear notch abuts the projecting lateral surface of the olecranon fossa. Finally, the effects of torque are further reduced in cercopithecines by the shortening and retroflexion of the medial epicondyle so that the flexor muscles can pass behind the humeral trochlea rather than to its medial side. This illustrates that the cercopithecine elbow is primarily adapted for cursorial-related activities, emphasizing the motions of flexion and extension.

The delto-pectoral crest or ridge, which is sometimes shelf-like, runs from about the midpoint of the shaft proximally, to terminate at the greater tuberosity. Fig. 35 depicts the delto-pectoral crest length index, which expresses the length of the delto-pectoral crest as a function of humerus length. The most striking feature of Fig. 35 is that there are no discernible patterns of separation among colobines and cercopithecines. The mean values for all cercopithecoids of the comparative sample (except for *Theropithecus*) fall between 44 and 48. The single exception is *T. gelada*, with a mean value of 53 and a range of about 48 to 57. The two fossil species *T. brumpti* and *T. oswaldi* have mean values of about 50.5 and 52, respectively. Thus, all three species of *Theropithecus* form a group and are significantly different from all other cercopithecoids in the study. This may be related to a manual foraging behavior such as "seed-eating" (Jolly, 1970). The only two fossil colobines in this analysis, *Mesopithecus* and *Paracolobus*, fall right at the means of *Colobus badius* and *Nasalis*, respectively.

A long history of functional/locomotor significance attributed to the length of the delto-pectoral crest in anthropoid primates dates back to the report by Napier and Davis (1959) on the forelimb of the fossil "ape" *Proconsul africanus*. They developed an index (their "Deltoid Index") which charts the distal extent of the deltoid crest and relates its relative length to the ability either to brachiate (arm-swing), as seen in the Great Apes and some (?) monkeys, or to move about in a wholly quadrupedal (usually terrestrial) fashion, as in many Old World monkeys. Because they could separate brachiators from non-brachiators, the former group having a much more distal extent for the insertion of the deltoid, Napier and Davis ascribed functional/locomotor significance to their "Deltoid Index." The index thus became an indicator of arboreality vs. terrestriality in anthropoids.

There are two major problems with the "Deltoid Index" and the functional/locomotor interpretation Napier and Davis ascribe to it. First, the function of the deltoid is considerably different in apes than in monkeys. In apes, such as the chimpanzee, it functions as an abductor of the forelimb—important in arm-swinging and hauling and hoisting behaviors. In monkeys, such as the baboon, the deltoid functions more as a flexor or protractor of the forelimb—important for anterior/

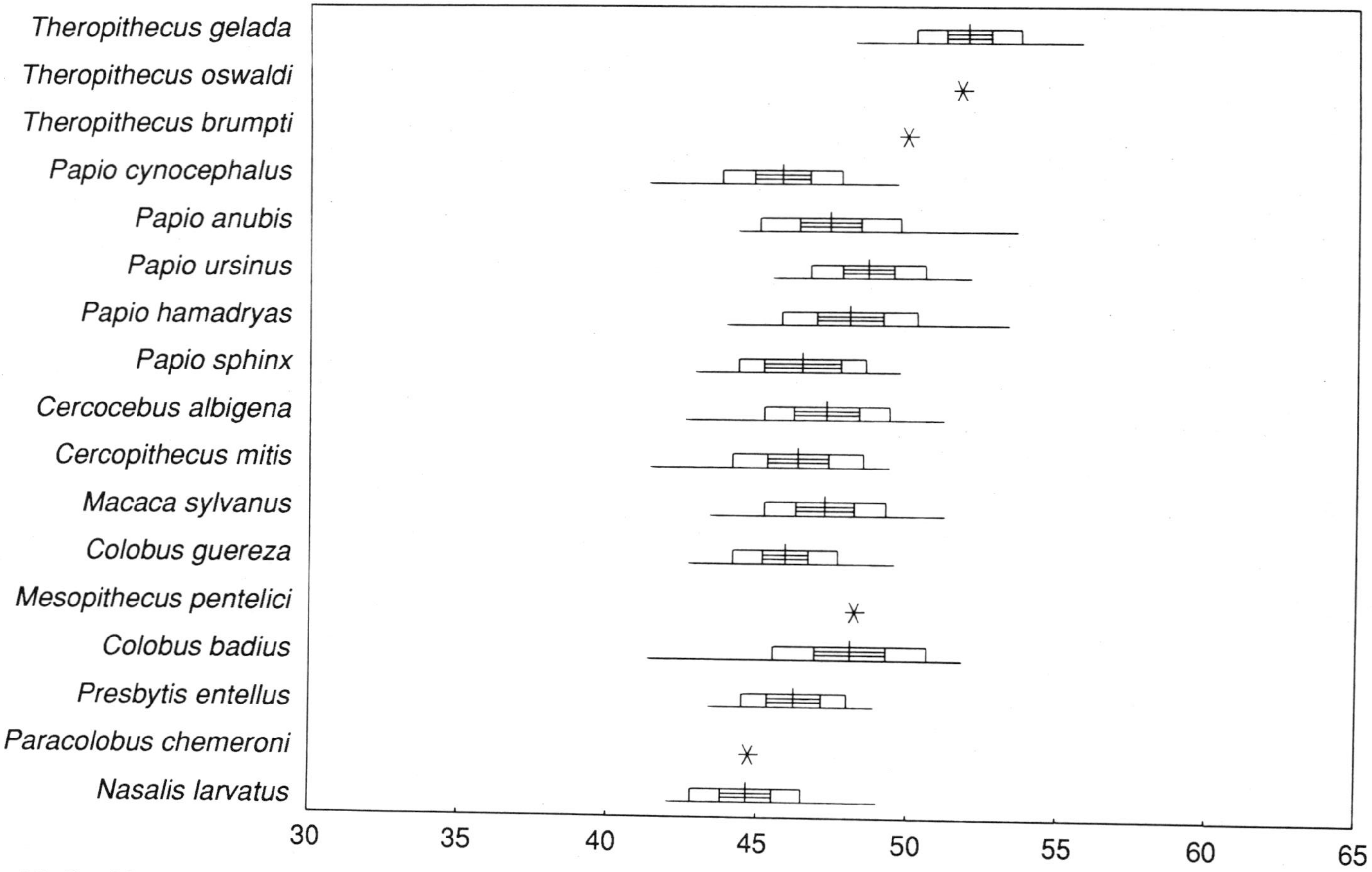

Figure 35. Graphic representation of the mean, two standard errors of the mean, one standard deviation, and total range for the delto-pectoral crest length index of species in the comparative sample. The delto-pectoral crest length index is calculated as follows: Delto-Pectoral Crest Length X 100/Humerus Length.

posterior limb excursions (in the parasagittal plane) such as quadrupedal running. This functional difference is closely connected with the differential shape of the thorax and placement of the scapula in apes vs. monkeys. Napier and Davis were actually comparing the insertion of a muscle in different primate groups and ascribing locomotor significance to its position, despite the non-homologous functions of the muscle. The second problem with the "Deltoid Index" concerns sample size and the choice of taxa. Had Napier and Davis based their index on larger samples (20+ individuals per species, instead of 3-8) and a wider array of cercopithecoid taxa, they would have discovered that there were no significant differences with respect to the insertion of the deltoid in cercopithecoids (see Fig. 35), and thus no intermediate locomotor categories (their "semi-brachiators"). Birchette (1982:157) arrives at a very similar conclusion and nicely sums up the implications of the "Deltoid Index" when he states: "At the very least, it seems clear that a straightforward, consistent relationship between deltoid crest length and arboreality does not exist among major primate taxa. . . ". My study is in full agreement with that statement.

For the proximal humerus, one variable, the greater tuberosity projection (Fig. 36), is very descriptive, and is presented here without the aid of an index. This variable expresses the position (proximal projection) of the greater tuberosity relative to the humeral head articular surface. Values below zero indicate a greater tuberosity that projects above the proximal sphere of the humeral head; values above zero, one that lies below the humeral head articular surface.

The most consistent way to interpret the dispersion of taxa in Fig. 36 is purely on the basis of locomotor/habitat differences. All of the wholly terrestrial, open-country-adapted papionins fall together in one tight group, while the remaining forest-adapted terrestrial/arboreal cercopithecines and all colobines sort together in another group, less well defined. Within the open-country-adapted papionins, the extinct forms *T. oswaldi* and *T. brumpti* are divergent, the latter falling far off the mean values for the other open-country-adapted species. This would be consistent with a more forested habitat preference for *T. brumpti* (see Eck and Jablonski, 1987; M.G. Leakey, 1993; Jablonski, 1993b). It is also worth noting the position of *Papio sphinx*, a large-bodied papionin by every other criterion, yet one whose habitat preference is rain forest or deciduous forest. All of the fossil colobines fall within the more forest-adapted group, and all have very similar tuberosity projection values. This latter interpretation differs from that of Birchette (1982), who found the projection of the greater tuberosity in *Paracolobus* to be quite unlike that of living colobines, though he considered *Rhinocolobus* to be more in line with living forms.

In cercopithecoids, the proximal surface of the greater tuberosity provides an area for the insertion of the supraspinatus muscle, which takes its origin from the scapula. As part of the rotator cuff complex, supraspinatus acts as a stabilizer of the shoulder joint: it helps to keep the head of the humerus in close approximation with the glenoid fossa. It also functions as a flexor of the humerus on the scapula, protracting the forelimb in an anterior direction in the parasagittal plane (see Birchette, 1982). Jolly (1965, 1967) refers to such motion as extension, but I will follow Birchette and use that term to refer to retraction of the forelimb in a posterior (rearward) direction. All

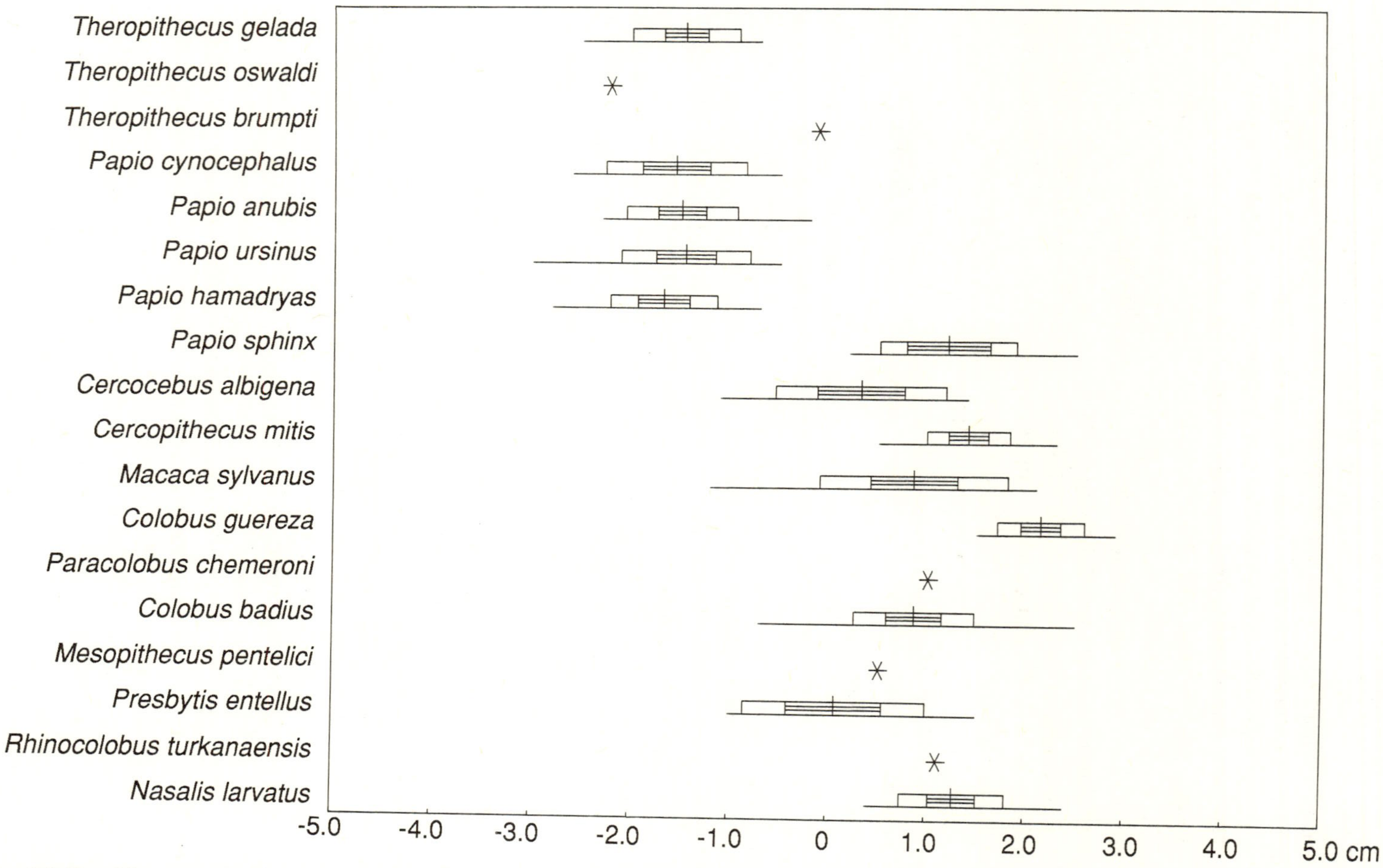

Figure 36. Graphic representation of the mean, two standard errors of the mean, one standard deviation, and total range for the greater tuberosity projection of the humerus of species in the comparative sample. The scale is in centimeters.

of the open-country-adapted papionins have a broad and proximally projecting greater tuberosity (Fig. 36). Jolly (1967:29) notes that "the expansion of the greater tuberosity in baboons raises the insertion of *M. supraspinatus*, thus increasing its moment about the joint." This change increases the mechanical advantage of supraspinatus by lengthening the perpendicular distance between its axis of rotation and the primary vector of muscle action (Birchette, 1982). As a flexor and protractor of the forelimb, supraspinatus acts together with the deltoid during the recovery phase of walking and running activities characteristic of the large-bodied, open-country papionins. In a ground-living quadrupedal primate, it would also act to resist passive extension at the glenohumeral joint during periods of little or no movement. On the opposite side of the scale, in the arboreal colobines, which have a small non-projecting greater tuberosity, the moment arm of supraspinatus is reduced, resulting in fast but weak flexion of the humerus in the recovery phase—a necessary adaptation for increased agility (Jolly, 1967).

The glenoid cavity shape index (Fig. 37) and the coraco-acromial projection index (Fig. 38) are both based on variables from the scapula. Fig. 37 expresses the depth of the glenoid fossa as a function of its length. The large-bodied papionins form a group characterized by a relatively long and deepened glenoid fossa. The other cercopithecines and the colobines, which have slightly less elongate and less deep glenoid fossae, are separated from the large-bodied papionins with respect to mean values, but their ranges overlap. The fossil colobines *Paracolobus* and *Rhinocolobus* both fall well within the range of extant colobines, with the former lying nearest the mean of *Colobus badius*.

The separation of taxa on the basis of the glenoid cavity shape index is minimal. Therefore, ascribing a functional significance to this index is not clear-cut. One possible explanation is that the elongate and deep glenoid fossa of the large-bodied papionins acts together with their oval-shaped humeral head to restrict or limit forelimb motion to the parasagittal plane. The glenoid of the cercopithecoid scapula faces ventrally and is oriented parallel to the body midline. Its shape in the large-bodied papionins very likely facilitates movement in an anterior/posterior direction while limiting medial/lateral (abduction/adduction) excursions of the forelimb. This more limited range of movement would be quite advantageous to a cursorially adapted terrestrial primate.

The coraco-acromial projection index (Fig. 38) measures the ventral projection of the acromion and coracoid past the glenoid fossa and charts it as a function of glenoid length. In an earlier study of the osteometrics of the anthropoid scapula, this index proved very useful in a systematic sorting of primate species (Ciochon and Corruccini, 1976). Fig. 38 shows a pattern often seen in previous univariate plots. Basically, three groups are distinguishable: the large-bodied papionins form one group, separated from a wholly colobine group; the small-bodied Papionini and Cercopithecini occupy an intermediate position. The single exception to this pattern is *Presbytis entellus*, which has the largest range of any species in the sample, overlapping all three groups. The only extinct species complete enough to be included in this plot is

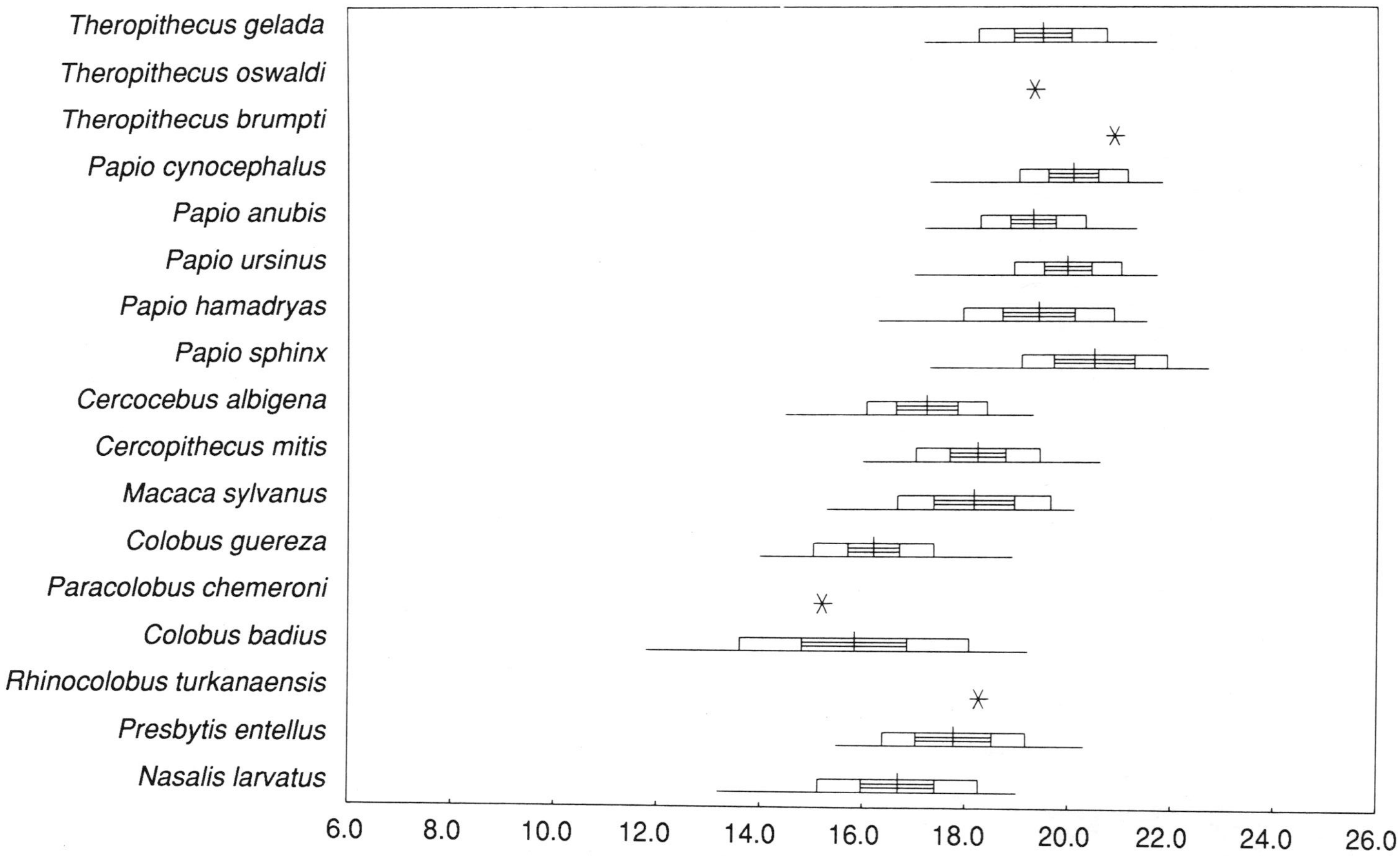

Figure 37. Graphic representation of the mean, two standard errors of the mean, one standard deviation, and total range for the glenoid cavity shape index of species in the comparative sample. The glenoid cavity shape index is calculated as follows: Glenoid Cavity Depth X 100/ Glenoid Cavity Length.

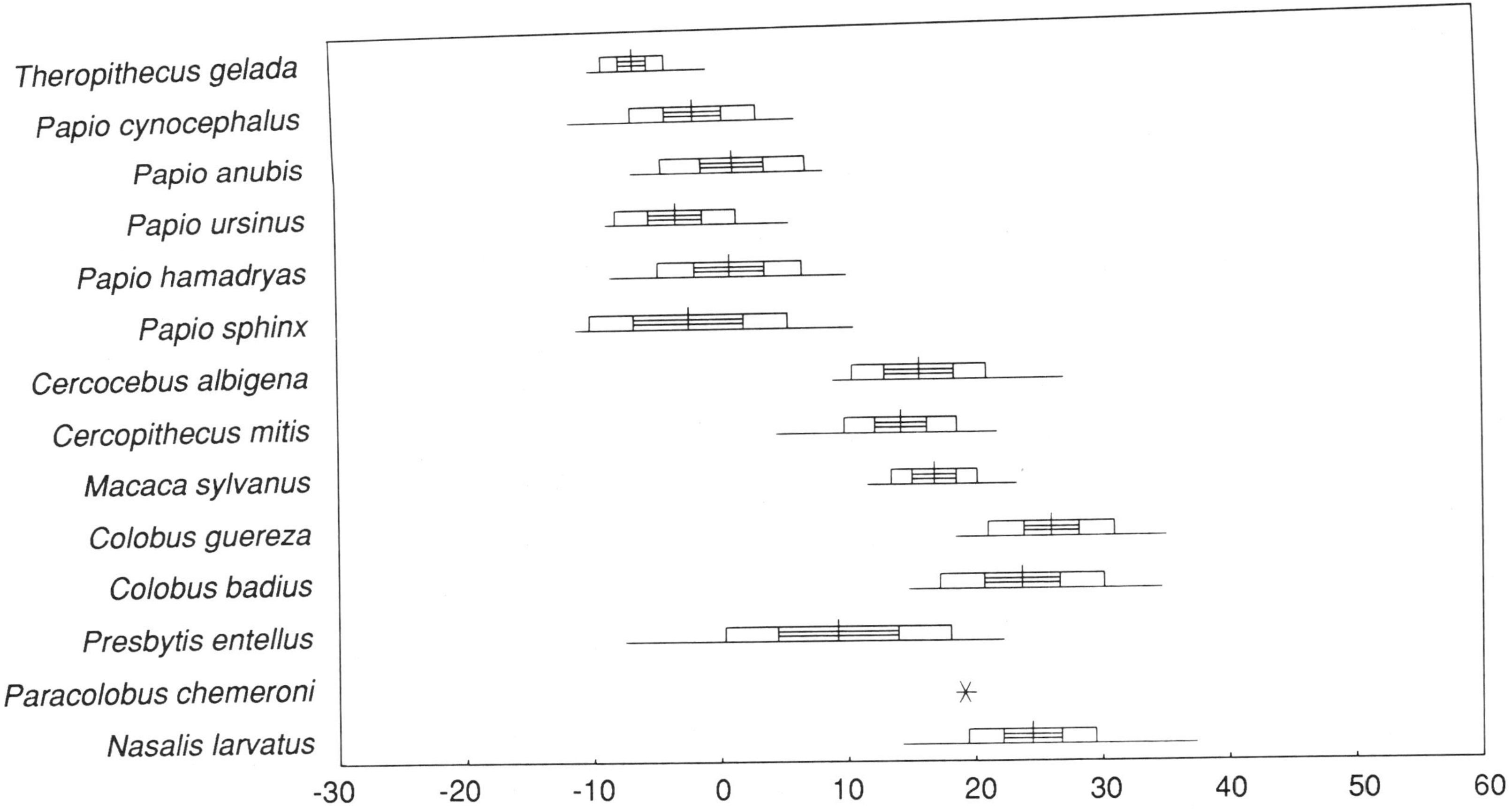

Figure 38. Graphic representation of the mean, two standard errors of the mean, one standard deviation, and total range for the coraco-acromial projection index of species in the comparative sample. The coraco-acromial projection index is calculated as follows: Coraco-Acromial Projection X 100/Glenoid Cavity Length.

Paracolobus, which falls at the upper range of the African colobine species and *Nasalis*.

The coraco-acromial projection index was originally designed to quantify the distance the acromion and coracoid project past the glenoid in the Hominoidea (apes and humans), and to compare this projection with that seen in a sampling of Old World and New World monkeys. All hominoids have a similar morphology of the gleno-humeral joint, related to their descent from a brachiating ancestor (Corruccini and Ciochon, 1976). Their ability to brachiate is directly related to an exceedingly wide range of potential movement at the shoulder joint (termed circumduction), which is associated with a projecting acromion and coracoid process, among other features.

One functional interpretation of this projection in hominoids is that it provides a greater mechanical advantage for the major forelimb abductor, deltoid, which takes its origin from the acromion. Hominoids have a mean range of 47 to 89 for the coraco-acromial projection index (Ciochon and Corruccini, 1977a), far surpassing that of any Old World monkeys. However, Fig. 38 reveals that most colobines and the more arboreal small papionins and *Cercopithecus* all have a more projecting acromion and coracoid than the terrestrial large-bodied papionins. In the latter group there is actually an absence of projection in more than half the individuals of each species, resulting in negative values for the index. In summary, the coraco-acromial projection index in cercopithecoids could also be a measure of the more advantageous placement for the abductors of the forelimb across the gleno-humeral joint, with colobines displaying the greatest range of abduction and large-bodied papionins the least. This index could also be an indicator of arboreality vs. terrestriality, since the one colobine species that overlaps nearly the entire range of the large-bodied papionins, *Presbytis entellus*, is also well known as the most terrestrially adapted of the living Colobinae.

Overview of Univariate Results

The preceding univariate analyses provide an overview of the patterns of variation seen in a selection of variables for the comparative sample of this study. This survey of some 30 variables, in raw form or with the use of an index, displays the systematic/phyletic and functional/locomotor diversity present in this sample. Specifically, a systematic sorting of colobines and cercopithecines is evident in many of the variables. Furthermore, within the cercopithecines, the large-bodied papionins often form a group, and the extinct and extant species of *Theropithecus* are even sometimes distinguishable as a subgroup. In general, the known fossil species usually fall within the range of living species, but occasionally deviate from the extant members of the sample. These results lend a good measure of confidence to the analysis of the 234 unknown fossil specimens presented in Chapter 7.

The interpretability of the univariate patterns of variation presented above also justifies more powerful statistical manipulations of the data base. Many of the preceding univariate plots displayed similar trends, but in a number of cases the same species were sorted in different manners. The application of multivariate analysis at this stage should allow a determination of how many distinctive variables are present and how they contribute to the final functional and phyletic sorting of the comparative sample.

6

RESULTS: MULTIVARIATE ANALYSIS OF THE COMPARATIVE SAMPLE

The results of the multivariate analysis for the entire comparative sample are presented in this chapter. The 85 continuous variables for the 20 extant and extinct species in the sample are analyzed in a series of 14 separate multivariate analyses, performed on: 1) the entire forelimb, 2) the shoulder joint, 3) the scapula, 4) the proximal humerus, 5) the elbow joint, 6) the distal humerus, 7) the proximal ulna, 8) the proximal radius, 9) the wrist joint, 10) the distal ulna, and 11) the distal radius. The primary multivariate method used for each of these anatomical units is principal components analysis (PCA). In addition, discriminate functions analysis (DFA) is also performed on 12) the shoulder, 13) the elbow, and 14) the wrist joint.

Results of the Multivariate Analysis

The results of the multivariate analyses presented below are each listed under a separate anatomical heading. For each analysis a figure is presented which plots principal component 1 (PC 1) vs. principal component 2 (PC 2), together with a table indicating vector loadings for all the measurements in the analysis. These vector loadings (or "component-loadings," as they are sometimes termed) express the degree of correlation between each measurement and the component. In essence, this is a measure of the relative influence that each variable exerts on the overall dispersion of species in the plot. The vector loadings have then been converted to correlation coefficients. The variables with correlation coefficients of +.700 to +.999 and –.700 to –.999 are those with 50% or more of the shared variance (see Blackith and Reyment, 1971). Since these variables account for most of the dispersion of species in the plots, they will be the ones most thoroughly discussed. Each of the 11 tables accompanying the plots also contains the correlation coefficients for the third and fourth components, as well as the percentage of contribution for the first four components. For the shoulder, elbow, and wrist joint, plots of discriminant axis 1 vs. discriminant axis 2 are also presented. This allows for a direct comparison of the results from the two primary multivariate techniques used in this study.

The Forelimb

All 85 continuous variables of the cercopithecoid forelimb were analyzed together in an effort to determine overall relationships in the extant sample and to discover which characters of the forelimb were the most effective for sorting subgroupings of the Cercopithecoidea. Fig. 39 presents the mean values for the extant sample, PC 1 vs. PC 2, based on an analysis of the entire forelimb. Note that PC 1, accounting for 37.4% of the variance, sorts colobines from cercopithecines, with *Nasalis* occupying the extreme colobine pole and *Papio ursinus* the extreme cercopithecine pole. No real pattern is discernible within the colobines, except to note that *Presbytis entellus* is the most cercopithecine-like of the colobines. Within the cercopithecines, the large-bodied papionins fall together along the extreme cercopithecine pole. The small-bodied papionins, along with *Cercopithecus* and *Papio sphinx*, occupy the intermediate position between the large-bodied Papionini and the Colobinae. Such a pattern is similar to that seen in many of the univariate plots presented in the preceding section.

Except for the separation of *Theropithecus* from the other large-bodied papionins, no significant patterns of taxa are evident along PC 2, which represents 15.3% of the sample's variance. However, when PC 3, representing 10.7% of the variance, is plotted against PC 1, an interesting, interpretable sorting of the taxa results (see Fig. 40), with three distinct groups in evidence. Colobines, large-bodied papionins, and small-bodied papionins, together with *Cercopithecus* and *Papio sphinx*, each form separate clusters. Such an easily interpretable pattern resulting from an analysis of the entire forelimb justifies further multivariate manipulations of the data set and sets the stage for assignment of the unknown fossil specimens, presented in the following chapter. In these assignments, the positions of taxa along the *first six components* will be utilized simultaneously. It is not possible to visually display more than two components, as in Figs. 39-40, but surely if the results are this interpretable when two components are plotted, the results should be even more definitive when the first six components are used.

Table 4 presents vector loadings for the first four components in the analysis of the entire forelimb. Discussing only those features with 50% or more shared variance for PC 1, it can be seen that 33 variables have values above +.7 or below –.7: 10 from the ulna, 8 from the radius, 10 from the humerus, and 5 from the scapula. Such an even spread of significant variables across all elements of the forelimb demonstrates that fundamental differences separating colobines from cercopithecines should be recognizable in all portions of the forelimb. This is an important finding, since many of the subsequent analyses deal only with a single element or portion of one.

An interpretation of the vector loadings in Table 4 indicates that the ulna of colobines is characterized by significantly greater radial notch heights (2)*, ulnar head articular surface widths (16), and degrees of midshaft curvature (23). In cercopithecines, on the other hand, the ulna is characterized by significantly greater

*The number of each variable discussed appears in parentheses throughout this chapter for easy cross-reference to the numbered variables listed in the tables of vector-loadings.

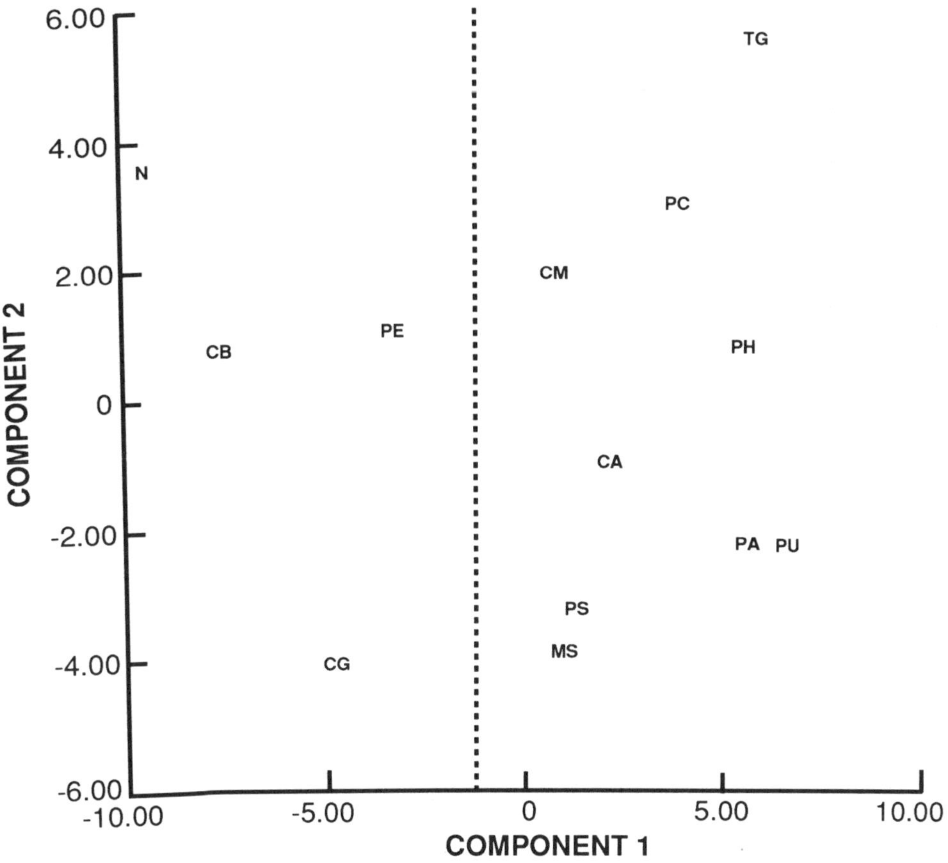

Figure 39. Plot of PC 1 vs. PC 2 for the extant sample, based on analysis of 85 variables of the cercopithecoid forelimb. PC 1 accounts for 37.4% of the total variance and PC 2 accounts for 15.3%. Vertical dashed line demarcates the separation of colobines (left) and cercopithecines (right). For key to symbols see "Taxon Abbrevations." The vector loadings for these components are presented in Table 4.

radial notch widths (1), trochlear notch/olecranon lengths (6), and widths (7), distal trochlear notch widths (10), coranoid process projections (11), trochlear/radial notch combined widths (13), and trochlear notch posterior widths (25).

The radius of colobines is characterized by significantly greater maximum medio-lateral and minimum antero-posterior head diameters (27 and 28), head maximum articular surface widths (29), maximum radial neck lengths (33), bicipital tuberosity widths (34), and ulnar notch articular facet widths (43). In cercopithecines, however, the radius is characterized by significantly greater extensor digitorum groove widths (42) and midshaft medio-lateral diameters (45).

The humerus of colobines has significantly greater proximal projections of the humeral head above the greater tuberosity (54), greater distal capitular and trochlear

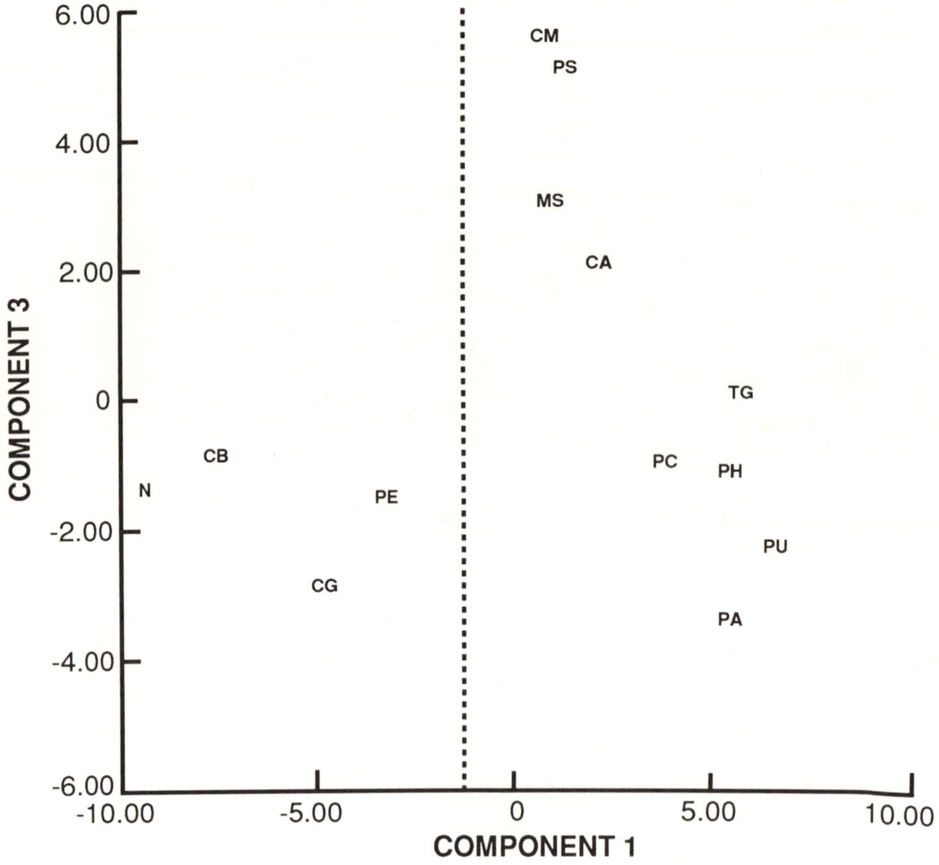

Figure 40. Plot of PC 1 vs. PC 3 for the extant sample, based on analysis of 85 variables of the cercopithecoid forelimb. PC 1 accounts for 37.4% of the total variance and PC 3 accounts for 10.7%. Vertical dashed line demarcates the separation of colobines (left) and cercopithecines (right). The vector loadings for these components are presented in Table 4.

gutter widths (62 and 70), and deeper olecranon fossae (66). In cercopithecines the humerus has significantly deeper bicipital grooves (53), and greater distal projections of the medial trochlea (60), distal trochlea widths (61), medial epicondyle/trochlear projections (65), and projections of the olecranon fossa lateral margin (67), and more pronounced midshaft curvatures (73).

The scapula of colobines is characterized by significantly longer coracoid processes (75) and greater projections of the coracoid and acromion past the glenoid (80). The cercopithecine scapula features significantly greater distances between the infraglenoid tubercle and the coracoid (84) and greater glenoid cavity depths (86) and scapula morphological lengths (87).

The most significant feature on PC 2 is the separation of *Theropithecus gelada* from the other cercopithecines and, to a lesser extent, the separation of *Nasalis* from

TABLE 4. Vector loadings for PC 1 through 4 for the cercopithecoid entire forelimb. Loadings are converted to correlation coefficients between each measurement and the component.

	Measurement	*Correlation Coefficients*			
		1	*2*	*3*	*4*
1	Radial Notch W	.709	.256	-.262	.200
2	Radial Notch H	-.799	.169	.118	.074
3	Troch Notch Med H	.583	.580	-.349	.244
4	Troch Notch Mid H	-.465	.620	-.426	.317
5	Coronoid/Olec H	.540	.227	-.474	.620
6	Troch Notch/Olec L	.856	.122	-.027	.381
7	Troch Notch/Olec W	.932	.035	-.019	.059
8	Prox Troch Notch W	-.355	-.510	-.484	.312
9	Mid Troch Notch W	-.351	-.604	-.045	.471
10	Dis Troch Notch W	.840	-.081	-.303	-.326
11	Coronoid Proc Proj	.701	-.424	.020	.299
12	Troch Notch Dp	-.550	.410	-.372	.358
13	Troch/Rad Notch W	.887	.089	-.222	.329
14	Styloid Process L	-.130	.474	-.062	.447
15	Styloid Process W	-.413	-.438	-.584	-.338
16	Ulna Head Artic W	-.846	.182	.307	-.229
17	Exten Carp Grv W	-.586	-.412	.236	.246
18	A-P Dis Ulna W	-.220	-.415	-.451	-.442
19	M-L Dis Ulna W	-.566	.025	-.487	-.047
22	Midshaft A-P Dia	-.021	-.830	.087	-.088
23	Midshaft Curv	-.801	-.284	.028	.119
24	Rad Notch Pos W	-.603	-.703	.104	-.088
25	Troch Notch Pos W	.716	-.371	.291	.203
26	Ulna L	-.536	.705	.017	-.051
27	Max M-L Head Dia	-.913	-.088	-.202	-.172
28	Min A-P Head Dia	-.851	.233	.158	-.350
29	Head Max Artic W	-.791	.410	.151	.052
30	Head Min Artic W	-.681	-.129	.142	.198
31	M-L Neck W	.313	.440	-.508	-.163
32	A-P Neck W	-.545	-.029	.265	-.482
33	Max Neck L	-.843	-.095	.261	.055
34	Bicipital Tuber W	-.752	-.035	.371	-.343
35	Styloid Process L	-.188	-.430	.466	-.280
36	Dis Artic Surf L	.493	.356	.189	-.303
37	Lunate Surface W	-.136	.766	.087	-.118

(continued on next page)

TABLE 4 Continued.

	Measurement	*Correlation Coefficients*			
		1	*2*	*3*	*4*
38	Max Dist Rad W	-.021	-.049	-.502	-.706
39	Scaphoid Surf W	.368	.060	-.805	-.304
40	Lister Tub Place	-.361	.197	.298	-.565
41	Lat Tub Dist Proj	.684	.322	.265	.455
42	Exten Digit Grv W	.769	-.204	-.041	-.133
43	Ulna Notch Fac W	-.892	.227	.053	-.080
44	Lunate Ridge Proj	-.597	.685	.246	-.083
45	Midshaft M-L Dia	.825	-.095	.448	-.199
46	Midshaft Curv	-.661	.180	.081	.369
47	Radius L	-.563	.704	.117	-.090
48	A-P Head Dia	.535	.360	-.197	-.177
49	M-L Head Dia	.056	-.337	.286	-.485
50	Head Max Prox Proj	-.446	.432	.065	-.186
51	Head Artic Sur W	.486	-.002	.448	-.162
52	Bicipital Grv W	.392	-.265	.560	.479
53	Bicipital Grv Dp	.811	.136	-.428	-.139
54	Grt Tub Prox Proj	-.704	-.141	.457	.417
55	Great Tuber Max W	.022	-.284	-.506	-.175
56	Less Tuber Max W	.687	.076	.155	.018
57	Less Tuber Place	.125	-.099	.177	.697
58	Biepicondylar W	-.220	-.487	-.378	-.704
59	M-L Artic Surf W	-.449	-.110	.093	-.591
60	M Troch Dist Proj	.810	-.134	.400	.227
61	Dist Trochlear W	.730	-.278	-.232	-.374
62	Dist Capitular W	-.785	.251	-.183	-.148
63	A-P Lat Troch Dia	-.552	.637	-.183	.384
64	Max Capitular H	.532	.630	-.435	.053
65	M Epicond/Troch Proj	.809	-.103	.302	-.162
66	Olecran Fossa Dp	-.706	.581	.268	-.172
67	Olec Lat Margin Proj	.889	-.066	-.054	-.307
68	Olecran Fossa W	-.319	-.216	-.581	-.063
69	Post Troch Artic W	-.647	-.376	-.396	.305
70	Troch Gutter W	-.736	-.229	-.412	.365
71	Midshaft M-L Dia	.189	-.538	.604	-.339
72	Delto Pect Crest L	-.248	.720	-.124	-.238
73	Midshaft Curv	.819	-.004	-.101	.246

(continued on next page)

TABLE 4 Continued.

	Measurement	Correlation Coefficients			
		1	*2*	*3*	*4*
74	Humerus L	-.697	.510	.057	-.082
75	Coracoid Proc L	-.761	.172	-.490	-.132
76	Coracoid Max W	-.240	-.512	-.169	-.618
77	Acromion W	.534	.430	-.153	-.365
78	Acromion L	.111	-.422	.240	.128
79	Grt Scap Notch Dp	.132	-.201	-.783	.149
80	Coraco-Acrom Proj	-.794	-.064	.215	.413
81	Supraglen Tub L	.138	.717	.294	.160
82	Glenoid Cavity W	.451	.573	.348	-.175
83	Glenoid Cavity L	.364	.561	.242	-.514
84	Infraglen/Corac Dist	.768	.095	.019	-.458
85	Infraglen/Acrom Dist	.533	.574	.154	-.429
86	Glenoid Cav Dp	.785	.352	.395	-.208
87	Scap Morphol L	.960	.135	-.153	.043
	Percentage of Variance	37.4%	15.3%	10.7%	10.4%

the other colobines. The vector loadings in Table 4 indicate that the following anatomical characters played a factor in these separations: greater midshaft antero-posterior diameter of the ulna (22), lengthened ulna (26), greater lunate articular surface width of the radius (37), lengthened radius (47), longer delto-pectoral crest of the humerus (72), and a lengthened supraglenoid tubercle of the scapula (81). The significance of a lengthened delto-pectoral crest for *T. gelada*, among other features, is discussed in the univariate analysis in Chapter 5.

The functional significance of many of the above anatomical features is discussed elsewhere in this study. The main point to be emphasized here is that multivariate analysis of the entire cercopithecoid forelimb produced a significant sorting of colobines and cercopithecines which can be related directly to specific morphological and functional features of the forelimb that characterize these two groups. Size is apparently not a significant factor in this multivariate analysis (or, for that matter, in subsequent analyses). The colobine with the largest body size, *Nasalis*, is at the opposite pole from the largest cercopithecine, *Papio ursinus* (see Fig. 39). Within the various subgroupings of the cercopithecoids (see Figs. 39-40), both small and large members cluster together. These results reinforce the view that the size/shape correction applied to this data set (see Chapter 4) functions properly.

Shoulder Joint

Results of the PCA of 22 variables describing the cercopithecoid shoulder joint are presented in Fig. 41. PC 1, accounting for 42% of the variance, provides a complete sorting of colobines and cercopithecines, with no overlap between the groups. *Colobus guereza* occupies the extreme colobine pole and *T. gelada* the extreme cercopithecine pole. Based on PC 1, *Nasalis* appears the most cercopithecine-like of the colobines and *Macaca sylvanus* is the most colobine-like of the cercopithecines. On PC 1, *Cercopithecus* and *Cercocebus* fall very near *Macaca*, while *Papio sphinx* is sorted with the other large-bodied papionins. This is in contrast to the position that *P. sphinx* occupied in plots of the entire forelimb (Figs. 39-40). As was seen there, PC 2, here accounting for 17.4% of the variance, separates *Nasalis* from the other colobines and *Theropithecus* from the other cercopithecines.

Table 5 presents vector loadings* for the first four components in the PCA of the shoulder joint. Referring only to those variables with 50% or more of the shared variance for PC 1, 10 features can be identified that contribute significantly to the sorting of colobines and cercopithecines. Colobines are characterized by a significantly greater projection of the proximal humeral head above (proximal to) the level of the greater tuberosity (54), and a scapula with greater projection of the coracoid process and acromion past the glenoid (80). In cercopithecines, there is a significantly greater diameter of the proximal humeral head in the antero-posterior plane (48), a deeper bicipital groove (53), and a wider lesser tuberosity (56). The scapula of cercopithecines is characterized by a glenoid cavity which is wider (82), longer (83), and deeper (86), and with greater infraglenoid tubercle/coracoid and infraglenoid tubercle/acromion distances (84 and 85). The vector loadings for PC 2 show that the lengthened coracoid process of the scapula (75) is a primary factor contributing to the separation of *Nasalis* and *Theropithecus*.

Fig. 42 presents the results of a DFA of the same 22 variables describing the cercopithecoid shoulder joint. Since DFA along with PCA will be used in the assignments of the unknown fossil specimens, it is employed here to test its concordance with PCA, the primary method used in this study. Both methods should yield basically similar results, though the spread of taxa along axis 2 might differ from that seen along component 2 because axes in DFA are not necessarily constructed perpendicular to one another, as required in PCA (see Chapter 4 for further discussion). Discriminant axis 1 in Fig. 42 separates colobines from cercopithecines in a pattern very similar to that seen in Fig. 41. One significant difference is that *Presbytis entellus* replaces *Nasalis* as the most cercopithecine-like of the colobines. The spread of taxa along discriminant axis 2 is the polar reverse of that seen in Fig. 41, though the interrelationships among taxa remain much the same. The large-bodied papionins form a group at the extreme cercopithecine pole, with *Papio sphinx* an outlier, while the small-bodied papionins and *Cercopithecus* are grouped nearer the center of the plot.

---

*Each variable appears by number and by name in the tables of vector loadings. When a variable is discussed in the text, its number also appears in parentheses to aid with its location in the tables.

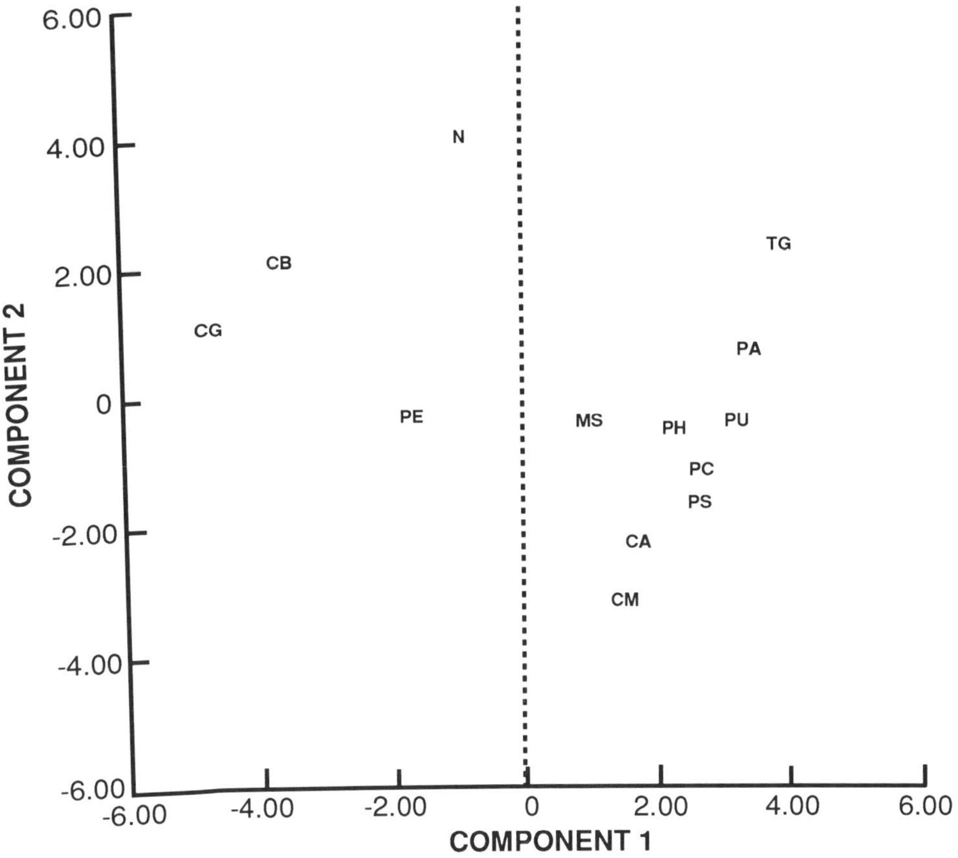

Figure 41. Plot of PC 1 vs. PC 2 for the extant sample, based on analysis of 22 variables of the cercopithecoid shoulder joint. PC 1 accounts for 42.0% of the total variance and PC 2 accounts for 17.4%. Vertical dashed line demarcates the separation of colobines (left) and cercopithecines (right). The vector loadings for these components are presented in Table 5.

Scapula

Results of the PCA of 12 variables describing the cercopithecoid scapula are presented in Fig. 43. In addition to the species in the extant sample, this analysis includes three species from the extinct comparative sample: *Paracolobus, Theropithecus oswaldi,* and *T. brumpti*. Because of the fragmentary and largely unassociated nature of the fossil cercopithecoid postcranial skeleton, no extinct forms could be included in the previous analyses. Their inclusion here and in following analyses is further justified on the basis that these same samples will be used in the series of multivariate analyses run for the purpose of assigning the unknown fossil specimens (see Chapter 7).

In Fig. 43, PC 1, accounting for 42.6% of the variance, separates cercopithecines

TABLE 5. Vector loadings for PC 1 through 4 for the cercopithecoid shoulder joint. Loadings are converted to correlation coefficients between each measurement and the component.

	Measurement	*Correlation Coefficients*			
		1	*2*	*3*	*4*
48	A-P Head Dia	.898	.214	-.036	-.161
49	M-L Head Dia	.520	.168	-.376	.561
50	Head Max Prox Proj	.311	.663	.327	-.005
51	Head Artic Sur W	.520	-.550	-.105	.495
52	Bicipital Grv W	.130	-.799	-.343	.172
53	Bicipital Grv Dp	.803	-.011	-.148	-.292
54	Grt Tub Prox Proj	-.933	-.075	.176	.046
55	Great Tuber Max W	.442	.542	-.399	-.105
56	Less Tuber Max W	.702	-.288	-.210	-.405
57	Less Tuber Place	-.217	-.632	-.300	.165
75	Coracoid Proc L	-.277	.780	.183	.262
76	Coracoid Max W	.265	.621	-.444	.475
77	Acromion W	.681	.087	.228	.168
78	Acromion L	.091	-.103	-.584	-.474
79	Grt Scap Notch Dp	.230	.515	-.529	-.289
80	Coraco-Acrom Proj	-.957	.123	.158	-.116
81	Supraglen Tub L	.403	-.204	.655	-.344
82	Glenoid Cavity W	.788	-.203	.370	.298
83	Glenoid Cavity L	.868	.206	.333	-.012
84	Infraglen/Corac Dist	.907	-.004	-.120	-.087
85	Ingraglen/Acrom Dist	.908	.088	.230	.022
86	Glenoid Cav Dp	.870	-.395	.190	.028
	Percentage of Variance	42.0%	17.4%	11.1%	8.0%

and colobines. The polar reverse in location of these two subfamily groupings, when compared to the pattern in previous figures, is simply a mathematical artifact of the statistical technique and has no bearing on the relationships of species. *Papio ursinus* lies at the extreme cercopithecine pole and *Colobus guereza* at the extreme colobine pole. *Macaca sylvanus* is the most colobine-like of the cercopithecines and *Presbytis entellus* the most cercopithecine-like colobine. Based on their positions along PC 1, all of the large-bodied papionins, including the two extinct species of *Theropithecus*, fall very near one another.

PC 2, accounting for 27.3% of the variance, sorts the three extinct species from the rest of the comparative sample, with *Paracolobus* occupying one pole and *T.*

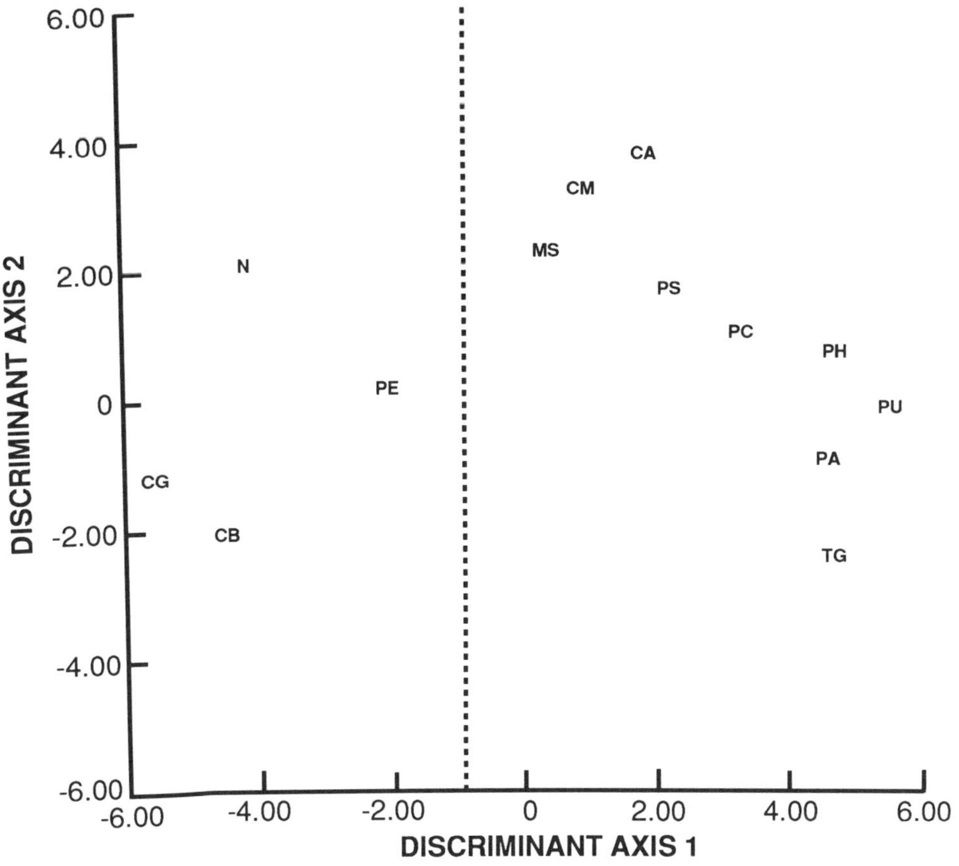

Figure 42. Plot of discriminant axes 1 vs. 2 for the extant sample, based on analysis of 22 variables of the cercopithecoid shoulder joint. Vertical dashed line demarcates the separation of colobines (left) and cercopithecines (right). Axis 1 accounts for 58.2% of the total variance and Axis 2 accounts for 12.2%.

brumpti and *T. oswaldi* the other. The significance of this sorting is that, though these extinct species can be assigned with ease to one of the two subfamilies of cercopithecoids, they do not bear a close relationship to any of the living genera within these subfamilies insofar as the scapula is concerned. PC 2 also rather curiously sorts *Papio cynocephalus* from the three baboon species and *M. sylvanus* from *Cercocebus albigena*; these patterns have not been observed in previous analyses.

Table 6 presents vector loadings for the first four components in the PCA of the scapula. The single significant feature related to the separation of colobines along PC 1 is the high degree of coraco-acromial projection (80). Cercopithecines, on the other hand, are characterized by a glenoid cavity that is both wide (82) and deep (86) and

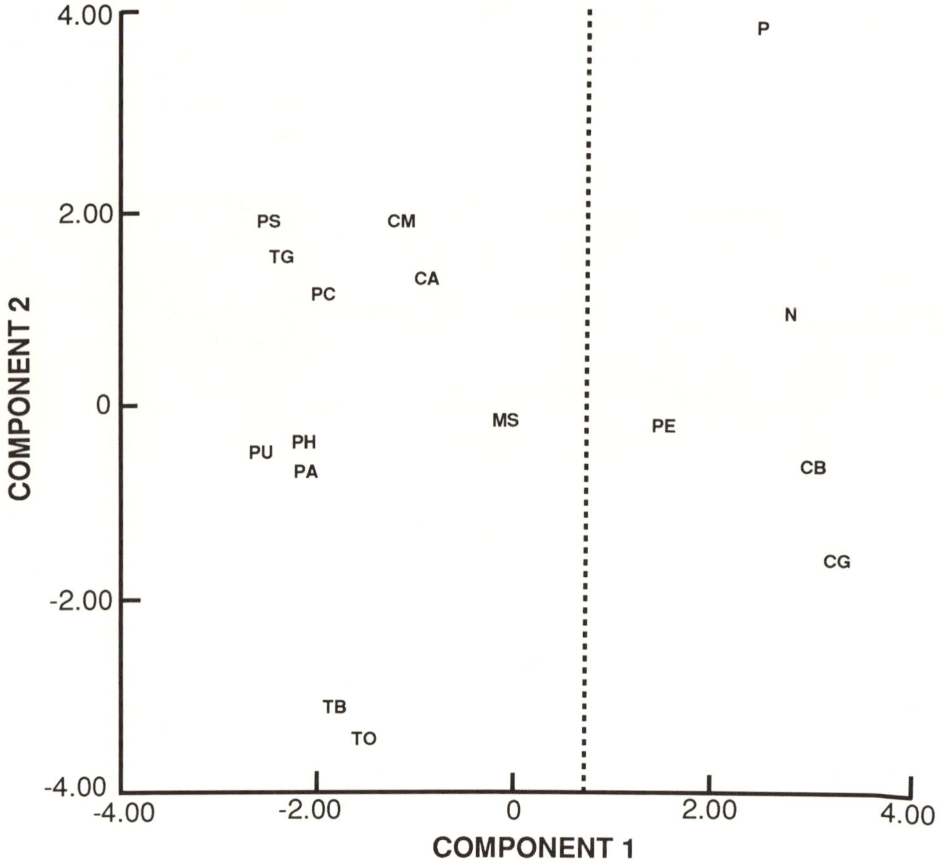

Figure 43. Plot of PC 1 vs. PC 2 for the extant sample, based on analysis of 12 variables of the cercopithecoid scapula. PC 1 accounts for 42.6% of the total variance and PC 2 accounts for 27.3%. Vertical dashed line demarcates the separation of colobines (right) and cercopithecines (left). The vector loadings for these components are presented in Table 6.

a coracoid process that is well distanced from the infraglenoid tubercle (84). The vector loadings for PC 2 indicate that those features contributing to the separation of *Paracolobus* are a significantly lengthened glenoid cavity (83), an acromion that is significantly far removed from the infraglenoid tubercle (85), and a lengthened supraglenoid tubercle (81). A major factor contributing to the separation of *T. oswaldi* and *T. brumpti* at the opposite pole of PC 2 is the presence in both these species of a significantly deepened greater scapular notch (79). All four of the anatomical features separating the fossil groups along PC 2 are significant departures from the patterns of variation seen in previous analyses of the extant sample; hence the fossil groups appear here as outliers in Fig. 43.

TABLE 6. Vector loadings for PC 1 through 4 for the cercopithecoid scapula. Loadings are converted to correlation coefficients between each measurement and the component

	Measurement	*Correlation Coefficients*			
		1	*2*	*3*	*4*
75	Coracoid Proc L	.603	-.054	.576	.504
76	Coracoid Max W	.650	.463	-.124	.312
77	Acromion W	-.651	.096	.629	-.063
78	Acromion L	-.495	.037	-.731	.398
79	Grt Scap Notch Dp	-.346	-.724	.174	.200
80	Coraco-Acrom Proj	.953	-.077	-.115	-.285
81	Supraglen Tub L	.051	.899	.093	-.016
82	Glenoid Cavity W	-.782	-.497	.125	-.010
83	Glenoid Cavity L	-.160	.945	.098	-.228
84	Infraglen/Corac Dist	-.941	.177	-.031	-.044
85	Infraglen/Acrom Dist	-.523	.759	.095	.310
86	Glenoid Cav Dp	-.948	.088	-.084	-.069
	Percentage of Variance	42.6%	27.3%	11.3%	6.5%

Proximal Humerus

Results of the PCA of 10 variables describing the cercopithecoid proximal humerus are presented in Fig. 44. This analysis includes all of the extant sample and nearly the full complement of the extinct sample. As in all previous analyses, PC 1, accounting for 45.4% of the variance, separates colobines and cercopithecines. The extreme colobine pole is occupied by *Colobus guereza* and the extreme cercopithecine pole by *Theropithecus oswaldi*. *Paracolobus* is the most cercopithecine-like colobine and *Macaca sylvanus* the most colobine-like cercopithecine. Within the cercopithecine cluster, all large-bodied papionins, both living and extinct, are grouped together. The small-bodied papionins, together with *Cercopithecus*, form another group closer to the center of the plot. The fossil species *T. oswaldi* and *T. brumpti* lie very near *T. gelada* on PC 1, though they are separated from the latter species on PC 2.

There are no apparent clusters within the Colobinae. The African Pliocene species *Paracolobus* lies nearest to *Presbytis entellus*. The fossil colobine from France, *Mesopithecus*, lies equidistant from *Colobus guereza* and *C. badius*. The East African fossil form *Rhinocolobus* is closest to *Nasalis*, but is actually quite far removed from all species in the analysis. PC 2, summarizing 23% of the variance, separates *Rhinocolobus* and *Nasalis* from the other colobines and, to a lesser extent, separates *T. brumpti* and *T. oswaldi* from the other cercopithecines. At the opposite

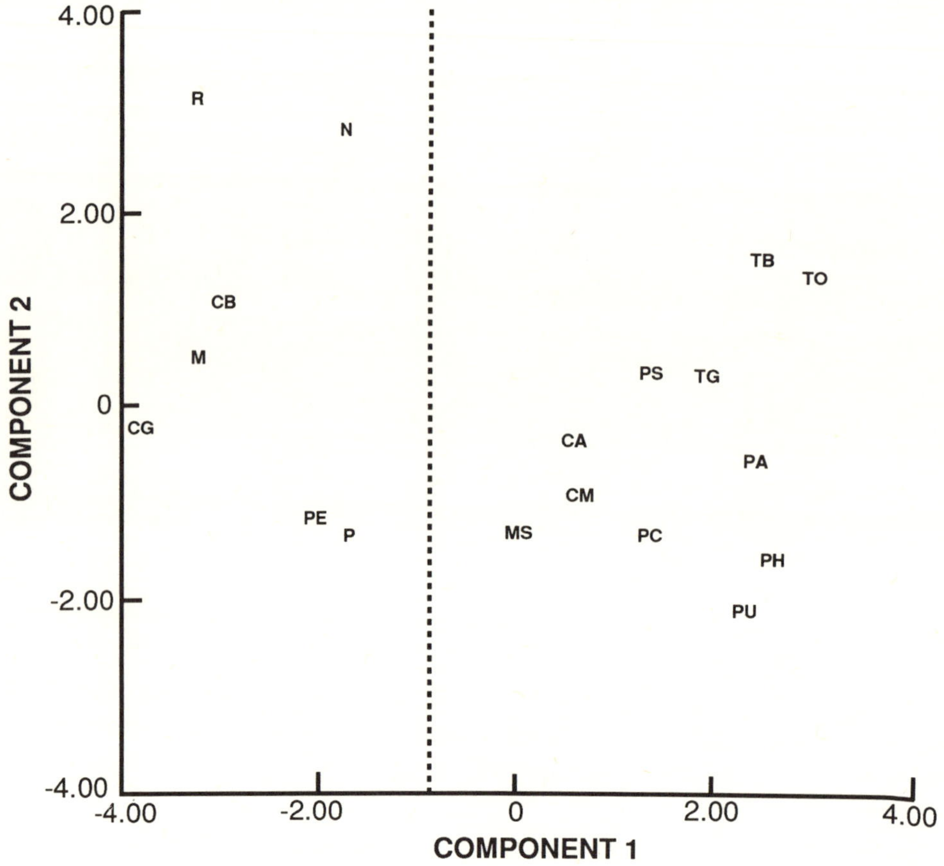

Figure 44. Plot of PC 1 vs. PC 2 for the extant sample, based on analysis of 10 variables of the cercopithecoid proximal humerus. PC 1 accounts for 45.4% of the total variance and PC 2 accounts for 23.0%. Vertical dashed line demarcates the separation of colobines (left) and cercopithecines (right). The vector loadings for these components are presented in Table 7.

pole of PC 2, *Presbytis entellus* and *Paracolobus* are somewhat separated from other colobines.

Table 7 presents vector loadings for the first four components in the PCA of the proximal humerus. The single significant feature related to the separation of colobines along PC 1 is the proximal projection of the humeral head above the plane of the greater tuberosity (54). In contrast, most cercopithecines have a greater tuberosity that projects above the humeral head. For further discussion of this feature, see Fig. 36 and Chapter 5. The significant features related to the separation of cercopithecines along PC 1 are an expanded diameter of the humeral head in the antero-posterior plane (48) and expanded maximum widths of both the greater tuberosity (55) and lesser tuberosity (56). Along PC 2, the main feature contributing

TABLE 7. Vector loadings for PC 1 through 4 for the cercopithecoid proximal humerus. Loadings are converted to correlation coefficients between each measurement and the component.

	Measurement	*Correlation Coefficients*			
		1	*2*	*3*	*4*
48	A-P Head Dia	.925	.229	.072	- .137
49	M-L Head Dia	.633	.500	-.254	.282
50	Head Max Prox Proj	.344	.364	.588	-.614
51	Head Artic Sur W	.438	-.366	.597	.502
52	Bicipital Grv W	.614	-.617	-.305	-.093
53	Bicipital Grv Dp	.581	-.351	-.573	-.273
54	Grt Tub Prox Proj	-.862	.317	-.153	-.013
55	Great Tuber Max W	.782	.547	.047	.032
56	Less Tuber Max W	.940	.036	-.053	.142
57	Less Tuber Place	.109	-.909	.274	-.137
	Percentage of Variance	45.4%	23.0%	13.0%	8.5%

to the separation of *Rhinocolobus* and *Nasalis* from the other colobines is a lesser tuberosity that is not far removed from the plane of the greater tuberosity (57).

Elbow Joint

Results of the PCA of 36 variables describing the cercopithecoid elbow joint are presented in Fig. 45. This analysis was performed only on species in the extant sample, because of the lack of fossil specimens that were both complete and associated (representing single individuals). PC 1, accounting for 47.8% of the variance, separates colobines and cercopithecines, with no overlap between the groups. *Nasalis* lies at the extreme colobine pole and *Papio anubis* at the extreme cercopithecine pole. Of the colobines, *Colobus guereza* is closest to the cercopithecine cluster, and *Papio sphinx* is closest to the colobine cluster. These two species did not occupy these positions in any of the preceding analyses. Along PC 1, all of the large-bodied papionins except *Papio sphinx* fall very near one another at the extreme cercopithecine pole. The next group, made up of the small-bodied papionins, the Cercopithecini, and *P. sphinx*, forms a nearly straight line close to the center of the plot. Within the Colobinae, *Colobus badius* and *C. guereza* are the two species most closely associated; *Nasalis* and *Presbytis entellus* are the outliers. PC 2, accounting for 18.9% of the variance, to some extent separates *T. gelada* from the other cercopithecines and serves to separate all of the colobine species about equally.

Table 8 presents vector loadings for the first four components in the PCA of the

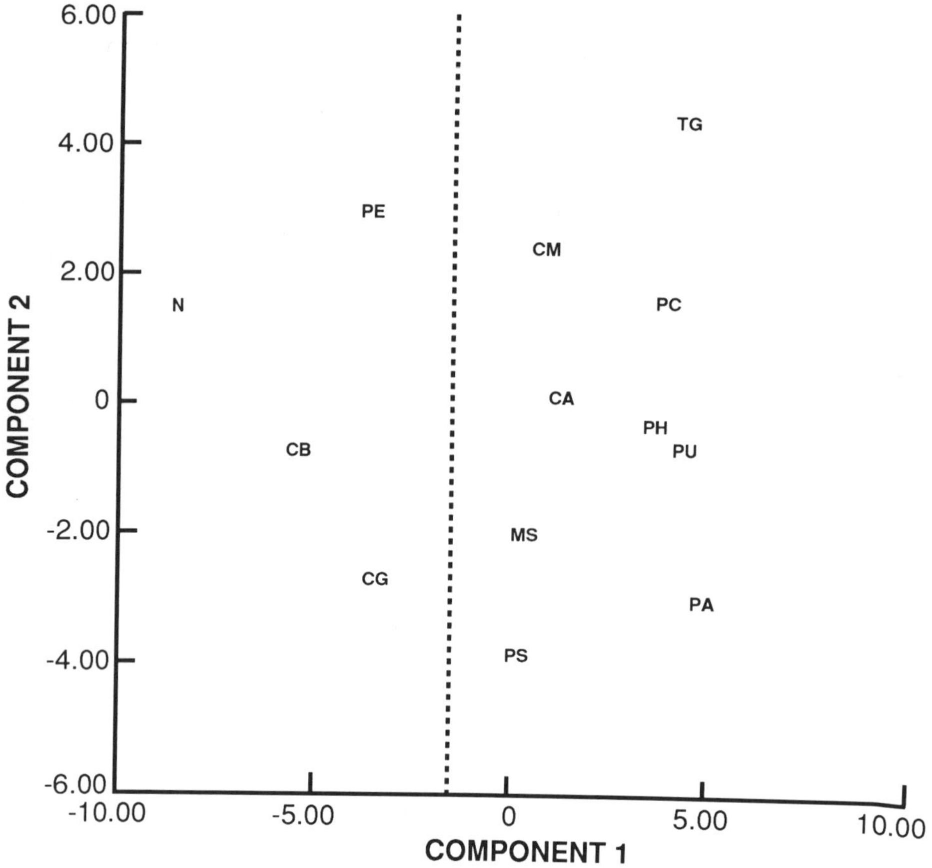

Figure 45. Plot of PC 1 vs. PC 2 for the extant and extinct species in the comparative sample, based on analysis of 36 variables of the cercopithecoid elbow joint. PC 1 accounts for 47.8% of the total variance and PC 2 accounts for 18.9%. Vertical dashed line demarcates the separation of colobines (left) and cercopithecines (right). The vector loadings for these components are presented in Table 8.

elbow joint. There are 17 variables that contribute significantly (with 50% or more of the shared variance) to the separation of colobines and cercopithecines along PC 1. Colobines have a distal humerus with a widened distal articular surface of the capitulum (62) and a greatly widened trochlear gutter (70). Cercopithecines, on the other hand, have a distal humerus with a medial trochlear ridge that projects posteriorly (60), a widened distal trochlear articular surface (61), a projecting medial trochlear ridge, when measured from the medial epicondyle (65), and a pronounced posteriorly projecting lateral margin of the olecranon fossa (67).

The vector loadings in Table 8 indicate that in colobines the proximal ulna has a radial notch articular surface that is significantly higher (2) than that seen in

TABLE 8. Vector loadings for PC 1 through 4 for the cercopithecoid elbow joint. Loadings are converted to correlation coefficients between each measurement and the component.

	Measurement	*Correlation Coefficients*			
		1	*2*	*3*	*4*
1	Radial Notch W	.679	.593	.002	-.254
2	Radial Notch H	-.787	.344	.085	-.157
3	Troch Notch Med H	.561	.771	.059	.184
4	Troch Notch Mid H	-.413	.801	-.147	.165
5	Coronoid/Olec H	.557	.602	-.479	.045
6	Troch Notch/Olec L	.833	.359	-.158	-.309
7	Troch Notch/Olec W	.897	.202	.085	-.224
8	Prox Troch Notch W	-.428	-.297	-.740	.293
9	Mid Troch Notch W	-.356	-.231	-.676	-.436
10	Dis Troch Notch W	.844	-.255	.119	.420
11	Coronoid Proc Proj	.680	-.034	-.260	-.606
12	Troch Notch Dp	-.540	.708	-.220	.168
13	Troch/Rad Notch W	.901	.321	-.223	-.116
24	Rad Notch Pos W	-.818	-.210	-.170	-.085
25	Troch Notch Pos W	-.672	-.472	-.155	-.387
27	Max M-L Head Dia	-.949	.089	.049	-.018
28	Min A-P Head Dia	-.835	.236	.442	-.041
29	Head Max Artic W	-.752	.540	.168	-.222
30	Head Min Artic W	-.665	.083	-.168	-.392
31	M-L Neck W	.222	.756	.235	.119
32	A-P Neck W	-.586	-.160	.577	-.114
33	Max Neck L	-.848	.051	.046	-.251
34	Bicipital Tuber W	-.781	-.154	.437	-.215
58	Biepicondylar W	-.365	-.649	.262	.260
59	M-L Artic Surf W	-.557	-.296	.559	.089
60	M Troch Dist Proj	.776	-.135	.021	-.379
61	Dist Trochlear W	.761	-.479	.102	.315
62	Dist Capitular W	-.813	.302	.240	.106
63	A-P Lat Troch Dia	-.507	.820	-.087	.036
64	Max Capitular H	.482	.728	.180	.401
65	M Epicond/Troch Proj	.774	-.208	.324	-.253
66	Olecran Fossa Dp	-.676	.509	.461	-.078
67	Olec Lat Margin Proj	.881	-.249	.253	.092
68	Olecran Fossa W	-.415	-.228	-.350	.711
69	Post Troch Artic W	-.679	-.089	-.566	.112
70	Troch Gutter W	-.735	.087	-.585	.046
	Percentage of Variance	47.8%	18.9%	11.2%	7.7%

cercopithecines, and a very wide posterior distance (24). The proximal ulna of cercopithecines has a widened, projecting, and posteriorly inclined olecranon process, as measured by trochlear notch/olecranon length (6) and width (7), a widened distal trochlear notch articular surface (10), and a significantly greater trochlear notch/radial notch combined maximum width (13).

The head of the proximal radius of colobines has a significantly wider diameter in both the medio-lateral (27) and antero-posterior (28) planes, and a maximally wide articular surface (29). The radius of colobines also has a significantly longer neck (33) and a widened bicipital tuberosity (34). No features of the cercopithecine proximal radius contribute significantly to the sorting of taxa along PC 1.

The variables contributing to the distribution of taxa along PC 2 for the distal humerus, and especially to the separation of *T. gelada*, are an expanded lateral diameter in the antero-posterior plane of the trochlear articular surface (63) and a capitulum that is maximally high (64) dorso-ventrally. The proximal ulna has a trochlear notch that is significantly higher on its medial side (3) and at its midline (4) and also significantly deeper (12). The significant feature for the proximal radius is an expansion of the neck width in the medio-lateral plane (31).

Fig. 46 presents the results of a DFA of the same 36 variables describing the cercopithecoid elbow joint. The distribution of taxa along discriminant axis 1 and 2 is remarkably similar to that seen in the PCA plot of the elbow joint (Fig. 45). The relative positions of all the taxa in their respective subgroupings are the same. The only apparent differences in Fig. 46 are that the colobine species are more spread out along axis 2 and the cercopithecine species are more clustered together. The position of *Nasalis* on axis 2 is also slightly different (compare Figs. 45 and 46). The striking concordance between these two different multivariate methods is an independent justification for the methodology used to assign the unknown fossil forelimb specimens in Chapter 7.

Distal Humerus

Results of the PCA of 13 variables describing the cercopithecoid distal humerus are presented in Fig. 47. In this analysis, all members of both the extant and extinct comparative sample are represented. This complete sample is due primarily to the favorable preservation of the distal end of the humerus in fossil assemblages. PC 1, accounting for 43.1% of the variance, fully separates cercopithecines and colobines. The colobine pole is on the right side of the plot in Fig. 47. *Papio anubis* lies at the extreme cercopithecine pole and *Nasalis* at the extreme colobine pole. *Cercocebus albigena* is the most colobine-like cercopithecine and *Mesopithecus* is the most cercopithecine-like colobine. All of the extant large-bodied papionins cluster together at the extreme cercopithecine pole. The small-bodied papionins and *Cercopithecus* are grouped together nearer the center of the plot. Curiously, *T. brumpti, T. oswaldi*, and European Pliocene *Paradolichopithecus geticus* fall with the small-bodied papionins on PC 1.

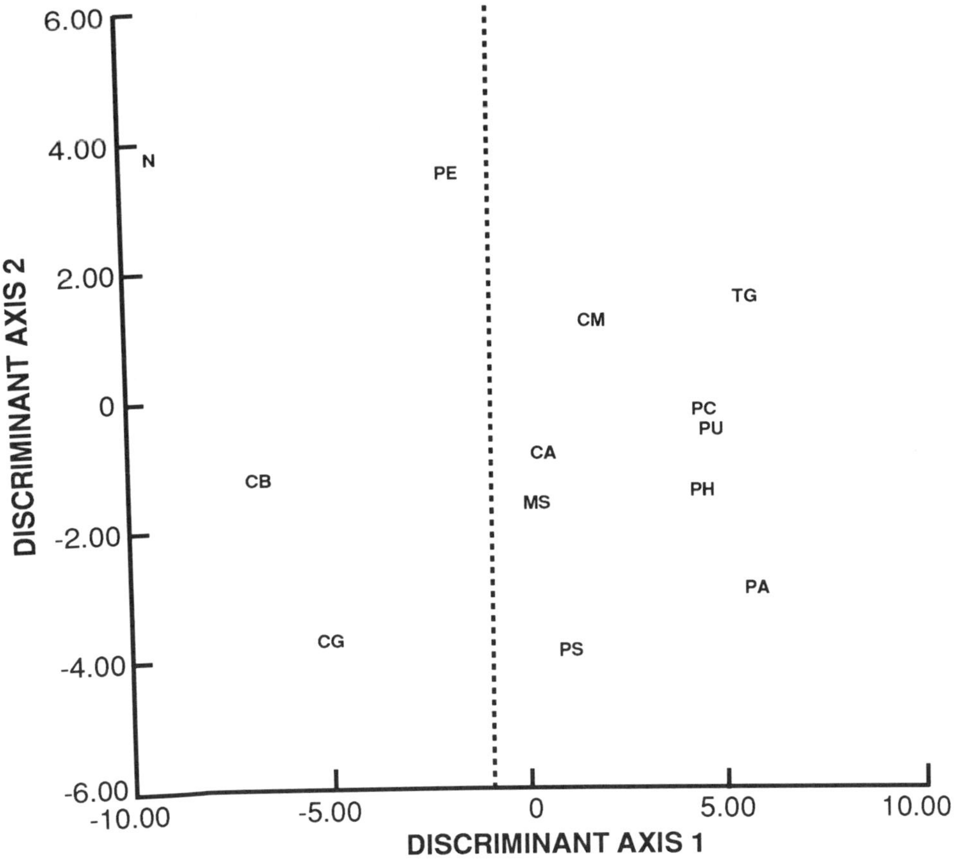

Figure 46. Plot of discriminant axes 1 vs. 2 for the extant sample, based on analysis of 36 variables of the cercopithecoid elbow joint. Vertical dashed line demarcates the separation of colobines (left) and cercopithecines (right). Axis 1 accounts for 58.1% of the total variance and Axis 2 accounts for 10.1%.

There are several interesting patterns within the Colobinae. *Rhinocolobus* is grouped with *Colobus guereza* and *C. badius*. *Presbytis entellus* lies nearest to *Paracolobus*. Finally, the Eurasian Mio-Pliocene species *Mesopithecus* and *Dolichopithecus* lie near the center of the plot, well separated from these other colobine groupings. The positioning of taxa along PC 2, which accounts for 17.9% of the variance, is not very informative. However, the colobine taxa are dispersed along PC 2 to a greater extent than are the cercopithecine taxa. This dispersed pattern of taxa may represent a greater morphological variability in the colobine distal humerus.

Table 9 presents vector loadings for the first four components in the PCA of the distal humerus. The functional significance of the features which separate colobines

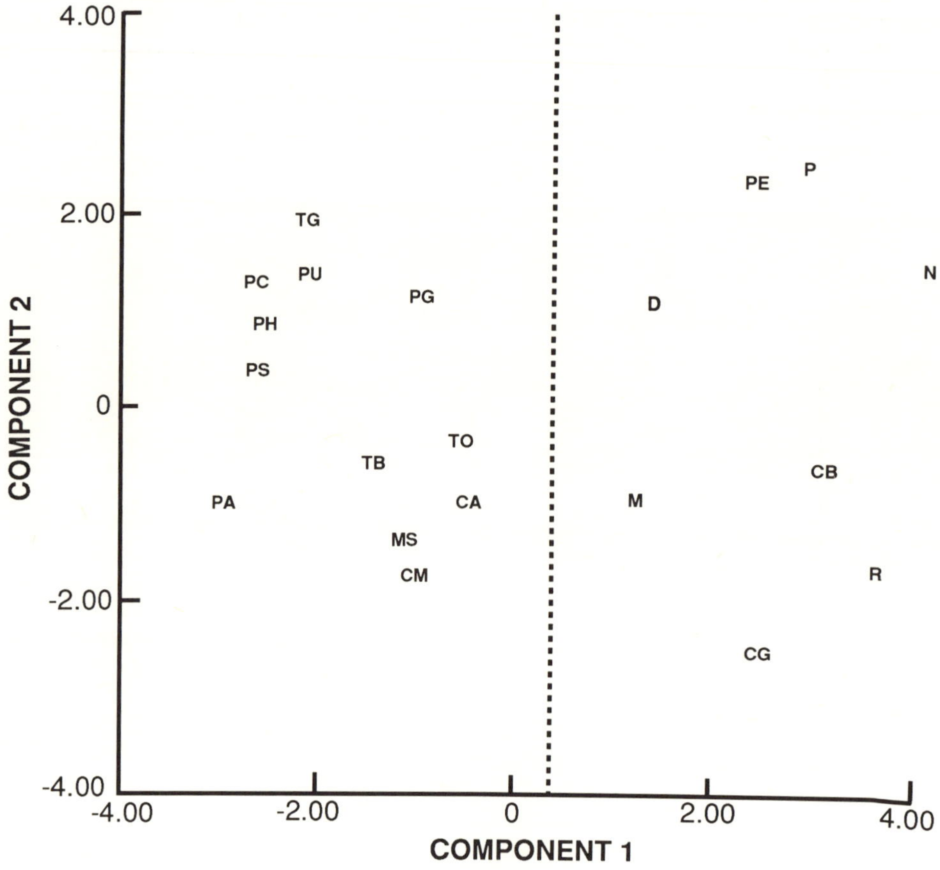

Figure 47. Plot of PC 1 vs. PC 2 for the extant and extinct species in the comparative sample, based on analysis of 13 variables of the cercopithecoid distal humerus. PC 1 accounts for 43.1% of the total variance and PC 2 accounts for 17.9%. Vertical dashed line demarcates the separation of colobines (right) and cercopithecines (left). The vector loadings for these components are presented in Table 9.

and cercopithecines (see section on the elbow joint) can be related to mobility potentials and weight-bearing adaptations at the elbow joint and is discussed further in the univariate section (see Chapter 5). Finally, the vector loadings for PC 2 show that a capitulum that is maximally wide (or high) dorso-ventrally (64) is the one significant feature that can be related to the dispersed pattern of taxa along PC 2.

Proximal Ulna

Results of the PCA of 15 variables describing the cercopithecoid proximal ulna are presented in Fig. 48. All 20 species of the comparative sample are represented. PC 1, accounting for 46.7% of the variance, separates colobines and cercopithecines,

TABLE 9. Vector loadings for PC 1 through 4 for the cercopithecoid distal humerus. Loadings are converted to correlation coefficients between each measurement and the component.

	Measurement	*Correlation Coefficients*			
		1	*2*	*3*	*4*
58	Biepicondylar W	.549	-.312	-.635	-.307
59	M-L Artic Surf W	.527	-.235	.309	-.637
60	M Troch Dist Proj	-.942	-.069	-.206	.012
61	Dist Trochlear W	.155	-.434	-.694	.111
62	Dist Capitular W	.736	.457	-.093	-.294
63	A-P Lat Troch Dia	.441	.695	-.033	.136
64	Max Capitular H	-.214	.814	-.151	.293
65	M Epicond/Troch Proj	-.750	.261	-.359	-.198
66	Olecran Fossa Dp	.267	.691	-.199	-.432
67	Olec Lat Margin Proj	-.906	-.030	-.225	-.060
68	Olecran Fossa W	.728	.111	-.180	.342
69	Post Troch Artic W	.791	-.091	-.366	.162
70	Troch Gutter W	.848	-.232	.177	.298
	Percentage of Variance	43.1%	17.9%	11.3%	9.0%

though the groups almost overlap. *Rhinocolobus* lies at the extreme colobine pole, while *T. gelada* and *Papio ursinus* occupy the extreme cercopithecine pole. *Presbytis entellus* and *Cercocebus albigena* nearly converge at the center of the plot as the most cercopithecine-like colobine and colobine-like cercopithecine, respectively. Within the Colobinae, *Mesopithecus* and *Dolichopithecus* lie very near *Presbytis entellus,* and *Paracolobus* is close to *Colobus guereza. Nasalis* is widely separated from the other colobines on PC 2, which accounts for 20% of the variance. Within the cercopithecines, the extant large-bodied papionins cluster together, with the exception of *Papio sphinx*. The extinct *Theropithecus* species *T. oswaldi* and *T. brumpti* are not grouped with *T. gelada* but instead fall nearest to *Papio sphinx,* as does *Paradolichopithecus.* The small-bodied papionins and *Cercopithecus* are very near one another on PC 1. Within the Cercopithecinae, PC 2 serves to separate the extant large-bodied papionins (minus *P. sphinx*) from the other species.

Table 10 presents vector loadings for the first four components in the PCA of the proximal ulna. In colobines, the trochlear notch is widened proximally (8), widened in the middle (9) and is quite deepened (12). The radial notch is more anteriorly placed, as indicated by its posterior width (24). The proximal ulna of cercopithecines has a widened radial notch articular surface (1), a greater maximum trochlear/radial notch combined width (13), an anteriorly placed trochlear notch, as judged from its

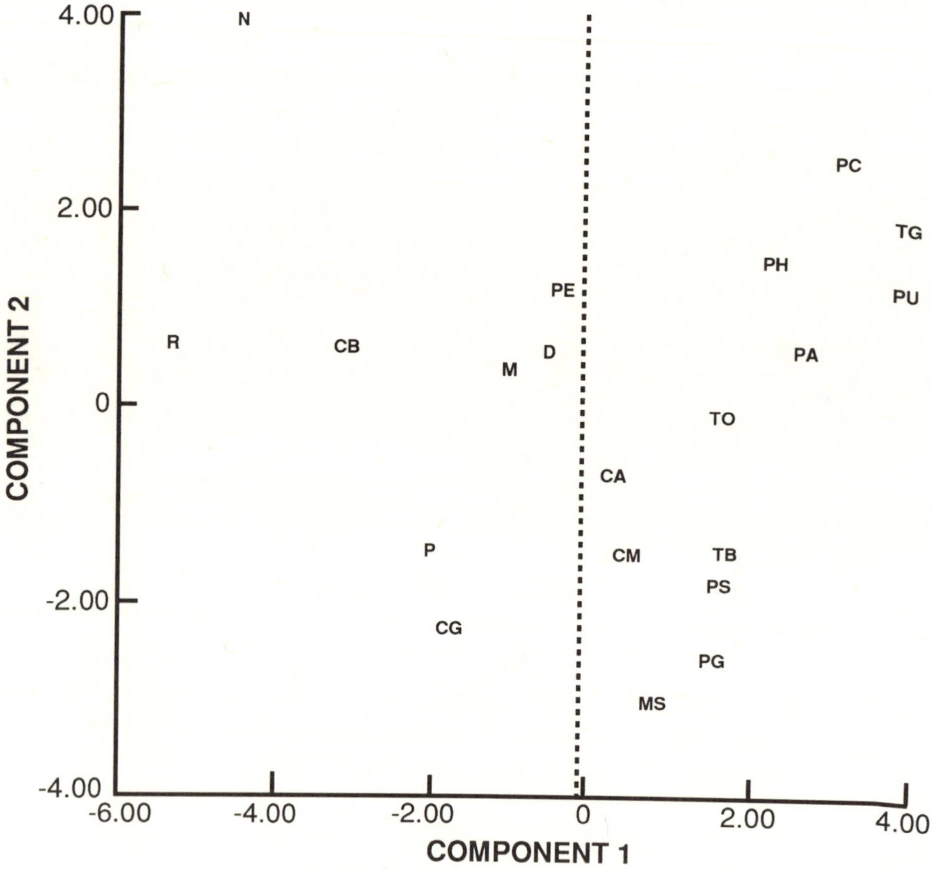

Figure 48. Plot of PC 1 vs. PC 2 for the extant and extinct species in the comparative sample, based on analysis of 15 variables of the cercopithecoid proximal ulna. PC 1 accounts for 46.7% of the total variance and PC 2 accounts for 20.0%. Vertical dashed line demarcates the separation of colobines (left) and cercopithecines (right). The vector loadings for these components are presented in Table 10.

posterior width (25), and an olecranon process that is significantly lengthened (6) and widened (7). The significant feature of PC 2, accounting in part for the separation of *Nasalis*, is a trochlear notch that is maximally high or "opened up" (3 and 4).

Proximal Radius

Results of the PCA of 8 variables describing the cercopithecoid proximal radius are presented in Fig. 49. This analysis included all available species in the comparative sample. PC 1, accounting for 41.2% of the variance, provides a basic sorting of colobines and cercopithecines. For the first time, however, one cercopithecine

TABLE 10. Vector loadings for PC 1 through 4 for the cercopithecoid proximal ulna. Loadings are converted to correlation coefficients between each measurement and the component.

	Measurement	*Correlation Coefficients*			
		1	*2*	*3*	*4*
1	Radial Notch W	.813	.267	-.353	.118
2	Radial Notch H	-.514	.384	-.620	-.370
3	Troch Notch Med H	.019	.692	.485	-.061
4	Troch Notch Mid H	-.034	.737	.135	.574
5	Coronoid/Olec H	.687	.349	.152	.035
6	Troch Notch/Olec L	.942	-.063	-.139	-.153
7	Troch Notch/Olec W	.926	-.043	-.091	-.011
8	Prox Troch Notch W	-.840	-.209	.401	.155
9	Mid Troch Notch W	-.813	-.418	.128	.093
10	Dis Troch Notch W	.164	-.248	.742	-.534
11	Coronoid Proc Proj	.296	-.800	.077	.253
12	Troch Notch Dp	-.826	.369	.038	-.033
13	Troch/Rad Notch W	.700	.003	.608	.147
24	Rad Notch Pos W	-.828	-.438	-.156	.156
25	Troch Notch Pos W	.723	-.619	-.210	.115
	Percentage of Variance	46.7%	20.0%	13.2%	6.3%

species, *Cercopithecus mitis,* is grouped with the colobines (center of Fig. 49). *Paracolobus* occupies the extreme colobine pole and *Papio ursinus* the extreme cercopithecine pole. *Presbytis entellus* is the most cercopithecine-like colobine, and *C. mitis* falls with the colobines and is thus the most colobine-like cercopithecine. Within the cercopithecines, all species of *Papio* except for *P. sphinx* cluster together at the extreme cercopithecine pole. Also, for the first time, all three species of *Theropithecus* sort together on PC 1, with *P. sphinx* joining these species. On PC 2, accounting for 24.8% of the variance, *T. oswaldi* is separated from the other *Theropithecus* species. Finally, the two small-bodied papionins fall near one another at the center of the plot.

Within the colobines, the two Eurasian Mio-Pliocene species *Mesopithecus* and *Dolichopithecus* fall near the extant *Colobus* species on PC 1, yet are separated from all the colobines on PC 2, which accounts for 24.8% of the variance. *Paracolobus*, the only other fossil colobine in the analysis, is not closely linked with any species. Finally, *Nasalis* falls near *Presbytis entellus* on both PC 1 and PC 2. Because *Cercopithecus mitis* falls with the colobines in this analysis, it can be concluded that the cercopithecoid proximal radius does not provide as reliable a sorting of the comparative sample as do the previous limb elements presented above. This

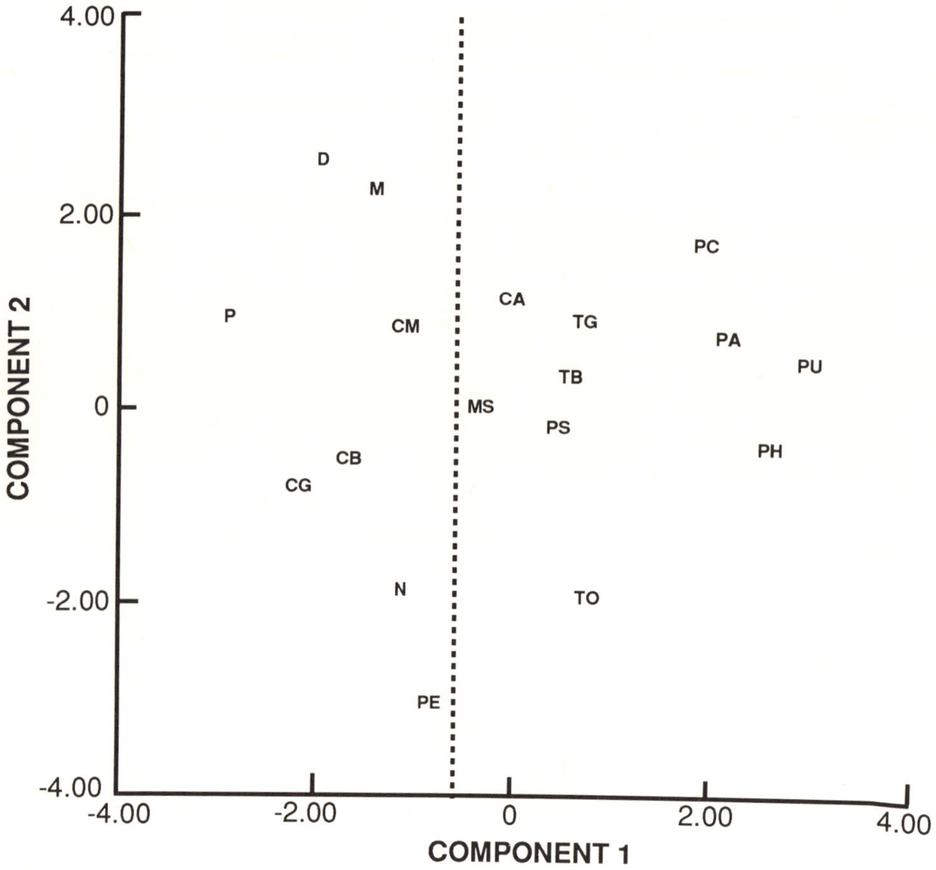

Figure 49. Plot of PC 1 vs. PC 2 for the extant and extinct species in the comparative sample, based on analysis of 8 variables of the cercopithecoid proximal radius. PC 1 accounts for 41.2% of the total variance and PC 2 accounts for 24.8%. Vertical dashed line demarcates the separation of colobines (left) and cercopithecines (right). The vector loadings for these components are presented in Table 11.

conclusion has implications for the taxonomic assignment of unknown cercopithecoid proximal radii presented in Chapter 7.

Table 11 presents vector loadings for the first four components in the PCA of the proximal radius. Colobines have a significantly more spherical radial head, indicated by relatively greater maximum (27) and minimum (28) radial head diameters. Colobines also exhibit a relatively greater minimum articular surface extent on the side of the radial head (30). The most notable feature contributing to the sorting of cercopithecines along PC 1 is a significantly greater medio-lateral radial neck width (31). The vector loadings for PC 2 show that the two features which contributed most to the separation of *Mesopithecus* and *Dolichopithecus* from the other colobines are significantly greater antero-posterior radial neck and bicipital tuberosity widths (32 and 34).

TABLE 11. Vector loadings for PC 1 through 4 for the cercopithecoid proximal radius. Loadings are converted to correlation coefficients between each measurement and the component.

	Measurement	*Correlation Coefficients*			
		1	*2*	*3*	*4*
27	Max M-L Head Dia	-.797	.217	-.086	.087
28	Min A-P Head Dia	-.856	.277	-.253	.073
29	Head Max Artic Ex	.030	-.631	.555	-.507
30	Head Min Artic Ex	-.779	-.108	.232	.548
31	M-L Neck W	.878	.037	.229	.169
32	A-P Neck W	.376	.781	-.215	-.143
33	Max Neck L	-.596	-.348	-.649	-.270
34	Bicipital Tuber W	-.223	.846	.260	-.247
	Percentage of Variance	41.2%	24.8%	12.8%	9.4%

Wrist Joint

Results of the PCA of 16 variables describing the cercopithecoid wrist joint are presented in Fig. 50. Only species of the extant sample are included in this analysis. PC 1, accounting for 31.7% of the variance, separates cercopithecines and colobines. The Colobinae are clustered on the right side of the plot. *Papio ursinus* lies at the extreme cercopithecine pole and *Nasalis* at the extreme colobine pole. *Macaca sylvanus* is the most colobine-like of the cercopithecines, and *Presbytis entellus* is the most cercopithecine-like colobine. Within the Cercopithecinae, all of the large-bodied papionins, along with *M. sylvanus*, are grouped about equidistant from one another along PC 1. On the other hand, PC 2, accounting for 29.4% of the variance, dramatically separates *Cercocebus* and *Cercopithecus* from all other cercopithecines (and from all colobines). Within the Colobinae, all species form a single cluster about equidistantly spaced from one another.

This analysis of the wrist joint differs from all previous analyses in one significant respect. For the first time, the amount of variance summarized by PC 2 is almost equal to that summarized by PC 1 (29.4% vs. 31.7%). This relationship is also reflected in the cluster shape of the taxa in Fig. 50. There is nearly as much separation within the Cercopithecinae as there is between the Cercopithecinae and the Colobinae. In essence, this means our ability to differentiate colobines from cercopithecines on the basis of wrist-joint morphology is not as precise as in analyses based on the shoulder and elbow joints and their components. This finding has implications for the assignments of the unknown fossil limb bones presented in Chapter 7.

Table 12 presents vector loadings for the first four components in the PCA of the wrist joint. The significant features relating to the separation of colobines along PC 1 are a distal ulna with a widened or expanded head articular surface (16) and a distal

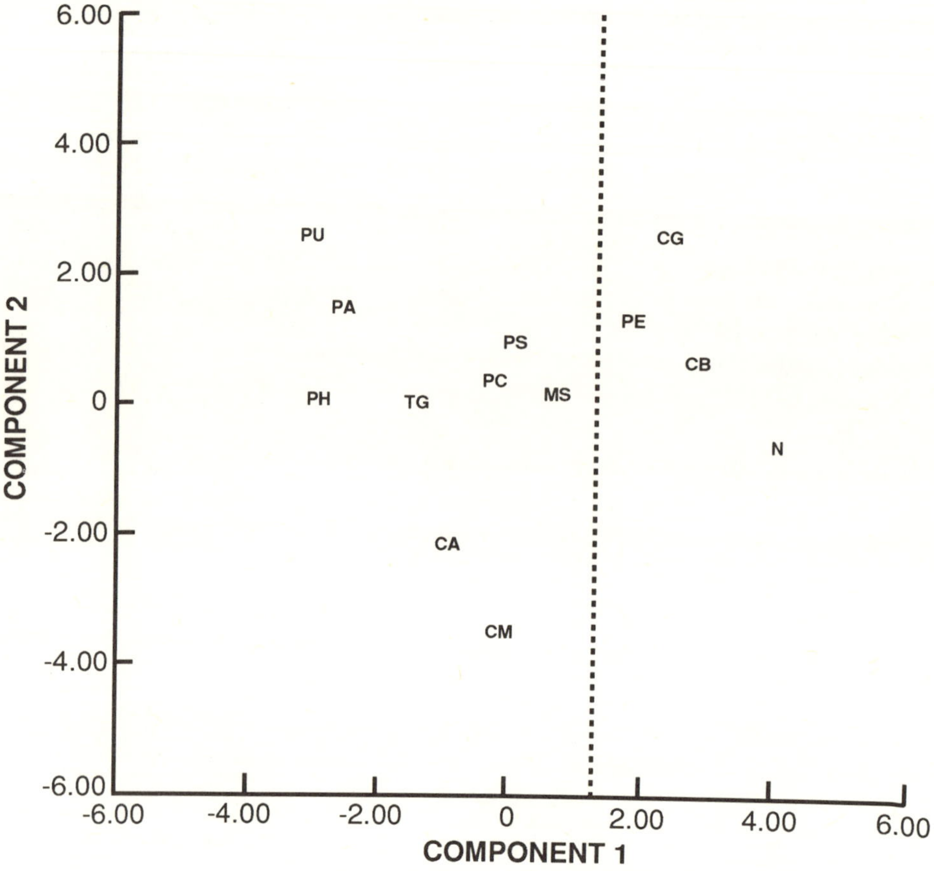

Figure 50. Plot of PC 1 vs. PC 2 for the extant and extinct species in the comparative sample, based on analysis of 16 variables of the cercopithecoid wrist joint. PC 1 accounts for 31.7% of the total variance and PC 2 accounts for 29.4%. Vertical dashed line demarcates the separation of colobines (right) and cercopithecines (left). The vector loadings for these components are presented in Table 12.

radius with a concomitantly widened ulnar notch articular surface (43). On the other hand, cercopithecines are characterized by a distal radius with a significantly longer distal articular surface (36), a distally projecting styloid process in relation to the placement of the lateral tubercle (41), and a greatly widened extensor digitorum groove (42). Along PC 2, the features contributing to the separation of *Cercocebus albigena* and *Cercopithecus mitis* from the other cercopithecines are a relatively narrow ulnar styloid process (15) and antero-posterior distal ulna (18).

Fig. 51 presents the results of a DFA of the same 16 variables describing the cercopithecoid wrist joint. The distribution of taxa along discriminant axis 1 is nearly identical to that seen along PC 1 in Fig. 50, but with the axes reversed. A species-by-species comparison of the taxon order along DF axis 1 vs. PC 1 shows a remarkable

TABLE 12. Vector loadings for PC 1 through 4 for the cercopithecoid wrist joint. Loadings are converted to correlation coefficients between each measurement and the component.

	Measurement	*Correlation Coefficients*			
		1	*2*	*3*	*4*
14	Styloid Process L	-.114	-.585	.069	.333
15	Styloid Process W	.433	.866	.124	-.029
16	Ulna Head Artic W	.857	-.369	-.023	-.288
17	Exten Carp Grv W	.471	-.122	.596	.479
18	A-P Dis Ulna W	.199	.927	.037	-.033
19	M-L Ulna W	.622	.556	-.316	.162
35	Styloid Process L	-.068	-.302	.871	-.242
36	Dis Artic Surf L	-.887	-.316	.039	-.043
37	Lunate Surface W	-.067	-.603	-.671	-.286
38	Max Dist Rad W	-.352	.646	-.059	-.597
39	Scaphoid Surf W	-.549	.586	-.220	-.086
40	Lister Tub Place	.014	-.506	.420	-.675
41	Lat Tub Dist Proj	-.707	-.442	-.169	.287
42	Exten Digit Grv W	-.903	.184	.091	.059
43	Ulna Notch Fac W	.837	-.253	-.045	-.063
44	Lunate Ridge Proj	.487	-.654	-.438	-.072
	Percentage of Variance	31.7%	29.4%	13.4%	9.5%

congruence, illustrating once again the complementary nature of the two multivariate techniques. The ordering of taxa along DF axis 2 (Fig. 51) differs somewhat from that seen along PC 2 (Fig. 50). The taxa along DF axis 2 are also more spread-out, especially with regard to *Colobus guereza*, which is actually separated from the other Colobinae. The differences between DF axis 2 and PC 2 could very well be due to the fact that, in DFA, axis 2 is not necessarily constructed perpendicular to axis 1, as is always the case in PCA. Thus the slight difference in results noted here may actually reflect nothing more than a mathematical difference in the techniques employed.

Distal Ulna

Results of the PCA of 6 variables describing the cercopithecoid distal ulna are presented in Fig. 52. This analysis includes both extant and extinct members of the comparative sample. PC 1, accounting for 36.8% of the variance, separates colobines and cercopithecines, though the two groups nearly overlap. *Colobus badius* occupies the extreme colobine pole and *Papio ursinus* the extreme cercopithecine pole. *Colo-*

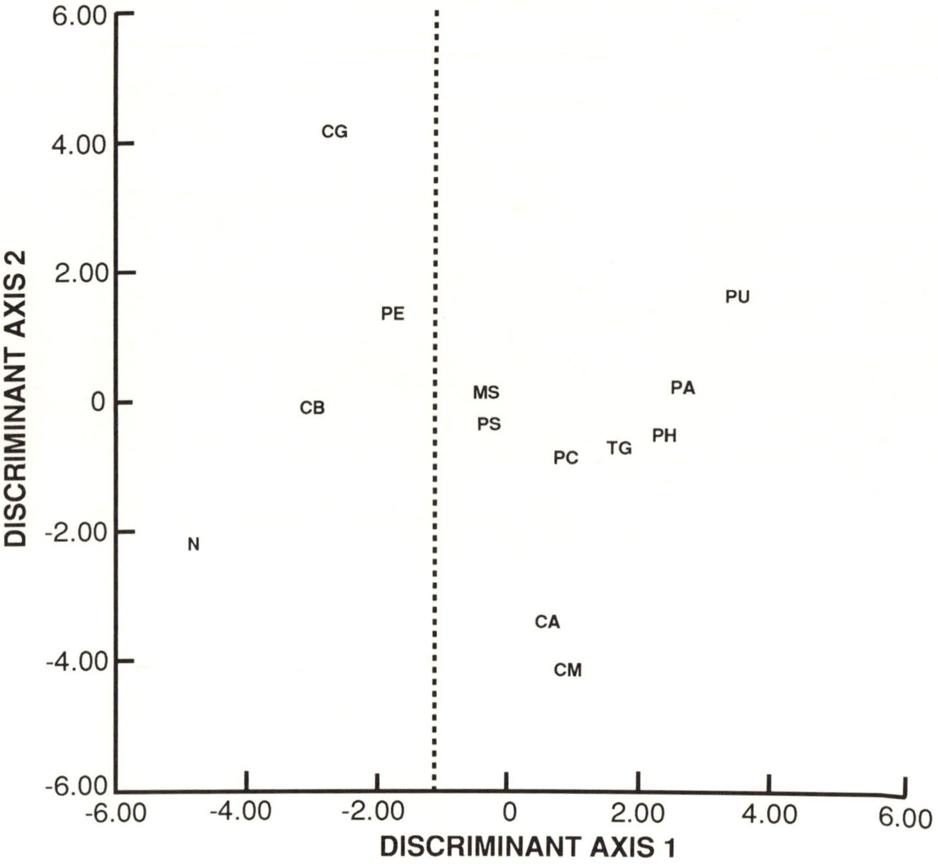

Figure 51. Plot of discriminant axes 1 vs. 2 for the extant sample, based on analysis of 16 variables of the cercopithecoid wrist joint. Vertical dashed line demarcates the separation of colobines (left) and cercopithecines (right). Axis 1 accounts for 42.3% of the total variance and Axis 2 accounts for 9.5%.

bus guereza is the most cercopithecine-like colobine and *Macaca sylvanus* is the most colobine-like of the cercopithecines. Based on the spread of taxa along PC 1, *Rhinocolobus* lies nearest to *C. badius, Mesopithecus* is near *Nasalis,* and *T. brumpti* and *T. oswaldi* are near *T. gelada.* However, when the position of these taxa along PC 2, accounting for 22.3% of the variance, is considered, many of these relationships change. *Rhinocolobus* is now quite distinct from all colobine species, and *T. oswaldi* and *T. brumpti* are separated at opposite ends of PC 2, with *T. gelada* equidistant yet far removed from both forms. The colobine grouping is more widely dispersed, whereas the cercopithecines are more tightly clustered, probably because the distal ulna is more morphologically diverse in extant and extinct colobines than in cercopithecines.

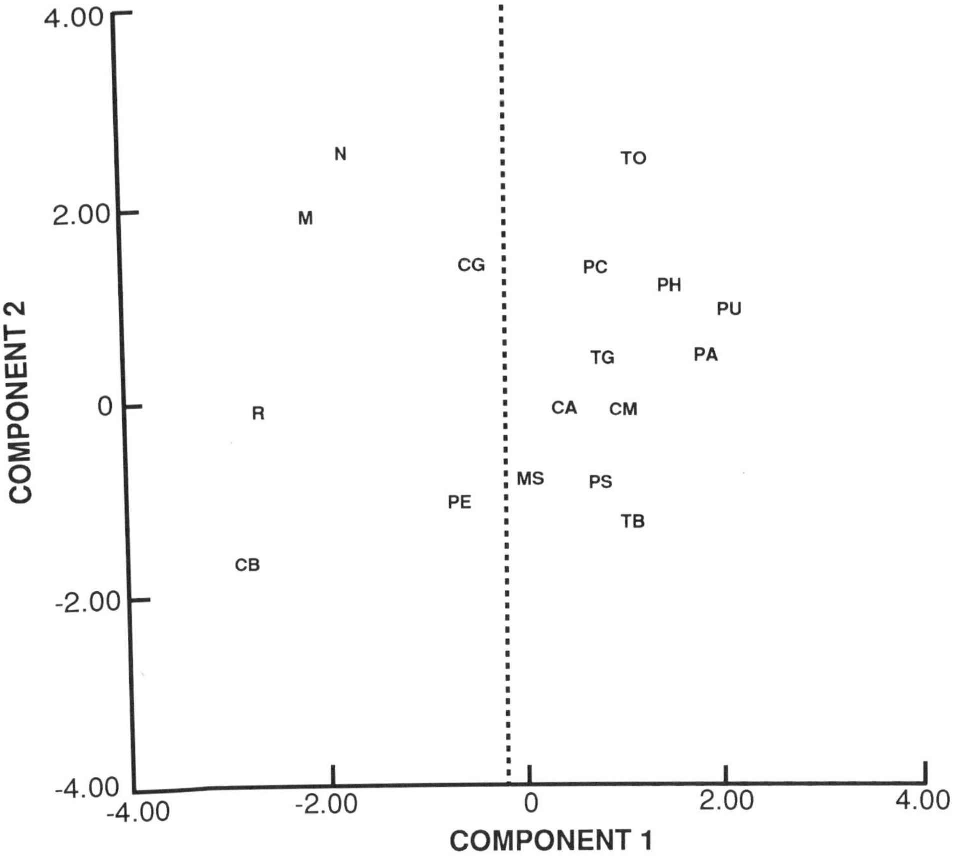

Figure 52. Plot of PC 1 vs. PC 2 for the extant and extinct species in the comparative sample, based on analysis of 6 variables of the cercopithecoid distal ulna. PC 1 accounts for 36.8% of the total variance and PC 2 accounts for 22.3%. Vertical dashed line demarcates the separation of colobines (left) and cercopithecines (right). The vector loadings for these components are presented in Table 13.

Table 13 presents vector loadings for the first four components in the PCA of the distal ulna. Of the 6 variables in this analysis, two can be significantly related to the separation of the Colobinae along PC 1. These are a relatively wide styloid process (15) and a distal ulnar that is greatly expanded in the antero-posterior dimension (18). Two features related to the separation of cercopithecines along PC 1 are a lengthened styloid process (14) and a widened extensor carpi ulnaris groove (17). These two cercopithecine features do not contribute as significantly (in a numerical sense) to the spread of taxa along PC 1 as do the two colobine features. As deduced from the vector loadings in Table 13, no single variable contributes significantly to the distribution of taxa along PC 2.

TABLE 13. Vector loadings for PC 1 through 4 for the cercopithecoid distal ulna. Loadings are converted to correlation coefficients between each measurement and the component.

	Measurement	*Correlation Coefficients*			
		1	*2*	*3*	*4*
14	Styloid Process L	.661	.556	.463	.017
15	Styloid Process W	-.777	.500	.054	.035
16	Ulna Head Artic W	-.296	-.681	.324	.583
17	Exten Carp Grv W	.655	-.248	-.694	-.048
18	A-P Dis Ulna W	-.807	.014	-.347	-.286
19	M-L Dis Ulna W	.014	-.502	.460	-.720
	Percentage of Variance	36.8%	22.3%	19.0%	15.8%

Distal Radius

Results of the PCA of 10 variables describing the cercopithecoid distal radius are presented in Fig. 53. This analysis includes all available species in the comparative sample. PC 1, accounting for 26.7% of the variance, separates colobines and cercopithecines, although the two groups nearly overlap at the center of the plot. *Nasalis* occupies the extreme colobine pole and *Cercopithecus* is at the extreme cercopithecine pole. The most cercopithecine-like colobine is *Mesopithecus,* and *Macaca sylvanus* is the most colobine-like cercopithecine. Within the Colobinae, *Dolichopithecus* and *Mesopithecus* fall very near Presbytis entellus on both PC 1 and PC 2. PC 2, accounting for 23.1% of the variance, separates *Colobus guereza* from the other colobines, including *C. badius. C. guereza* and *C. badius* were also quite far removed from one another in the PCA of the distal ulna (Fig. 52), suggesting that different wrist-joint adaptations characterize these two species.

Within the Cercopithecinae, all of the large-bodied papionins are clustered together, with *T. gelada* at the top of the cluster and *T. brumpti* and *T. oswaldi* at the bottom (Fig. 53). PC 2 serves to separate the two fossil *Theropithecus* species from the living *T. gelada. Cercocebus* and *Cercopithecus* lie at the extreme cercopithecine pole, far removed from *Macaca sylvanus*, which is near the center of the plot. This is an unusual occurrence not seen in any of the previous analyses. If these results from the distal radius seem a little atypical, it should be noted that the variance summarized by PC 1 vs. PC 2 for this element (26.7% vs. 23.1%) is nearly identical, and the sum total variance for both components is a good deal less than in any of the previous analyses. These factors indicate that the amount of significant taxonomic information in the distal radius is less than that seen in other parts of the forelimb.

Table 14 presents vector loadings for the first four components in the PCA of the

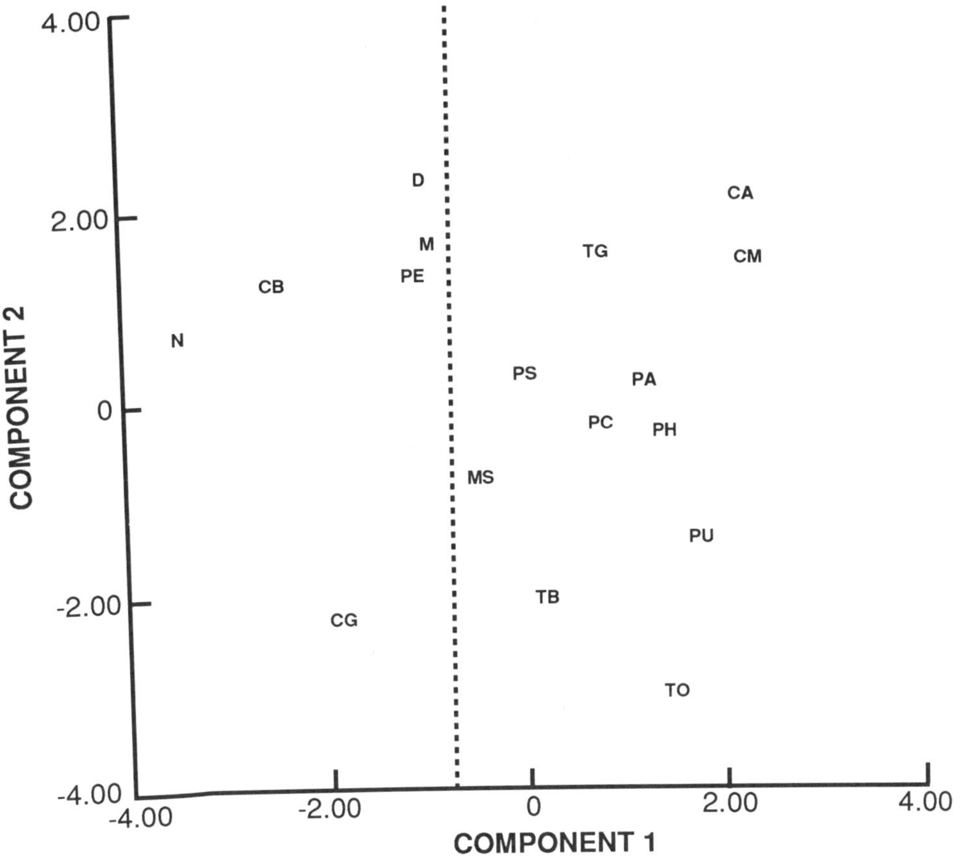

Figure 53. Plot of PC 1 vs. PC 2 for the extant and extinct species in the comparative sample, based on analysis of 10 variables of the cercopithecoid distal radius. PC 1 accounts for 26.7% of the total variance and PC 2 accounts for 23.1%. Vertical dashed line demarcates the separation of colobines (left) and cercopithecines (right). The vector loadings for these components are presented in Table 14.

distal radius. Three of the 10 variables in this analysis contribute significantly to the separation of the colobines and cercopithecines along PC 1. Colobines are characterized by significantly greater maximum distal radial and ulnar notch articular facet widths (38 and 43). Cercopithecines have a significantly more projecting radial styloid process when measured in relation to the relative placement of the lateral tubercle (41). The major variable to contribute significantly to the sorting of taxa along PC 2 is a distally projecting lunate articular facet ridge (44), which accounts in part for the separation of *Cercocebus* and *Dolichopithecus*. To a lesser extent, the separation of *T. oswaldi* and *T. brumpti* on PC 2 can be related to a lengthened radial styloid process (35).

TABLE 14. Vector loadings for PC 1 through 4 for the cercopithecoid distal radius. Loadings are converted to correlation coefficients between each measurement and the component.

	Measurement	*Correlation Coefficients*			
		1	*2*	*3*	*4*
35	Styloid Process L	-.073	-.686	-.311	-.348
36	Dis Artic Surf L	.539	.239	-.654	.290
37	Lunate Surface W	-.225	.625	.237	-.003
38	Max Dist Rad W	-.700	-.195	-.331	-.201
39	Scaphoid Surf W	-.117	-.624	.297	.018
40	Lister Tub Place	-.461	-.226	-.400	.720
41	Lat Tub Dist Proj	.803	-.181	.371	.039
42	Exten Digit Grv W	.364	.460	-.551	-.497
43	Ulna Notch Fac W	-.836	.108	.040	-.237
44	Lunate Ridge Proj	-.353	.805	.175	.111
	Percentage of Variance	26.7%	23.1%	14.1%	10.8%

Overview of the Multivariate Results

The preceding 14 multivariate analyses of the cercopithecoid forelimb and all its components show a remarkable degree of uniformity. Beyond any doubt, it is possible to consistently sort colobines and cercopithecines on the basis of forelimb morphology. Furthermore, intra-colobine and intra-cercopithecine relationships are also remarkably consistent. The numerous multivariate plots presented in the preceding discussion serve to document and emphasize this point. This is an important finding which lays a solid foundation for the analysis and *correct* assignment of the 234 unknown fossil specimens presented in Chapter 7.

A total of 85 continuous variables that metrically and functionally describe the forelimb were analyzed in multivariate fashion in this chapter. Of these 85 variables describing the scapula, humerus, ulna, and radius, more than 50% contributed significantly to the sorting of taxa in the 14 multivariate analyses. The variables were designed to metrically express the size and shape of joint surfaces, muscle and ligament attachments, and overall shaft dimensions and robusticity. The fact that such a large number of variables contributed significantly to the taxonomic sorting of the extant and extinct species in the comparative sample indicates that the bony surfaces being quantified actually have important functional and phyletic significance.

Finally, if we compare the univariate and multivariate results presented in this chapter, it is possible to see how the latter results amplify the former. Many of the trends barely noticeable in the univariate results become fully defined in the multivariate results. For instance, in the univariate analyses based upon the distal humerus (e.g., humerus trochlea projection index), the cercopithecines are a cohesive group and *Paradolichopithecus* and the extinct *Theropithecus* species are indistinguishable from the other large-bodied papionins (see Fig. 34 in Chapter 5). With regard to the colobines, the ranges for 3 of the 4 extant species and for all of the extinct species overlap with those of the cercopithecines. However, in the multivariate analyses of the distal humerus (see Fig. 47), the colobines and cercopithecines are fully separated. Furthermore, *Paradolichopithecus* and the extinct *Theropithecus* species fall closer to the small-bodied papionins and are separated from the large-bodied papionins which form a cluster. Thus, in this example, it is clear that the multivariate analyses yield more interpretable results.

7

RESULTS: ANALYSIS OF UNKNOWN FOSSIL LIMB BONES

Introduction to the Techniques

Multivariate analysis was used in Chapter 6 to assess the relationships among extant and extinct species in the comparative sample. Now it will be used to assign isolated forelimb elements of unknown taxonomic affinity to known taxonomic groups, based on statistical comparisons with species in the comparative sample.

DFA has long been used to discriminate between groups and to assign individuals of unknown affinity. Indeed, DFA was initially developed for such taxonomic purposes (Fisher, 1936). One of the first applications of the technique to the primate fossil record was by Day (1967), who successfully interpolated a fossil hominoid phalanx of unknown affinity into a discriminant analysis of hominid and pongid phalanges, ultimately assigning it to the Hominidae, a view which corroborated previous anatomical analysis (Day and Napier, 1966).

In my study, DFA was employed, together with two specialized applications of PCA and a standard version of SCA, to assign 234 isolated cercopithecoid forelimb elements to groups of known taxonomic affinity. These four complementary techniques were used in order to obtain a more accurate assessment of the taxonomic affinities of these fossil limb specimens.

An example using individuals of already known taxonomic affinity will first be given to illustrate these complementary techniques. Tables 15-17 present the taxonomic assignments of 10 individuals of known affinity. These specimens were introduced into the analyses as if they were part of the unknown group. Their actual affinity appears in the far left column, and their assigned affinity in the far right column. Three separate joint complexes of each of the 10 test cases were analyzed, in an effort to examine the concordance among analyses of different regions of the forelimb. Table 15 displays the results based on the elbow joint, Table 16 on the shoulder joint, and Table 17 on the wrist joint.

Four independent multivariate techniques were used in the analysis of each test case shown in Tables 15-17 and for all of the fossil unknowns: SCA, PCA Euclidean Distance, PCA Membership Quotient, and Stepwise DFA. The methodologies underlying these analyses are documented in detail in Chapter 4. To recap the techniques briefly: SCA measures the straight-line distance between two points in the original variable space, while PCA Euclidean Distance measures the straight-line distance between individual points in principal components space. PCA Membership Quotient is also a distance measure, but it takes into account the dispersion of individuals about the mean, assigning the unknown to known groups on the basis of a numerical value: 0.00 to 1.00 implying group membership, and values greater than 1.00 implying non-membership. Stepwise DFA maximizes the between-group separations and computes both a distance statistic (Mahalanobis Distance) and a membership quotient (Posterior Probability) ranging from 0.00 (no chance of membership) to 1.00 (certain membership).

A rank ordering of the three taxa closest to the unknown specimen is given for each test case in Tables 15-17 and for all the subsequent fossils analyzed below. This provides a maximum amount of information for making the taxonomic assignments. For the PCA Membership Quotient and DFA Posterior Probability such a ranking is not always useful, since the unknown is usually a member of the taxon in rank order 1, but exceptions do exist. For the SCA and PCA Euclidean Distance, however, the unknown can be equidistant from two different groups, so an assignment based on rank order 2 is sometimes nearly as valid as an assignment based on rank order 1.

As is evident from Tables 15-17, each technique does not always yield the same rank order of taxa. In about half of the test cases, the first ranking is different in one of the four techniques. This is not surprising when analyzing taxa that are not actually members of any species in the comparative sample. An additional problem of concordance arises in evaluation of the unassigned specimens with respect to other extinct taxa, because some of these extinct taxa could only be included in two of the techniques, SCA and PCA Euclidean Distance. The other techniques, PCA Membership and DFA, require larger sample sizes than are available for many of the extinct species. As a result, the probable "assignments" for a given fossil can be different, depending on which technique is emphasized. It is in cases such as these, where full concordance is not evident, that a subjective element is introduced into the final taxonomic assignments. The various techniques have to be subjectively weighed as to their robustness and reliability, and the number of individuals comprising each species in the extinct comparative sample has to be taken into account. Thus in the final step of the assignment procedure a subjective, almost philosophical, decision often has to be made.

The Test Cases

Tables 15-17, on the known test cases and their final taxonomic assignments, show how these techniques are applied in assigning the fossil unknowns presented in the next section. For instance in Table 15 the assignment of A/H 5 to *Theropithecus gelada* is very straightforward. All the techniques yield the same assignment, and with a posterior probability of 1.000 there is no question that A/H 5 is a member of the *T. gelada* taxon. The same conclusion can be drawn for 16, *Papio ursinus*.

Assignment of the next case, 120388, presents a different problem. The drill *Papio leucophaeus* is not included in the comparative sample. The analyses indicate that 120388 is a member of the genus *Papio* with affinities to *P. sphinx* and *P. hamadryas*. A decision is thus made to assign 120388 to *Papio* species indet. In reality, *P. leucophaeus* is the sister species of *P. sphinx*, and this relationship is suggested by the posterior probability score in Table 15. In the next test case, OM 4284, *Erythrocebus patas* is placed nearest *Cercopithecus mitis* by two of the techniques and closest to *Cercocebus albigena* by the others. Note, however, that for the SCA and PCA Euclidean distance the second rank order is *Cercopithecus mitis*. On the basis of these analyses, OM 4284 is assigned to cf. *Cercopithecus*. In reality, *Cercopithecus* is the sister genus of *Erythrocebus*; in fact, the latter is sometimes included in the former in some classifications.

For 81250, *Cercocebus galeritus*, the assignment is straightforward. With three of the four techniques, assignment in Table 15 is to *C. albigena*; in SCA assignment to CA is in rank order 2, a short distance behind rank order 1 assignment. The decision for the final assignment is therefore *Cercocebus* species indet. Note that 81250 is not actually a member of CA in the PCA membership quotient, nor does it have a particularly strong posterior probability score. In the next test case, 20898, *Macaca maura* is assigned to *M. sylvanus* by three of the four techniques. Note, however, that it is judged a possible member of two groups in the PCA membership quotient. This is not unusual, since *M. maura* is not part of the comparative sample. The final taxonomic assignment of 20898 is judged to be *Macaca* species indet. A similar situation exists for 137555, *Colobus polykomos*. In all four analyses, it is closest to *C. guereza*, but it is not a member according to the PCA membership quotient nor does it have a strong posterior probability score. Assignment is thus made to *C.* cf. *guereza*. In reality, its sister species is *C. guereza*.

In the next example, OM 3100, *C. guereza* is assigned without question to the group to which it belongs. Note the high posterior probability score. A similar situation occurs with 68684: based on a concordance of all four analyses, it can be correctly assigned to *Nasalis larvatus*. However, note that it is not a member of this group according to the PCA membership quotient (it has a value greater than 1.00). In this case, this particular technique may not be as finely tuned as the DFA posterior probability. Finally, the last test case in Table 15 is 26887, *Rhinopithecus roxellanae*. This rare colobine, restricted to southern China and northern Vietnam, is assigned to *Colobus badius* by three techniques and to *Nasalis larvatus* by the DFA routine. Note that none of the absolute distances, especially Mahalanobis D^2, are very close, indicating that this colobine is quite far removed. It is, therefore, classified as Colobinae genus indet., which is its correct allocation with respect to the taxa in the extant comparative sample.

Similar analyses were performed on the shoulder and wrist joints of these same test cases (Tables 16 and 17). There is significant overall similarity to the results in Table 15, though in Tables 16-17 assignments based on the shoulder and wrist joints do show a few departures from those based on the elbow joint. From an overview of the multivariate results presented in Chapter 6, analyses of the elbow joint seem to produce a more "taxonomically significant" sorting of the comparative sample than those based on the shoulder and wrist joints. This finding is corroborated by previous

TABLE 15. Taxonomic assignments, based on the cercopithecoid elbow joint, of 10 known specimens, to test the effectiveness of the morphometric techniques employed in this study.

Known Taxon	*Identification Number*	*Body Part*	*Present Location*	*Collection Locality*	*Rank Order*	*SCA Similarity Coefficient*	
Theropithecus gelada	A/H 5	Elbow joint	UCB-AH	Debre Birhan, Ethiopia	1	TG	.682
					2	PC	.803
					3	PU	.840
Papio ursinus	16	Elbow joint	UCB-AW	Victoria Falls, Zambia	1	PU	.762
					2	PC	.784
					3	PH	.843
Papio leucophaeus	120388	Elbow joint	AMNH-M	West Africa	1	PS	.773
					2	PH	.930
					3	PU	1.015
Erythrocebus patas	OM 4284	Elbow joint	KNM-O	Tigoni Primate Center., Kenya	1	CA	.779
					2	CM	.882
					3	PU	.850
Cercocebus galeritus	81250	Elbow joint	AMNH-M	Akenge, Zaire	1	CM	1.009
					2	CA	1.032
					3	MS	1.059
Macaca maura	20898	Elbow joint	MVZ	Celebes (Sulawesi)	1	CA	.774
					2	MS	.911
					3	CA	.952
Colobus polykomos	137555	Elbow joint	MVZ	West Africa	1	CG	.784
					2	CB	.813
					3	PE	.927
Colobus guereza	OM 3100	Elbow joint	UCB-AH	Thomson's Falls, Kenya	1	CG	.839
					2	CB	1.166
					3	PE	1.205
Nasalis larvatus	68684	Elbow joint	FMNH-M	Sandakan Dist., North Borneo	1	N	.886
					2	CB	1.368
					3	PE	1.707
Rhinopithecus roxellanae	26887	Elbow joint	USNM-M	Wen Chuan, Szechuan, China	1	CB	1.074
					2	N	1.246
					3	CG	1.466

TABLE 15. Continued.

PCA Euclidean Distance		*PCA Membership Quotient*		*DFA Mahalanobis Distance*		*DFA Posterior Probability*		*Taxonomic Assignment*
TG	3.153	TG	.985	TG	32.45	TG	1.000	*Theropithecus gelada*
PC	3.277	PC	1.150	PU	53.06	PU	.000	
PU	4.455	PU	1.221	PC	71.36	PC	.000	
PU	2.555	PU	.703	PU	60.16	PU	.838	*Papio ursinus*
PC	3.187	PC	.742	PH	64.17	PH	.113	
PH	3.980	PA	.942	PC	65.82	PC	.049	
PH	4.108	PH	.803	PS	55.81	PS	.998	*Papio* species indet.
PS	5.679	PS	1.055	PH	68.66	PH	.002	
PU	5.943	PU	1.400	PU	78.28	PU	.000	
CA	3.571	CM	.806	CM	61.78	CM	.986	Cercopithecinae cf. *Cercopithecus*
CM	3.726	CA	.920	PU	70.34	PU	.014	
PU	3.995	PU	.981	PA	84.78	PA	.000	
CA	5.493	CA	1.465	CA	86.44	CA	.507	*Cercocebus* species indet.
MS	5.748	MS	1.480	PS	86.52	PS	.487	
PS	5.969	PS	1.551	MS	95.44	MS	.006	
MS	3.493	MS	.808	MS	69.78	MS	.967	*Macaca* species indet.
CA	4.127	CA	.853	CA	76.57	CA	.032	
CM	4.877	PH	1.033	CM	85.41	CM	.001	
CG	3.560	CG	1.088	CG	50.42	CG	.691	*Colobus* cf. *guereza*
CB	4.204	CB	1.123	CB	52.03	CB	.309	
PE	4.566	PE	1.301	PE	93.79	PE	.000	
CG	4.075	CG	.687	CG	79.79	CG	1.000	*Colobus guereza*
CB	7.015	PE	1.590	PE	127.59	PE	.000	
PE	7.714	CB	1.715	CB	127.63	CB	.000	
N	3.995	N	1.310	N	87.60	N	1.000	*Nasalis larvatus*
CB	7.625	PE	2.333	CB	108.03	CB	.000	
PE	9.068	CB	2.409	CG	172.72	CG	.000	
CB	5.500	CB	1.790	N	128.47	N	.998	Colobinae genus indet.
N	6.950	N	2.180	CB	141.18	CB	.002	
CG	7.452	PE	2.500	CG	212.33	CG	.000	

TABLE 16. Taxonomic assignments, based on the cercopithecoid shoulder joint, of 10 known specimens, to test the effectiveness of the morphometric techniques employed in this study.

Known Taxon	*Identification Number*	*Body Part*	*Present Location*	*Collection Locality*	*Rank Order*	*SCA Similarity Coefficient*	
Theropithecus gelada	A/H 5	Shoulder joint	UCB-AH	Debre Birhan, Ethiopia	1	TG	.539
					2	PC	.696
					3	PH	.868
Papio ursinus	16	Shoulder joint	UCB-AW	Victoria Falls, Zambia	1	PU	.690
					2	PC	.756
					3	PH	.813
Papio leucophaeus	120388	Shoulder joint	AMNH-M	West Africa	1	PS	.715
					2	PH	.742
					3	PU	.912
Erythrocebus patas	OM 4284	Shoulder joint	KNM-0	Tigoni Primate Center, Kenya	1	MS	.819
					2	CA	.962
					3	CM	.997
Cercocebus galeritus	81250	Shoulder joint	AMNH-M	Akenge, Zaire	1	MS	.947
					2	CA	.992
					3	PH	1.093
Macaca maura	20898	Shoulder joint	MVZ	Celebes (Sulawesi)	1	MS	.627
					2	CA	.710
					3	PH	.790
Colobus polykomos	137555	Shoulder joint	MVZ	West Africa	1	CG	.850
					2	CB	1.068
					3	PE	1.113
Colobus guereza	OM 3100	Shoulder joint	UCB-AH	Thomson's Falls, Kenya	1	CG	.663
					2	CG	1.097
					3	PE	1.146
Nasalis larvatus	68684	Shoulder joint	FMNH-M	Sandakan Dist., North Borneo	1	N	.723
					2	CB	1.140
					3	PE	1.201
Rhinopithecus roxellanae	26887	Shoulder joint	USNM-M	Wen Chuan, Szechuan, China	1	N	1.166
					2	CB	1.465
					3	PE	1.475

TABLE 16. Continued.

PCA Euclidean Distance		*PCA Membership Quotient*		*DFA Mahalanobis Distance*		*DFA Posterior Probability*		*Taxonomic Assignment*
TG	2.211	TG	.694	TG	17.90	TG	.993	*Theropithecus*
PC	2.369	PC	.741	PC	27.73	PC	.007	*gelada*
PS	3.586	PS	.768	PH	35.97	PH	.000	
PC	2.726	PC	.751	PC	22.84	PC	.934	*Papio*
PU	3.750	PU	.781	PU	29.36	PU	.036	cf. *cynocephalus*
PH	3.839	PS	.932	PH	29.69	PH	.030	
PU	3.217	PU	.518	PH	30.03	PH	.583	*Papio*
PH	3.396	PS	.635	PS	30.91	PS	.375	species indet.
PS	3.991	PH	.710	PU	35.29	PU	.042	
CA	4.244	PS	1.088	CM	43.53	CM	.994	*Cercopithecinae*
CM	4.284	CM	1.363	MS	54.71	MS	.004	genus indet.
MS	4.390	CA	1.511	CA	57.11	CA	.001	
CA	4.190	CA	1.264	CA	45.50	CA	.907	*Cercocebus*
MS	4.674	CM	1.299	PC	50.34	PC	.084	species indet.
CM	5.352	MS	1.344	MS	55.25	MS	.007	
MS	3.197	MS	.582	CA	29.95	CA	.921	*Cercopithecinae*
CA	3.511	PS	.669	MS	35.73	MS	.051	cf. *Macaca*
CM	3.748	CA	.840	PE	37.82	PE	.018	
CG	4.074	PE	1.390	CG	22.15	CG	.997	*Colobus*
CB	4.626	CG	1.535	CB	35.73	CB	.001	cf. *guereza*
PE	5.335	CM	1.839	PE	36.35	PE	.001	
CG	2.083	CG	.563	CG	30.97	CG	1.000	*Colobus*
CB	4.234	PE	1.191	N	63.41	N	.000	*guereza*
PE	4.989	CB	1.461	CB	74.39	CB	.000	
N	3.208	N	.928	N	24.53	N	1.000	*Nasalis*
CB	5.422	PE	1.447	CG	48.25	CG	.000	*larvatus*
CG	5.522	CB	1.745	CB	49.94	CB	.000	
N	5.981	PE	1.412	N	103.39	N	1.000	*Colobinae*
CB	6.928	N	1.677	CB	124.53	CB	.000	cf. *Nasalis*
PE	7.464	CB	2.117	PE	144.01	PE	.000	

TABLE 17. Taxonomic assignments, based on the cercopithecoid wrist joint, of 10 known specimens, to test the effectiveness of the morphometric techniques employed in this study.

Known Taxon	*Identification Number*	*Body Part*	*Present Location*	*Collection Locality*	*Rank Order*	*SCA Similarity Coefficient*	
Theropithecus gelada	A/H 5	Wrist joint	UCB-AH	Debre Birhan, Ethiopia	1	TG	.607
					2	CA	.814
					3	PA	.921
Papio ursinus	16	Wrist joint	UCB-AW	Victoria Falls, Zambia	1	PU	.676
					2	PC	.679
					3	PA	.778
Papio leucophaeus	120388	Wrist joint	AMNH-M	West Africa	1	PH	.679
					2	PU	.778
					3	PS	.800
Erythrocebus patas	OM 4284	Wrist joint	KNM-0	Tigoni Primate Center, Kenya	1	CM	1.181
					2	CA	1.190
					3	PA	1.198
Cercocebus galeritus	81250	Wrist joint	AMNH-M	Akenge, Zaire	1	CA	.611
					2	PC	.624
					3	MS	.667
Macaca maura	20898	Wrist joint	MVZ	Celebes (Sulawesi)	1	MS	.784
					2	CA	.821
					3	CM	1.173
Colobus polykomos	137555	Wrist joint	MVZ	West Africa	1	CG	.824
					2	PE	1.282
					3	CM	1.312
Colobus guereza	OM 3100	Wrist joint	UCB-AH	Thomson's Falls, Kenya	1	CG	.714
					2	PE	1.295
					3	CB	1.311
Nasalis larvatus	68684	Wrist joint	FMNH-M	Sandakan Dist., North Borneo	1	N	.820
					2	CB	1.086
					3	MS	1.208
Rhinopithecus roxellanae	26887	Wrist joint	USNM-M	Wen Chuan, Szechuan, China	1	CB	1.083
					2	PE	1.210
					3	N	1.232

Table 17. Continued.

PCA Euclidean Distance		*PCA Membership Quotient*		*DFA Mahalanobis Distance*		*DFA Posterior Probability*		*Taxonomic Assignment*
TG	2.343	TG	.799	TG	9.52	TG	.884	*Theropithecus*
CA	3.021	CA	.925	PA	14.86	PA	.061	*gelada*
PH	3.325	PC	1.051	PH	15.24	PH	.051	
PC	2.660	PA	.629	PU	14.45	PU	.638	*Papio*
PU	2.858	PU	.689	PH	17.55	PH	.135	cf. *ursinus*
PH	3.182	PC	.708	PC	18.33	PC	.092	
PS	2.816	PS	.739	PH	15.05	PH	.471	*Papio*
PA	2.822	PH	.839	PS	15.75	PS	.331	species indet.
PH	2.842	PA	.854	PA	16.94	PA	.184	
CM	5.438	CA	.655	CA	48.20	CA	.396	Cercopithecinae
PC	5.566	CM	.689	CM	49.27	CM	.232	genus indet.
CA	5.730	PC	.723	PC	49.56	PC	.201	
CA	1.866	CA	.732	CA	9.26	CA	.505	*Cercocebus*
MS	2.410	PC	.782	MS	11.09	MS	.202	*albigena*
PS	2.531	PS	.818	PS	11.72	PS	.147	
MS	2.657	MS	.958	CA	14.73	CA	.832	Cercopithecinae
CM	3.046	CM	1.056	MS	17.99	MS	.163	cf. *Macaca*
CA	5.411	CA	1.536	PH	25.76	PH	.003	
CG	3.166	CG	1.000	CG	19.74	CG	1.000	*Colobus*
PE	5.012	PE	1.518	PE	41.32	PE	.000	*guereza*
CM	5.438	CB	1.550	CA	45.73	CA	.000	
CG	1.925	CG	.696	CG	21.51	CG	1.000	*Colobus*
PE	5.664	PE	1.549	CG	54.04	CB	.000	*guereza*
CB	5.698	CB	1.595	PE	58.91	PE	.000	
N	1.988	N	.559	N	23.64	N	1.000	*Nasalis*
CB	4.679	CB	.938	CB	42.08	CB	.000	*larvatus*
PE	5.365	PE	1.248	PS	63.49	PS	.000	
CB	3.148	CB	.760	N	42.08	N	.480	Colobinae
PE	3.897	PE	.926	CB	43.21	CB	.274	genus indet.
CA	4.522	N	1.174	CA	44.71	CA	.129	

multivariate studies (compare McHenry and Corruccini, 1975 vs. 1983), and the pattern is also evident from the test-case assignments presented here.

A final comment should be made on the application of these techniques for assignments of the unknown fossils. These techniques are, for the most part, designed to sort forelimb elements at the generic level. In the test cases presented in Tables 15-17, most assignments are at the species level because these are all extant cercopithecoids. Such specific assignments are not usually significant when applied to fossils that are 2-4 million years old. Most living Old World monkey species simply do not have a great time range. One exception to this is in the assignments to species level of *Theropithecus oswaldi* and *T. brumpti*; both are wholly extinct species, contemporaneous with many of the unknown fossil specimens. Another exception is the assignment of forelimb specimens from the South African Transvaal caves, which often contain many samples of a single well-identified cercopithecoid species.

In summary, then, assignments to the generic level of the unknown fossils presented in the following tables is the goal of this study. In some instances, an assignment to subfamily is the best the techniques can produce. In any case, the identifications of the forelimb postcranial bones presented below are a significant advance over any previous attempts.

The Fossil Cercopithecoid Sample

Fossil monkeys are among the most common faunal elements of the African Plio-Pleistocene (Freedman, 1957; Eck, 1977; Delson, 1984; Pickford, 1987a). More than 30 taxa have been identified (see Fig. 54) from dozens of localities spanning a time frame from 6 million to 200,000 years B.P. (Fig. 55). As usual for the vertebrate fossil record, associations between cercopithecoid cranial and postcranial remains are rare but not unknown. Examples of some fossil taxa with associated cranial and postcranial remains are *Theropithecus oswaldi* (Leakey and Leakey, 1973; Isaac, 1977), *Paracolobus chemeroni* (R. E. F. Leakey, 1969; Birchette, 1982), and *Rhinocolobus turkanaensis* (M. G. Leakey, 1982). (See discussion of the extinct comparative sample in Chapter 2 for further examples.) These few examples aside, most African Plio-Pleistocene cercopithecoid postcranial specimens are isolated skeletal elements, usually preserving only the proximal or distal portions. In studies of cercopithecoid evolution, these remains have routinely been overlooked in favor of the more "taxonomically significant" cranial remains.

The 234 unknown forelimb elements analyzed in this study are derived from Plio-Pleistocene localities in East and South Africa and a few localities in the circum-Mediterranean region. These specimens are analyzed and compared in a series of multivariate analyses with the 13 extant and 7 extinct species in the comparative sample. Tables 18-44 list all pertinent information regarding the fossil unknowns, their final taxonomic assignments, and site-by-site summaries of the assignments.

By and large, the multivariate techniques were of sufficient robusticity to allocate all unknown specimens; however, the level of assignment differs, for a variety of reasons. The taxonomic assignments of the 234 unknowns vary from species to subfamily. The most reliable assignments were made to those fossil taxa that were the best known postcranially, such as *Theropithecus oswaldi, T. brumpti,* and *Paracolobus chemeroni*, or to specimens from geologically recent horizons (i.e., middle Pleistocene) that could be directly assigned to extant taxa. A problem complicating the assignment of some bones is the differential representation of elements in the comparative sample. For instance, in the case of *Rhinocolobus turkanaensis*, the ulna, humerus, and scapula are known, but the radius is not. Therefore, it is possible to assign a distal humerus to that taxon, but not a proximal radius.

A more serious problem complicating the assignment of some fossils is the complete lack of associated cranial and postcranial remains for certain taxa. For example, the genus *Parapapio* has long been identified on the basis of cranial remains from a number of localities. Postcranial remains certainly exist, yet their identification and assignment is difficult without a known reference sample for comparison. However, by analyzing postcrania from those localities with a predominance of cranially identified *Parapapio* (e.g., Sterkfontein) it is possible to gain suggested evidence of this taxon's forelimb anatomy that can be used to make assignments at other localities. Finally, the identification of bones from the geologically oldest horizons (i.e., early Pliocene) is complicated by the fact that specimens from these levels may share little affinity with *any* extant or extinct members of the comparative sample. Thus their taxonomic assignment may be less precise.

These multivariate techniques, as outlined above, assign unknown limb remains to known taxonomic groups. These assignments are in part determined by the existence of cranially identified remains at the locality in question (see Fig. 54). There is a good possibility that at some localities there may be postcranial remains for taxa which have not yet been identified cranially. This is especially true at those localities where only a few specimens have been collected; sampling error can be a major factor. In this study, assignments of postcranial unknowns at localities where no cranially identified specimens exist for the assigned taxon have sometimes been made, resulting in a possible extension of the known range for some Plio-Pleistocene cercopithecoid paleospecies.

It is worth noting that the African fossil record of cercopithecoid postcranials is by nature a fragmentary one. Many of the sampled localities, especially from the Omo Shungura Formation, are predominantly fluviatile—and long bones break up during deposition. In developing the research design for this project, such factors were taken into account. The techniques employed for the assignment of fossil unknowns are based only on measurements from the proximal or distal ends of the forelimb elements, not on complete bones. In fact, those few examples of complete forelimb long bones are analyzed twice, once for the proximal end and again for the distal end, in order to conform to the research design used for the rest of the more fragmentary sample. Furthermore, only fully intact ends for which it was possible to collect a complete set of measurements are included in this study. Therefore, the fragmentary nature of the fossils was not a factor in the sorting and assignments of the fossil unknowns.

Sites	*Parapapio antiquus*	*Parapapio broomi*	*Parapapio jonesi*	*Parapapio whitei*	*Parapapio* or *Cercocebus* sp.	*Parapapio* or *Papio* sp.	*Cercocebus* sp.	*Papio (Papio) angusticeps*	*Papio (Papio) baringensis*	*Papio (Papio) izodi*	*Papio (Papio) robinsoni*	*Papio (Dinopithecus) ingens*	*Papio (D) quadratirostris*	*Papio* sp. indet.	*Gorgopithecus major*	*Theropithecus brumpti*	*Theropithecus darti*	*Theropithecus oswaldi*	*Theropithecus* sp. indet.	*Cercopithecus* sp.	*Macaca* sp.	Cercopithecinae indet.	*Cercopithecoides kimeui*	*Cercopithecoides williamsi*	*Paracolobus chemeroni*	*Paracolobus mutiwa*	*Rhinocolobus turkanaensis*	Colobinae sp. "A"	*Colobus* sp.	Colobinae indet.
Omo upper Usno Fm. (BS/WS)						X							X			?				X							X			
Omo Shungura Fm., mbr. A													?						X								X			
Omo Shungura Fm., mbr. B					X	X							X			X				X						X	X	X		
Omo Shungura Fm., mbr. C					?	X							X			X		X								X	X	X		
Omo Shungura Fm., mbr. D					?	?							X			X		?								X	X			
Omo Shungura Fm., mbr. E					?	X							X			X		?								X	X			
Omo Shungura Fm., mbr. F					?	X							X			X		X								X	X			
Omo Shungura Fm., lower mbr. G					?	?							X			X		X		X						X	X			
Omo Shungura Fm., upper mbr. G																?		X												
Omo Shungura Fm., mbr. H																		X												
Hadar Fm., SH mbr.																	X										X	X		X
Hadar Fm., DD mbr.																	X										X			
Hadar Fm., lower KH mbr.																	X													
Hadar Fm., upper KH mbr.			X														X													
Afar Leadu beds																	X											X		
Afar Ahmado beds			X														X										?			
Kanjera																		X												
Olorgesailie																		X												
Chesowanja loc.																			?											
Lothagam-3 Beds																			X			X								
Chemeron Fm., lower Beds									X																X					
Koobi Fora sub-KBS						X	X											X					X	X		X	X			
Koobi Fora sub-Burgi						X										?	X	X					X					X		X
Olduvai Bed I					X	X								X								X							X	
Olduvai Bed II						?																	X						X	
Olduvai Bed III and above						?								X									X						X	
Taung fauna loc.	X					X				X																				
Makapan mbrs. 2-3		X	X	X			X										X							X						
Makapan mbr. 4		X	X	X													X							X						X
Sterkfontein, mbr. 4		X	X	X	X					X	X													X						
Sterkfontein mbr. 5						X																								
Swartkrans mbr. 1							X				X	X			X			?												
Kromdraai A loc.					X			X			?				X															
Kromdraai B loc.							X	X			X				X									X						
Bolt's Farm, Pit 6											X													X						
Bolt's Farm, Pit 23		X									X													X						
Tamar Hat, (Algeria)																					X									

Figure 54. The distribution of Plio-Pleistocene cercopithecoids at fossil localities throughout East and South Africa, based on analysis of cranial remains and compiled from the work of Delson (1984) and other sources. The horizontal column on the left lists all identified taxa and the diagonal column at the top lists all fossil localities with cranial *and* postcranial remains that were included in this study. Key: X= taxon present at locality; ? = taxon possibly present at locality, but identification is uncertain. The presence or absence of a taxon is based solely on the identification of cranial remains. Chart modified after Table 1 in Delson (1984).

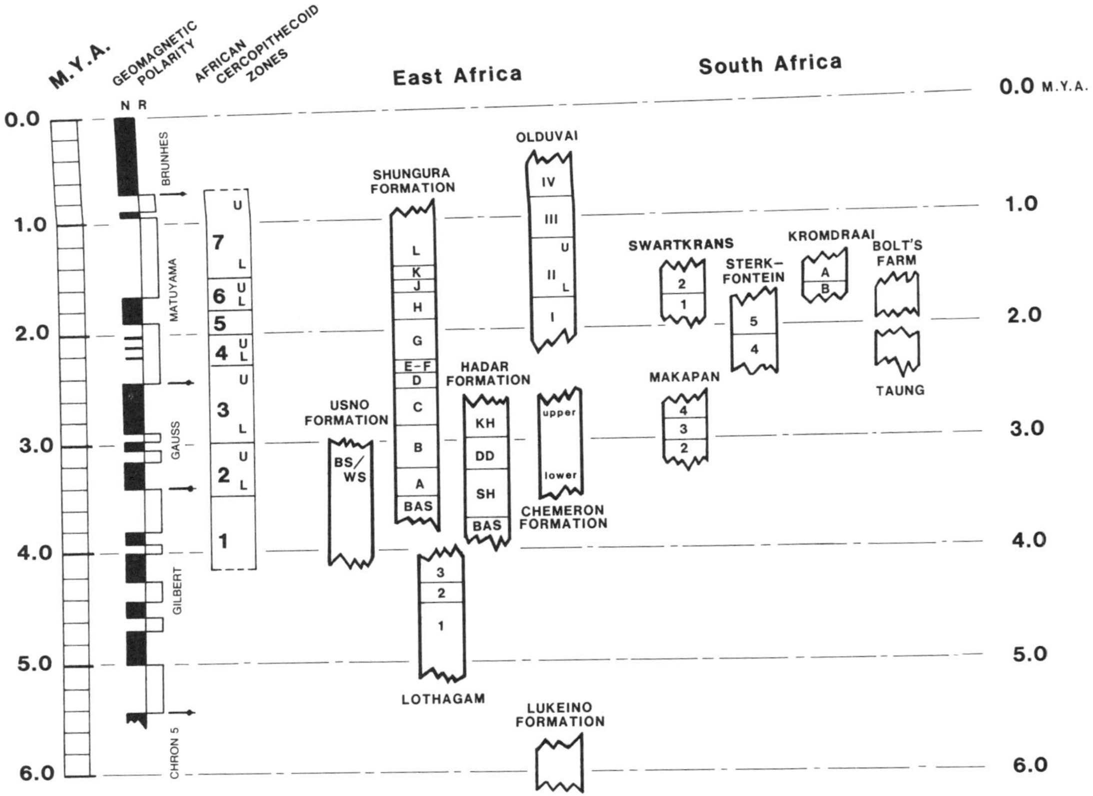

Figure 55. Correlation of African Plio-Pleistocene localities with cranial and postcranial cercopithecoid remains, based on a geochronologic calibration of radiometric and paleomagnetic dates (F. C. Howell, 1978; Walter and Aronson, 1982; de Heinzelin, 1983; Brown et al., 1985; Feibel et al., 1989) and a biozonation scheme of African cercopithecoid cranial remains proposed by Delson (1984). Nearly all fossil localities listed in Figure 54 can be dated and cross-correlated with the aid of this figure. Note that localities of the Koobi Fora Formation and the Nachukui Formation are not listed here, since fossil postcranial unknowns from those localities have not been included in this study.

Keeping in mind the several limitations noted above, it is possible to directly compare the percentages for assignments of postcranial unknowns based on these multivariate methodologies with like percentages for identified cranial remains. Of all the African Plio-Pleistocene sites, such data exists for only one series of localities—those from the Usno and Shungura formations of the lower Omo basin in southern Ethiopia. Discussion of the Omo fossil Cercopithecoidea will be followed by a site-by-site review of all the other localities where postcranial remains have been identified.

The Lower Omo Basin, southern Ethiopia

The fossiliferous beds of the Omo Usno and Shungura formations span a time range from about 3.5 Ma to about 1.0 Ma (see Fig. 55). More than 3000 cranial remains of cercopithecoids, the great majority isolated teeth, have been identified from all of the well-dated stratigraphic levels. Postcranial remains of cercopithecoids are considerably rarer, in part because the fluviatile mode of deposition of the Omo beds caused the fragmentation of many limb bones before burial. Some 326 forelimb postcranial elements have been recovered, giving approximately a 1:10 ratio of forelimb bones recovered to identified cranial remains. Of these 326 forelimb elements, only 203 could be included in this study, primarily due to the broken and abraded nature of a number of the rejected specimens.

All pertinent information regarding each forelimb specimen (identification number, body-part description, locality, stratigraphic level, etc.), together with its taxonomic assignment, appears in Tables 18-30 of this chapter and Table 47 in the Appendix. The specimens listed in Table 47 are from *Theropithecus brumpti* in the extinct comparative sample (see Chapter 2) whereas the specimens in Tables 18-30 represent the unknown Omo forelimb postcranial sample now newly identified. A summary of all the taxonomic assignments for the Omo fossil forelimb material, organized by stratigraphic level, is given in Table 31.

Since the multivariate techniques employed in this study used a comparative sample of 20 known species, some of the taxonomic assignments based on the postcrania are more inclusive than those based on the crania. For instance, there is an indication that three paleospecies of *Theropithecus* may have existed throughout the Omo sequence. Certainly *T. brumpti* is a distinct lineage ranging back to the upper Usno Formation—about 3.25 Ma. *T. oswaldi* appears distinct by member E in the Shungura Formation (2.3 Ma) and could range back into member C. If the identifications from member C are substantiated, this will prove to be the oldest dated occurrence of *T. oswaldi*. Finally, there is a third lineage of *Theropithecus* that is not yet identified to species. It is found throughout the sequence and could either represent a variant of one of the two identified species or a third species distinct from both *oswaldi* and *brumpti* and possibly ancestral to either *T. oswaldi* or *T. gelada*. Since *T. darti* has recently been identified as the most common cercopithecoid in the Hadar Formation in northeastern Ethiopia (Eck, 1993), it will likely prove to be this third lineage of *Theropithecus* I have identified in the Omo sequence.

Jablonski (1986) has recently analyzed the hand elements (metacarpals and phalanges) of the L865 partial skeleton of *T. brumpti* from unit 4, member E, originally identified on the basis of cranial remains. The distal humerus from the L865 skeleton, analyzed in this study, is also allocated to *T. brumpti* on the basis of morphometric techniques (see Table 23). A distal humerus from L4 in unit 3, member E, can be just as confidently assigned to *T. oswaldi* (Table 23), using the same methodology. Thus, the separation and distinctiveness of *T. brumpti* and *T. oswaldi* by member E (2.3 Ma) in the Omo sequence is documented here for the first time. For these closely related species to coexist in the same region of southern Ethiopia, they probably partitioned their environment, with *T. brumpti* occupying the gallery forest, riverine habitat (the proximal community), as suggested by Jablonski (1986), and *T. oswaldi* occupying the open-country savanna habitat near the gallery forests (the distal community), as suggested by Dechow and Singer (1984).

Another interesting point revealed in Table 31 is the apparent existence of *Parapapio* throughout the Omo sequence. Though this taxon has been identified from other localities in East and South Africa, its existence at Omo has not as yet been documented on the basis of cranio-dental remains. However, many isolated teeth from the Usno and Shungura Formations are identified only as small papionins. Furthermore, since *Parapapio* has been identified from the Hadar Formation (see next section) and Laetoli (M. G. Leakey and Delson, 1987), its existence in the Usno and lower Shungura Formations should be expected, since all of these stratigraphic levels are temporally equivalent and geographically proximate.

An unexpected outcome of this study is the apparent rarity of the genus *Papio* throughout the Omo sequence. This very common African savanna-dwelling species can be identified postcranially only back as far as member E (2.3 Ma), though cranial identifications have shown a greater antiquity. Recent reassessments by Eck and Jablonski (1984, 1987) of the cranial and dental evidence for *Papio* in the Plio-Pleistocene of eastern Africa have resulted in the transference of many specimens once identified as *Papio* to the *Theropithecus* species-group. The postcranial evidence from the Omo Group deposits analyzed here supports this contention.

Analysis of the colobines shows the existence of distinct *Paracolobus* and *Rhinocolobus* lineages back through member C, but there is also a surprising number of specimens assigned to Colobinae genus indet. These may prove to be a yet undiagnosed colobine, Colobine gen. et sp. indet, represented by 17 isolated teeth in the Shungura Formation (see M.G. Leakey, 1987). It is unlikely they represent *Cercopithecoides*, an extinct colobine known from elsewhere in East and South Africa (M. G. Leakey, 1982) but not yet identified on the basis of dental remains from the Omo sequence (M. G. Leakey, 1987). A final note should be made on the identification of *Colobus* from members A, C, and G of the Shungura and the Usno Formations. Though this genus has not yet been identified on the basis of cranial evidence, it very possibly exists throughout the sequence. Alternately, these few specimens could also represent Colobine gen. et sp. indet or atypical specimens of another already-identified colobine taxon.

TABLE 18. Taxonomic assignments of previously unidentified cercopithecoid forelimb elements from the Lower Omo Basin, southern Ethiopia.

Identification Number	*Body Part*	*Present Location*	*Collection Locality*	*Rank Order*	*SCA Similarity Coefficient*	
W7-473b	Right	ENM	Usno Fm.	1	TB	.712
(same individual	prox		White Sands	2	MS	.724
as W7-473a)	ulna			3	PS	.781
B7-215	Right	ENM	Usno Fm.	1	TB	1.091
	prox		Brown Sands	2	TG	1.207
	ulna			3	TO	1.211
B-420	Left	ENM	Usno Fm.	1	TB	.819
	prox		Brown Sands	2	MS	.833
	ulna			3	PS	.920
B-818a	Right	ENM	Usno Fm.	1	CB	1.075
	prox		Brown Sands	2	CA	1.177
	ulna			3	D	1.196
W7-473a	Right	ENM	Usno Fm.	1	TB	.561
(same individual	prox radius		White Sands	2	MS	.673
as W7-473b)	w/shaft			3	TO	.681
B7-118	Right	ENM	Usno Fm.	1	CM	.769
	prox radius		Brown Sands	2	CA	.911
	w/shaft			3	CB	1.067
B7-267	Right	ENM	Usno Fm.	1	MS	.368
	prox radius		Brown Sands	2	PS	.414
	w/shaft			3	TB	.458
W7-366	Left	ENM	Usno Fm.	1	PE	.967
	distal		White Sands	2	CB	.968
	humerus			3	N	1.015
W-1034	Left	ENM	Usno Fm.	1	TO	.863
	distal		White Sands	2	CA	.867
	humerus			3	TB	.950
W8-825	Left	ENM	Usno Fm.	1	CA	.381
	distal		White Sands	2	MS	.455
	humerus			3	CM	.519

Table 18. Continued.

PCA Euclidean Distance		*PCA Membership Quotient*		*DFA Mahalanobis Distance*		*DFA Posterior Probability*		*Taxonomic Assignment*
PS	2.096	TB	.582	TO	15.37	TO	.863	*Theropithecus*
TB	2.353	MS	.624	MS	19.09	MS	.135	species indet.
MS	2.486	PS	.811	TB	27.94	TB	.002	
TB	3.597	TB	.963	TB	31.74	TB	.951	*Theropithecus*
TO	3.838	TO	1.166	TO	38.52	TO	.032	cf. *brumpti*
TG	3.958	PC	1.295	PS	40.62	PS	.011	
TB	2.308	TB	.619	TB	28.84	TB	.946	*Theropithecus*
MS	2.457	MS	.860	MS	35.62	MS	.032	cf. *brumpti*
PS	3.304	PS	1.169	PS	36.43	PS	.021	
CB	4.283	PE	1.318	CB	26.94	CB	.976	Colobinae
PE	4.462	CA	1.367	PE	34.46	PE	.023	cf. *Colobus*
CA	4.501	CB	1.716	CA	42.97	CA	.000	
MS	1.811	TB	.607	TB	5.06	TB	.310	*Theropithecus*
TB	2.212	MS	.716	MS	5.61	MS	.235	species indet.
CM	2.231	CG	.734	CM	6.18	CM	.177	
CM	1.899	CM	.743	CM	15.97	CM	.566	*Cercopithecus*
CA	2.529	CA	.913	CA	17.35	CA	.284	species indet.
CB	2.864	TB	1.015	TB	18.67	TB	.147	
PS	1.136	PS	.358	MS	1.77	MS	.358	Cercopithecinae
MS	1.228	MS	.359	CA	3.99	CA	.118	cf. *Parapapio*
TB	1.434	TB	.368	TB	4.08	TB	.112	
CB	2.136	CB	.784	PE	28.81	PE	.613	Colobinae
PE	2.643	PE	.902	CB	30.16	CM	.311	genus indet.
P	3.218	TO	1.126	CA	36.48	CA	.013	
TO	2.454	TB	.802	CA	20.82	CA	.868	Cercopithecinae
CA	2.644	TO	1.166	TO	25.92	TO	.068	cf. *Theropithecus*
TB	2.708	CA	1.215	TB	26.23	TB	.058	
CA	1.066	CA	.432	CA	5.04	CA	.819	Cercopithecinae
CM	1.152	CM	.467	CM	9.46	CM	.088	cf. *Cercocebus*
MS	1.295	MS	.653	MS	9.88	MS	.073	

TABLE 19. Taxonomic assignments of previously unidentified cercopithecoid forelimb elements from the Lower Omo Basin, southern Ethiopia.

Identification Number	*Body Part*	*Present Location*	*Collection Locality*	*Rank Order*	*SCA Similarity Coefficient*	
B-525	Right distal humerus	ENM	Usno Fm. Brown Sands	1	MS	.484
				2	CA	.488
				3	TB	.522
O18-135	Left distal humerus	ENM	Usno Fm. White Sands	1	M	.641
				2	D	.781
				3	PE	.823
O18-488	Right distal humerus	ENM	Shungura Fm. member A3	1	CG	.648
				2	M	.821
				3	P	.850
P732-1	Left prox ulna	ENM	Shungura Fm. member Bl	1	CB	1.152
				2	CA	1.163
				3	PE	1.217
Ll-497	Left prox ulna	ENM	Shungura Fm. member B11	1	TB	.807
				2	CA	1.176
				3	TO	1.184
L367-3	Left prox ulna	ENM	Shungura Fm. member B12	1	TB	.888
				2	TO	1.236
				3	CA	1.248
L793-2	Left prox radius	ENM	Shungura Fm. member B2	1	CG	.704
				2	CM	1.004
				3	CB	1.016
Ll-1397	Left prox radius	ENM	Shungura Fm. member B10	1	PS	.610
				2	TB	.671
				3	PC	.703
Ll-322	Right prox radius	ENM	Shungura Fm. member B11	1	MS	.305
				2	TB	.338
				3	CB	.454
Ll-324	Right distal radius	ENM	Shungura Fm. member B11	1	CA	.991
				2	PU	1.096
				3	MS	1.198

TABLE 19. Continued.

PCA Euclidean Distance		*PCA Membership Quotient*		*DFA Mahalanobis Distance*		*DFA Posterior Probability*		*Taxonomic Assignment*
MS	1.622	MS	.382	CA	6.50	CA	.548	Cercopithecinae
CA	1.679	CA	.630	MS	8.14	MS	.241	cf. *Parapapio*
TB	1.683	TB	.682	TB	8.85	TB	.170	
M	2.042	CA	.656	PE	23.80	PE	.670	Colobinae
D	2.514	PE	.766	CA	26.76	CA	.169	genus indet.
PE	2.894	TB	.982	CM	26.55	CM	.152	
CG	1.474	CG	.686	CG	25.36	CG	.856	Colobinae
M	2.347	CB	.890	CM	29.36	CB	.116	cf. *Colobus*
P	2.642	TO	1.025	CA	33.39	CA	.016	
CB	4.062	CA	1.001	PE	33.52	PE	.882	Colobinae
PE	4.239	PE	1.355	CA	37.74	CA	.107	genus indet.
CA	4.305	CB	1.552	CG	43.93	CG	.005	
TB	2.896	TB	.776	TB	17.31	TB	.999	*Theropithecus*
PS	4.214	TO	1.324	CM	33.13	CM	.000	*brumpti*
TO	4.247	PE	1.540	CA	34.53	CA	.000	
TB	3.511	TB	.719	TB	26.83	TB	1.000	*Theropithecus*
PG	4.634	TO	1.331	TO	43.28	TO	.000	*brumpti*
CA	4.837	MS	1.603	CA	45.58	CA	.000	
CG	2.174	CG	.875	CM	8.24	CM	.887	Colobinae
CM	3.031	CM	.890	CG	12.98	CG	.083	genus indet.
CA	3.088	PE	1.145	CA	16.13	CA	.017	
TB	1.757	TB	.544	PS	7.33	PS	.287	Cercopithecinae
PS	2.149	PS	.603	TB	7.73	TB	.234	cf. *Theropithecus*
PC	2.315	PC	.661	TG	8.44	TG	.164	
MS	.976	MS	.250	MS	1.65	MS	.269	Cercopithecinae
CA	1.124	CA	.320	CA	2.07	CA	.218	cf. *Parapapio*
TB	1.515	CB	.509	TB	2.28	TB	.196	
MS	2.938	MS	1.039	PU	18.38	PU	.716	Cercopithecinae
PU	3.684	PA	1.062	PH	21.38	PH	.160	cf. *Parapapio*
CA	3.893	CA	1.151	MS	23.37	MS	.059	

TABLE 20. Taxonomic assignments of previously unidentified cercopithecoid forelimb elements from the Lower Omo Basin, southern Ethiopia.

Identification Number	*Body Part*	*Present Location*	*Collection Locality*	*Rank Order*	*SCA Similarity Coefficient*	
Ll-319	Right distal humerus	ENM	Shungura Fm. member B11	1 2 3	PG M CA	.822 .824 .840
O3.0-1410	Left distal humerus w/shaft	ENM	Shungura Fm. member B12	1 2 3	PE N CB	1.075 1.179 1.203
L556-3	Right prox ulna	ENM	Shungura Fm. member C?	1 2 3	TB TO MS	1.005 1.120 1.142
L107-4	Left prox ulna w/shaft	ENM	Shungura Fm. member C	1 2 3	P CG MS	1.779 1.815 1.904
L183-23	Right prox ulna	ENM	Shungura Fm. member C5	1 2 3	PE TB MS	.898 1.013 1.023
L373-3	Left prox ulna w/shaft	ENM	Shungura Fm. member Cl	1 2 3	N CB CG	.934 .951 1.187
O157-308	Right prox ulna	ENM	Shungura Fm. member C6	1 2 3	MS TB CM	.700 .758 .863
L32-283	Right prox radius	ENM	Shungura Fm. member C7	1 2 3	CM CG CA	.839 1.066 1.117
L292-7	Left prox humerus w/shaft	ENM	Shungura Fm. member C7	1 2 3	R N TB	.941 1.254 1.425
L899-1	Left prox humerus	ENM	Shungura Fm. member C8	1 2 3	TO TB TG	.892 1.033 1.299

TABLE 20. Continued.

PCA Euclidean Distance		*PCA Membership Quotient*		*DFA Mahalanobis Distance*		*DFA Posterior Probability*		*Taxonomic Assignment*
CA	1.992	CA	.887	CA	29.51	CA	.616	Cercopithecinae
M	2.477	TB	1.185	PH	30.62	PH	.353	cf.*Cercocebus*
TB	2.410	PS	1.196	TB	36.16	TB	.022	
N	3.680	PE	1.369	PE	28.07	PE	.835	Colobinae
PE	3.722	N	1.426	CB	32.07	CB	.112	genus indet.
CB	3.901	TO	1.606	TG	33.90	TG	.045	
TB	3.494	TB	.907	TB	30.02	TB	.976	*Theropithecus*
PS	3.728	TO	1.276	PH	39.05	PH	.011	*brumpti*
TO	3.907	MS	1.501	MS	40.61	MS	.005	
CG	6.660	CG	1.550	CG	154.18	CG	.988	Colobinae
P	6.670	MS	2.033	MS	162.98	MS	.012	cf. *Paracolobus*
MS	6.941	TB	2.448	TB	173.90	TB	.000	
TB	2.587	TB	.787	TB	23.64	TB	.552	*Theropithecus*
M	3.042	PE	.889	PE	25.36	PE	.234	cf. *brumpti*
TO	3.224	PC	1.326	MS	27.47	MS	.081	
CB	2.735	CB	.954	N	32.33	N	.593	Colobinae
N	2.796	N	1.145	CG	33.55	CG	.332	genus indet.
D	3.937	PE	1.393	CB	36.27	CB	.083	
TB	1.932	TB	.451	MS	17.24	MS	.300	Cercopithecinae
MS	2.245	MS	.813	CM	17.28	CM	.293	cf. *Theropithecus*
PS	2.283	PS	1.012	TB	18.56	TB	.155	
CM	2.719	CM	1.045	CM	5.66	CM	.744	Cercopithecinae
CG	3.397	TB	1.177	CA	9.37	CA	.088	cf. *Cercopithecus*
CA	3.544	PS	1.193	PE	11.09	PE	.049	
R	2.672	N	1.252	N	52.06	N	.903	*Rhinocolobus*
N	3.838	TB	2.013	TB	57.15	TB	.071	*turkanaensis*
TB	4.453	PS	2.083	CB	59.15	CB	.026	
TO	2.972	TB	1.087	TG	31.44	TG	.908	*Theropithecus*
TB	3.507	PU	1.804	PH	36.93	PH	.058	cf. *oswaldi*
TG	4.315	PH	1.810	PC	39.14	PC	.019	

TABLE 21. Taxonomic assignments of previously unidentified cercopithecoid forelimb elements from the Lower Omo Basin, southern Ethiopia.

Identification Number	*Body Part*	*Present Location*	*Collection Locality*	*Rank Order*	*SCA Similarity Coefficient*	
L304-10	Left prox humerus	ENM	Shungura Fm. member C	1	TB	1.251
				2	TO	1.262
				3	PA	2.059
L759-1	Right distal humerus	ENM	Shungura Fm. member C5	1	PS	1.066
				2	TG	1.125
				3	TO	1.137
O18-61	Right distal humerus	ENM	Shungura Fm. member C6	1	CA	.636
				2	TO	.652
				3	TB	.692
O18-79	Right distal humerus	ENM	Shungura Fm. member C6	1	CG	1.084
				2	M	1.258
				3	D	1.311
L869-5	Right distal humerus w/shaft	ENM	Shungura Fm. member C7	1	TB	.772
				2	CM	.950
				3	MS	1.003
L632-1	Left distal humerus	ENM	Shungura Fm. member C8	1	M	.621
				2	D	.625
				3	CA	.677
P397-1	Right complete ulna (prox)	ENM	Shungura Fm. member D	1	TB	.875
				2	PS	.905
				3	PE	.910
P397-1	Right complete ulna (distal)	ENM	Shungura Fm. member D	1	TB	.726
				2	TO	1.190
				3	PS	1.329
O153-277	Right prox radius	ENM	Shungura Fm. member D3	1	TB	.987
				2	CM	.990
				3	CB	.991
L11-19	Right prox radius	ENM	Shungura Fm. member D5	1	CM	.858
				2	TB	.998
				3	CA	1.026

TABLE 21. Continued.

PCA Euclidean Distance	*PCA Membership Quotient*	*DFA Mahalanobis Distance*	*DFA Posterior Probability*	*Taxonomic Assignment*
TB 3.731	TB 1.320	TB 24.54	TB 1.000	*Theropithecus*
TO 3.733	PU 2.661	PS 55.56	PS .000	*brumpti*
PA 6.432	PS 2.697	PU 58.83	PU .000	
TG 3.232	TO 1.291	TO 81.70	TO .886	*Theropithecus*
MS 3.244	PS 1.367	N 85.80	N .114	cf. *oswaldi*
CA 3.268	PE 1.406	CB 103.80	CB .000	
MS 1.114	TO .645	TB 12.87	TB .325	Cercopithecinae
CA 1.690	TB .727	MS 12.96	MS .311	cf. *Theropithecus*
TO 2.146	CA .947	CA 13.12	CA .286	
CG 2.662	CG 1.552	CG 33.74	CG .998	Colobinae
M 4.068	CB 1.675	CB 47.41	CB .001	cf. *Colobus*
CB 4.144	TB 2.131	TB 48.46	TB .001	
TB 2.532	TB .786	TB 23.18	TB .827	*Theropithecus*
MS 3.051	MS 1.021	CM 27.48	CM .096	*brumpti*
CM 3.179	TO 1.381	MS 27.98	MS .075	
D 1.861	CA .910	CA 15.12	CA .518	Colobinae
M 2.144	PE .927	PE 16.72	PE .233	genus indet.
CA 2.244	TO 1.003	MS 16.95	MS .207	
TB 3.002	TB .533	TB 28.66	TB .751	*Theropithecus*
PE 3.190	PS .724	PS 31.00	PS .232	*brumpti*
PS 3.381	PE .878	PE 39.58	PE .004	
TB 3.170	PS 1.123	PS 13.36	PS .692	*Theropithecus*
TO 3.290	PA 1.312	PA 17.36	PA .093	*brumpti*
PS 3.546	PE 1.458	PE 17.56	PE .085	
TB 2.918	TB .865	TB 14.73	TB .595	*Theropithecus*
CB 3.017	PS 1.140	CM 17.50	CM .149	*brumpti*
CM 3.159	PE 1.166	CB 17.50	CB .149	
CM 2.427	TB .897	TB 11.16	TB .798	*Theropithecus*
CA 2.852	CM .973	CM 14.70	CM .136	*brumpti*
TB 2.975	CB 1.146	CA 17.20	CA .039	

TABLE 22. Taxonomic assignments of previously unidentified cercopithecoid forelimb elements from the Lower Omo Basin, southern Ethiopia.

Identification Number	*Body Part*	*Present Location*	*Collection Locality*	*Rank Order*	*SCA Similarity Coefficient*	
O5.0-Cl	Left prox radius	ENM	Shungura Fm. member D	1	CM	1.111
				2	TB	1.300
				3	N	1.313
O59-(1968)	Right distal humerus	ENM	Shungura Fm. member D1	1	CG	1.071
				2	TB	1.149
				3	CB	1.193
L551-2	Left distal humerus	ENM	Shungura Fm. member D1	1	P	.635
				2	PU	.718
				3	CA	.729
L226-2	Left distal humerus	ENM	Shungura Fm. member D2	1	TB	.674
				2	CA	.811
				3	CG	.998
L161-43	Left distal humerus w/shaft	ENM	Shungura Fm. member D3	1	TB	.783
				2	PA	.787
				3	CM	.805
L11-18	Left distal humerus	ENM	Shungura Fm. member D5	1	PE	.862
				2	CB	.881
				3	CA	.906
L11-21	Left distal humerus	ENM	Shungura Fm. member D5	1	TB	.883
				2	MS	.971
				3	CA	1.088
L4-18	Right prox ulna w/shaft	ENM	Shungura Fm. member E3	1	TB	.741
				2	D	.794
				3	TO	.857
L4-17b	Left prox ulna	ENM	Shungura Fm. member E3	1	TO	1.079
				2	PS	1.102
				3	CM	1.165
L236-la	Right prox ulna	ENM	Shungura Fm. member E4	1	CM	.923
				2	CB	.944
				3	CA	.974

TABLE 22. Continued.

PCA Euclidean Distance		*PCA Membership Quotient*		*DFA Mahalanobis Distance*		*DFA Posterior Probability*		*Taxonomic Assignment*
CM	3.638	CM	1.380	TB	12.54	TB	.430	Cercopithecinae
TB	4.101	TB	1.404	N	13.81	N	.229	cf. *Theropithecus*
CG	4.298	N	1.448	CA	14.30	CA	.178	
CB	2.442	TO	1.384	TB	40.34	TB	.384	Cercopithecinae
TB	3.542	TB	1.410	CB	41.05	CB	.270	cf. *Theropithecus*
CG	3.712	CB	1.495	CG	41.23	CG	.246	
P	2.288	PE	.695	PE	13.54	PE	.911	Colobinae
MS	2.368	CA	.784	CA	19.35	CA	.050	cf. *Paracolobus*
PE	2.590	TB	.814	PH	21.57	PH	.016	
TB	2.125	TB	.870	TB	11.39	TB	.587	*Theropithecus*
CA	2.774	TO	.950	CA	12.19	CA	.393	*brumpti*
CG	2.851	CA	1.274	MS	19.40	MS	.011	
PA	2.481	TB	.658	TB	19.87	TB	.701	*Theropithecus*
TB	2.527	PA	.811	PC	24.02	PC	.088	*brumpti*
CM	2.782	TO	.867	TO	24.33	TO	.076	
PE	2.713	PE	1.004	PE	25.19	PE	.588	Colobinae
CB	2.871	CB	1.016	CA	26.78	CA	.267	genus indet.
CA	2.987	TO	1.037	CB	29.63	CB	.064	
TB	2.380	TB	.645	TB	29.96	TB	.495	*Theropithecus*
MS	2.623	MS	1.088	MS	31.50	MS	.229	*brumpti*
CA	3.125	TO	1.118	PS	31.76	PS	.201	
D	2.372	TB	.559	TB	22.27	TB	.507	*Theropithecus*
TB	2.559	CA	.950	TO	22.48	TO	.457	cf. *brumpti*
CA	2.426	TO	.955	MS	29.65	MS	.013	
PG	3.652	TO	1.144	PS	38.85	PS	.658	Cercopithecinae
TO	3.912	TB	1.239	TO	42.12	TO	.118	cf. *Theropithecus*
PS	4.075	PS	1.386	CM	42.97	CM	.084	
M	3.277	PE	1.215	CB	27.78	CB	.304	Colobinae
CB	3.327	TB	1.308	MS	28.07	MS	.263	genus indet.
CM	3.352	MS	1.356	CM	28.22	CM	.243	

TABLE 23. Taxonomic assignments of previously unidentified cercopithecoid forelimb elements from the Lower Omo Basin, southern Ethiopia.

Identification Number	*Body Part*	*Present Location*	*Collection Locality*	*Rank Order*	*SCA Similarity Coefficient*	
O57.4-164	Left prox ulna	ENM	Shungura Fm. member E4	1	P	1.129
				2	D	1.147
				3	R	1.247
O151-685	Left prox ulna	ENM	Shungura Fm. member E3	1	TB	.626
				2	CA	.819
				3	CM	.917
L185-11	Right prox ulna	ENM	Shungura Fm. member E3	1	TO	.821
				2	TB	.826
				3	CM	.871
L236-lb	Right prox radius	ENM	Shungura Fm. member E4	1	MS	.772
				2	CB	.820
				3	CM	.940
O57.5-121	Right prox radius	ENM	Shungura Fm. member E5	1	PS	.757
				2	PC	.863
				3	CG	.898
L187-6	Right prox humerus	ENM	Shungura Fm. member E5	1	PS	.800
				2	PH	.810
				3	TB	.850
L4-38	Left distal humerus	ENM	Shungura Fm. member E3	1	PS	.803
				2	TO	.913
				3	MS	.939
L101-14	Right distal humerus	ENM	Shungura Fm. member El	1	PS	.910
				2	PH	.921
				3	MS	.975
L865-1c	Right distal humerus	ENM	Shungura Fm. member E4	1	TB	.761
				2	MS	.825
				3	CM	1.013
O207-1762	Left scalpula fragment	ENM	Shungura Fm. member E3	1	PS	.853
				2	PC	.960
				3	PU	1.035

TABLE 23. Continued.

PCA Euclidean Distance		*PCA Membership Quotient*		*DFA Mahalanobis Distance*		*DFA Posterior Probability*		*Taxonomic Assignment*
P	3.920	CB	1.781	CB	41.35	CB	.746	Colobinae
D	4.356	PE	1.783	CA	44.49	CA	.155	cf. *Paracolobus*
R	4.437	TB	1.966	CM	44.65	CM	.087	
TB	1.915	TB	.418	TB	15.80	TB	.877	*Theropithecus*
PE	2.812	PE	.859	CA	20.82	CA	.071	*brumpti*
CA	3.017	TO	1.202	CM	21.82	CM	.043	
PS	2.259	TB	.742	TO	21.94	TO	.686	*Theropithecus*
TO	2.312	TO	.999	TB	25.02	TB	.145	cf. *oswaldi*
PG	2.453	MS	1.182	CM	26.61	CM	.066	
CB	2.363	CM	.796	CB	7.12	CB	.663	Colobinae
CM	2.632	CB	.851	CM	9.69	CM	.185	genus indet.
M	2.731	MS	1.029	MS	12.21	MS	.052	
PS	.757	PS	.709	PS	7.12	PS	.447	Cercopithecinae
CG	.863	PC	.931	CG	8.61	CG	.212	cf. *Papio*
PC	.898	TB	.984	PC	8.92	PC	.181	
PS	2.616	PS	.863	PS	9.48	PS	.910	*Papio*
MS	2.620	PH	.887	TB	15.37	TB	.048	species indet.
PH	2.726	TB	1.176	PH	17.35	PH	.018	
TO	2.508	TO	.880	TO	18.29	TO	.825	*Theropithecus*
MS	2.584	TB	.960	CA	23.38	CA	.065	*oswaldi*
TB	2.835	MS	1.020	TB	23.74	TB	.054	
PS	2.645	PS	.744	PH	25.23	PH	.364	*Papio*
MS	2.898	PH	1.017	PS	25.60	PS	.303	species indet.
PH	3.270	CA	1.110	CA	26.82	CA	.164	
TB	1.838	TB	.713	TB	30.18	TB	.492	*Theropithecus*
MS	1.890	MS	.744	MS	30.30	MS	.462	*brumpti*
TO	2.629	TO	.904	CA	36.78	CA	.018	
PC	1.582	PS	.827	PS	12.04	PS	.327	*Papio*
PS	1.719	PC	.865	PH	12.54	PH	.255	species indet.
CM	1.965	CA	.876	CM	12.96	CM	.207	

TABLE 24. Taxonomic assignments of previously unidentified cercopithecoid forelimb elements from the Lower Omo Basin, southern Ethiopia.

Identification Number	*Body Part*	*Present Location*	*Collection Locality*	*Rank Order*	*SCA Similarity Coefficient*	
L867-15	Right prox radius	ENM	Shungura Fm. member F2	1	CM	.376
				2	CA	.441
				3	MS	.477
L238-51	Right prox radius	ENM	Shungura Fm. member F3	1	PH	.597
				2	TO	.610
				3	PU	.703
O130-1908	Left distal radius	ENM	Shungura Fm. member F3	1	TB	1.291
				2	MS	1.437
				3	PC	1.489
L28-29	Left complete humerus (prox)	ENM	Shungura Fm. member Fl	1	CG	.675
				2	TG	.708
				3	PC	.789
O33-404	Left prox humerus	ENM	Shungura Fm. member Fl	1	TB	.566
				2	TG	1.042
				3	TO	1.101
L867-13	Right prox humerus	ENM	Shungura Fm. member F2	1	TO	.799
				2	TB	.860
				3	PH	1.409
L238-257	Left prox humerus	ENM	Shungura Fm. member F3	1	TB	.818
				2	TG	1.236
				3	MS	1.313
O199-1273	Right prox humerus	ENM	Shungura Fm. member F3	1	PE	.666
				2	PS	.723
				3	CB	.730
O108-4	Right prox humerus	ENM	Shungura Fm. member F	1	TB	.693
				2	TO	1.241
				3	TG	1.602
L28-29	Left complete humerus (distal)	ENM	Shungura Fm. member Fl	1	CG	.741
				2	CA	.801
				3	M	.868

TABLE 24. Continued.

PCA Euclidean Distance		*PCA Membership Quotient*		*DFA Mahalanobis Distance*		*DFA Posterior Probability*		*Taxonomic Assignment*
CA	1.690	TB	.461	CM	2.62	CM	.314	Cercopithecinae
MS	1.706	MS	.591	MS	2.37	MS	.298	genus indet.
TB	1.946	CA	.688	CA	3.62	CA	.159	
TO	1.840	TO	.532	TO	6.44	TO	.220	*Theropithecus*
PH	2.001	PH	.533	PH	6.76	PH	.187	*oswaldi*
PU	2.247	PU	.675	PU	7.55	PU	.126	
TB	4.253	PC	1.671	PC	27.31	PC	.895	*Theropithecus*
MS	5.026	MS	1.717	MS	32.49	MS	.067	cf. *brumpti*
PC	5.070	PA	1.748	TG	34.97	TG	.019	
TG	1.795	CG	.934	PS	11.40	PS	.372	Colobinae
CG	2.151	PS	.942	CG	12.50	CG	.214	genus indet.
PH	2.215	TG	1.025	MS	12.70	MS	.194	
TB	1.470	TB	.573	TB	6.70	TB	1.000	*Theropithecus*
TG	3.374	PS	1.215	PS	23.30	PS	.000	*brumpti*
TO	3.495	PU	1.455	PU	31.54	PU	.000	
TO	2.450	TB	.771	TB	20.19	TB	1.000	*Theropithecus*
TB	2.754	PU	1.678	PH	36.41	PH	.000	cf. *oswaldi*
TG	4.621	PS	1.688	PU	40.70	PU	.000	
TB	2.259	TB	.776	TB	19.78	TB	1.000	*Theropithecus*
TG	3.708	PS	1.426	TG	40.47	TG	.000	*brumpti*
PH	3.887	PH	1.673	PS	42.70	PS	.000	
PS	2.151	PE	.714	PE	11.35	PE	.785	Colobinae
PE	2.240	PS	.824	PS	15.66	PS	.091	genus indet.
CB	2.347	CB	.977	CG	17.19	CG	.042	
TB	2.374	TB	.666	TB	12.67	TB	1.000	*Theropithecus*
TO	4.311	PS	1.751	PS	42.45	PS	.000	*brumpti*
PS	5.231	PU	2.106	TG	51.47	TG	.000	
CG	1.799	CG	.566	CA	20.28	CA	.627	Colobinae
M	1.987	TG	.891	CG	21.64	CG	.317	genus indet.
CA	2.564	TB	.907	MS	26.27	MS	.031	

TABLE 25. Taxonomic assignments of previously unidentified cercopithecoid forelimb elements from the Lower Omo Basin, southern Ethiopia.

Identification Number	*Body Part*	*Present Location*	*Collection Locality*	*Rank Order*	*SCA Similarity Coefficient*	
L52-54	Right distal humerus	ENM	Shungura Fm. member F3	1	D	.640
				2	M	.703
				3	CA	.907
O130-2	Left distal humerus	ENM	Shungura Fm. member F3	1	TB	.755
				2	PA	.804
				3	PH	.885
L238-102	Left distal humerus	ENM	Shungura Fm. member F3	1	TB	.859
				2	CA	1.087
				3	MS	1.109
L238-46	Right distal humerus	ENM	Shungura Fm. member F3	1	TG	.731
				2	PU	.734
				3	PS	.825
L504-15	Left prox ulna	ENM	Shungura Fm. member G4	1	PC	.929
				2	PH	1.055
				3	M	1.058
L74-32c	Right prox ulna	ENM	Shungura Fm. member G5	1	TB	1.087
				2	TO	1.095
				3	PS	1.112
L74-31b	Right prox ulna	ENM	Shungura Fm. member G5	1	TB	.692
				2	PS	.726
				3	MS	.777
O222-2765	Right prox ulna	ENM	Shungura Fm. member G7	1	PH	.650
				2	PC	.711
				3	PU	.778
L776-2	Right prox ulna	ENM	Shungura Fm. member G5	1	TO	.732
				2	PG	.774
				3	PS	.849
F305-1	Left prox ulna w/shaft	ENM	Shungura Fm. member G9	1	PH	.582
				2	PC	.671
				3	M	.701

TABLE 25. Continued.

PCA Euclidean Distance		*PCA Membership Quotient*		*DFA Mahalanobis Distance*		*DFA Posterior Probability*		*Taxonomic Assignment*
D	1.497	TB	1.052	CA	21.56	CA	.877	Cercopithecinae
M	1.738	TO	1.113	TB	26.68	TB	.068	genus indet.
CG	2.213	CA	1.221	MS	28.98	MS	.021	
PA	2.518	TB	.687	TB	17.92	TB	.693	*Theropithecus*
TB	2.760	PA	.900	PH	20.80	PH	.164	*brumpti*
PH	3.100	TO	1.036	PU	21.56	PU	.113	
TB	2.789	TB	.867	TB	28.37	TB	.997	*Theropithecus*
MS	3.079	TO	1.336	CA	40.56	CA	.002	*brumpti*
CA	3.083	CA	1.436	CM	42.97	CM	.001	
PU	2.123	TO	.808	TG	19.55	TG	.845	*Theropithecus*
TG	2.306	PU	.946	TB	23.99	TB	.092	species indet.
TO	2.515	PS	.996	TO	27.25	TO	.018	
PC	3.321	PU	.881	PC	34.60	PC	.682	*Papio*
PU	3.440	PC	1.083	TB	36.45	TB	.270	species indet.
M	3.615	TB	1.269	PH	41.19	PH	.025	
TO	1.855	TO	.691	TO	36.67	TO	.955	*Theropithecus*
PS	2.640	TB	1.132	PC	43.82	PC	.027	*oswaldi*
TB	2.756	PE	1.189	PA	44.84	PA	.016	
TB	1.813	TB	.522	PS	20.49	PS	.632	*Theropithecus*
PS	2.013	MS	.821	TB	22.14	TB	.277	cf. *brumpti*
MS	2.462	PS	.842	MS	24.62	MS	.080	
PH	2.043	PH	.616	PH	21.49	PH	.450	*Papio*
PU	2.051	TB	.692	PC	23.07	PC	.204	species indet.
TG	2.396	PC	.745	TB	23.24	TB	.188	
PG	2.313	TO	.547	TO	25.48	TO	.861	*Theropithecus*
TO	2.807	TB	.614	PS	29.48	PS	.117	*oswaldi*
PS	2.995	PS	.689	PH	34.33	PH	.010	
PH	1.910	PH	.708	PH	15.41	PH	.868	*Papio*
M	1.972	PC	.772	PC	20.64	PC	.063	species indet.
PC	2.229	PE	.882	TB	21.60	TB	.039	

TABLE 26. Taxonomic assignments of previously unidentified cercopithecoid forelimb elements from the Lower Omo Basin, southern Ethiopia.

Identification Number	*Body Part*	*Present Location*	*Collection Locality*	*Rank Order*	*SCA Similarity Coefficient*	
L626-71	Right prox ulna	ENM	Shungura Fm. member G12	1	CM	1.012
				2	CA	1.036
				3	TB	1.099
L627-237	Left prox ulna	ENM	Shungura Fm. member G13	1	TO	.780
				2	PG	.806
				3	TB	.859
L895-1f	Right prox ulna	ENM	Shungura Fm. member G15	1	M	1.025
				2	D	1.029
				3	MS	1.045
F513-1	Left prox ulna	ENM	Shungura Fm. member G27	1	TB	.797
				2	MS	.931
				3	CM	.932
O75N-1866	Right distal ulna w/shaft	ENM	Shungura Fm. member G12	1	N	.725
				2	CB	.776
				3	CG	.796
L65-27	Left complete radius (prox)	ENM	Shungura Fm. member Gl	1	M	.852
				2	CA	.956
				3	MS	.981
O93-576	Left prox radius w/shaft	ENM	Shungura Fm. member G6	1	TB	1.254
				2	PA	1.427
				3	PC	1.444
O75-814	Left prox radius w/shaft	ENM	Shungura Fm. member G8	1	CB	.630
				2	CA	.674
				3	MS	.735
O50-4449	Right prox radius	ENM	Shungura Fm. member G8	1	P	.718
				2	PS	.728
				3	M	.759
L615-9	Right prox radius	ENM	Shungura Fm. member G9	1	PC	.414
				2	PS	.561
				3	MS	.686

TABLE 26. Continued.

PCA Euclidean Distance		*PCA Membership Quotient*		*DFA Mahalanobis Distance*		*DFA Posterior Probability*		*Taxonomic Assignment*
PG	1.902	TB	.819	CM	42.50	CM	.803	Cercopithecinae
TB	2.319	TO	1.327	CA	46.24	CA	.124	genus indet.
CM	2.472	CM	1.392	MS	47.46	MS	.068	
TO	1.387	TB	.905	TO	19.97	TO	.765	*Theropithecus*
PG	2.558	TO	.985	PS	24.34	PS	.086	*oswaldi*
TB	3.126	MS	1.097	TB	24.91	TB	.065	
M	2.213	PE	1.008	CB	35.29	CB	.902	Colobinae
D	3.216	MS	1.046	MS	40.74	MS	.059	genus indet.
MS	3.333	TB	1.054	CA	42.90	CA	.020	
TB	2.562	TB	.549	TB	20.94	TB	.598	*Theropithecus*
CM	3.236	TO	1.049	TO	23.30	TO	.184	*brumpti*
TO	3.359	MS	1.150	CA	24.86	CA	.085	
N	1.965	PS	.732	N	4.07	N	.279	Colobinae
CG	2.196	N	.736	CB	4.32	CB	.246	genus indet.
MS	2.216	PA	.805	MS	5.67	MS	.125	
M	2.493	MS	.923	CA	9.71	CA	.531	Cercopithecinae
CA	3.029	CA	1.065	MS	10.33	MS	.389	cf. *Parapapio*
MS	3.045	TB	1.080	TB	14.77	TB	.042	
TB	3.720	TB	1.144	TB	27.60	TB	.781	*Theropithecus*
PC	4.138	PA	1.383	PE	31.03	PE	.141	*brumpti*
PA	4.344	PS	1.463	CB	34.04	CB	.031	
CA	1.925	CB	.672	CB	6.40	CB	.387	Colobinae
CB	1.976	CG	.704	CG	6.70	CG	.334	cf. *Colobus*
MS	2.297	CA	.745	MS	8.47	MS	.138	
P	2.318	PS	.657	CA	6.02	CA	.389	Colobinae
M	2.353	CA	.796	PS	7.42	PS	.193	cf. *Paracolobus*
PS	2.380	MS	.927	MS	7.79	MS	.160	
PC	1.333	PC	.429	PC	1.53	PC	.584	Cercopithecinae
PS	1.845	PS	.528	PA	4.60	PA	.126	cf. *Papio*
MS	2.168	TB	.662	PS	7.03	PS	.037	

TABLE 27. Taxonomic assignments of previously unidentified cercopithecoid forelimb elements from the Lower Omo Basin, southern Ethiopia.

Identification Number	*Body Part*	*Present Location*	*Collection Locality*	*Rank Order*	*SCA Similarity Coefficient*	
F305-2	Left prox radius	ENM	Shungura Fm. member G9	1	CM	.847
				2	MS	.850
				3	CA	.863
O75N-689	Right prox radius	ENM	Shungura Fm. member G10	1	PU	1.031
				2	PA	1.082
				3	PH	1.145
O249-4987	Left prox radius	ENM	Shungura Fm. member G13	1	TB	.680
				2	CB	.817
				3	CA	.883
O242-4790	Right prox radius	ENM	Shungura Fm. member G13	1	TB	.431
				2	MS	.433
				3	PS	.523
L627-376	Left prox radius	ENM	Shungura Fm. member G13	1	CA	.726
				2	P	.849
				3	CB	.864
L895-lg	Left prox radius	ENM	Shungura Fm. member G15	1	CB	.420
				2	CA	.456
				3	CM	.481
L65-27	Left complete radius (distal)	ENM	Shungura Fm. member Gl	1	CM	1.000
				2	PC	1.001
				3	M	1.072
L726-9	Right distal radius	ENM	Shungura Fm. member G5	1	PH	.779
				2	D	.965
				3	PA	1.018
L25-132	Left distal radius	ENM	Shungura Fm. member G13	1	MS	.506
				2	PC	.644
				3	PA	.695
L74-32b	Right prox humerus w/shaft	ENM	Shungura Fm. member G5	1	TO	.889
				2	TB	1.175
				3	N	1.224

TABLE 27. Continued.

PCA Euclidean Distance		*PCA Membership Quotient*		*DFA Mahalanobis Distance*		*DFA Posterior Probability*		*Taxonomic Assignment*
CA	2.433	MS	.702	CM	7.97	CM	.453	Cercopithecinae
CG	2.564	CA	.947	CA	8.86	CA	.291	cf. *Parapapio*
CM	2.601	CM	.954	MS	10.97	MS	.101	
PU	3.263	PU	.990	PU	12.77	PU	.441	*Papio*
PA	3.486	PA	1.018	PA	12.82	PA	.431	species indet.
PH	3.761	PH	1.134	PH	15.58	PH	.108	
TB	2.119	TB	.567	TB	9.84	TB	.498	*Theropithecus*
PS	2.625	PS	.850	CA	11.33	CA	.236	*brumpti*
CA	2.648	CB	.972	CB	13.24	CB	.091	
TB	1.071	TB	.325	TB	2.39	TB	.294	*Theropithecus*
MS	1.462	MS	.494	MS	3.32	MS	.184	*brumpti*
PS	1.765	PS	.555	PC	3.40	PC	.177	
CA	2.298	CA	.866	CA	6.85	CA	.534	Cercopithecinae
P	2.434	MS	.888	CM	7.98	CM	.305	cf. *Parapapio*
CB	2.788	CB	1.040	MS	11.29	MS	.058	
PE	.997	CB	.406	CM	1.97	CM	.411	Colobinae
CG	1.701	CG	.657	CB	3.63	CB	.179	genus indet.
CM	1.950	CM	.723	CA	5.17	CA	.083	
CM	3.326	PC	1.010	CM	46.02	CM	.919	Cercopithecinae
PC	3.555	CM	1.054	PC	51.99	PC	.046	cf. *Parapapio*
M	3.805	CA	1.271	PH	54.44	PH	.014	
D	2.867	PA	.675	CM	17.90	CM	.673	Cercopithecinae
PH	3.190	PH	1.016	CA	21.65	CA	.103	genus indet.
PA	3.419	PS	1.133	PH	21.96	PH	.088	
PH	1.583	PC	.671	PC	7.23	PC	.344	*Papio*
MS	2.089	PA	.722	PA	7.98	PA	.236	species indet.
PC	2.123	PH	.753	PH	8.53	PH	.180	
TO	2.693	TB	.865	TB	26.97	TB	.994	*Theropithecus*
TB	3.929	PS	1.441	PS	38.43	PS	.003	cf. *oswaldi*
N	4.115	PU	1.633	TG	41.83	TG	.001	

TABLE 28. Taxonomic assignments of previously unidentified cercopithecoid forelimb elements from the Lower Omo Basin, southern Ethiopia.

Identification Number	*Body Part*	*Present Location*	*Collection Locality*	*Rank Order*	*SCA Similarity Coefficient*	
L434-8	Left	ENM	Shungura Fm.	1	R	.861
	prox		member G7	2	TO	1.256
	humerus			3	TB	1.373
O75S-461	Right	ENM	Shungura Fm.	1	TO	.564
	prox		member G7	2	TB	.771
	humerus			3	PH	1.267
O195-1354	Left	ENM	Shungura Fm.	1	TG	.694
	prox		member G7	2	PC	.833
	humerus			3	TB	.838
O75N-688	Right	ENM	Shungura Fm.	1	PS	.541
	prox		member G12	2	PE	.841
	humerus			3	MS	.884
O75N-693	Left	ENM	Shungura Fm.	1	TB	1.055
	prox		member G12	2	PS	1.063
	humerus			3	PC	1.134
O75N-C204	Right	ENM	Shungura Fm.	1	TO	.868
	prox		member G12	2	TB	1.126
	humerus			3	PH	1.132
O75N-C205	Left	ENM	Shungura Fm.	1	TO	.953
	prox		member G12	2	TB	.985
	humerus			3	PS	1.052
O75N-C202	Left	ENM	Shungura Fm.	1	PC	.553
	prox		member G12	2	PU	.555
	humerus			3	PH	.576
O214-4156	Right	ENM	Shungura Fm.	1	PS	1.028
	prox		member G27	2	TB	1.041
	humerus			3	PH	1.076
F501-1	Right	ENM	Shungura Fm.	1	MS	.708
	prox		member G28	2	CM	.723
	humerus			3	M	.744

TABLE 28. Continued.

PCA Euclidean Distance		*PCA Membership Quotient*		*DFA Mahalanobis Distance*		*DFA Posterior Probability*		*Taxonomic Assignment*
R	1.766	CB	.601	CB	47.38	CB	.997	*Rhinocolobus*
TO	3.660	TB	1.891	PS	59.95	PS	.002	*turkanaensis*
TB	3.697	PS	2.041	TB	60.74	TB	.001	
TO	1.789	TB	.624	TB	11.22	TB	1.000	*Theropithecus*
TB	2.234	PU	1.490	PU	39.51	PU	.000	cf. *oswaldi*
PH	3.971	PS	1.609	PS	42.51	PS	.000	
TG	1.816	PS	.882	TG	17.83	TG	.748	*Theropithecus*
MS	2.257	TG	.944	TB	20.41	TB	.205	cf. *brumpti*
PC	2.316	PC	1.045	PU	23.83	PU	.037	
PS	1.513	PS	.506	PS	8.84	PS	.900	*Papio*
PE	2.463	PE	.843	TB	13.32	TB	.096	species indet.
PH	2.581	PU	1.094	PE	20.67	PE	.002	
PH	2.887	TB	1.123	TB	17.98	TB	1.000	*Theropithecus*
TB	2.994	PS	1.149	PE	36.81	PE	.000	*brumpti*
PS	3.072	PU	1.380	TG	49.56	TG	.000	
TO	2.497	TB	.927	TB	21.58	TB	1.000	*Theropithecus*
TB	3.160	PU	1.309	PU	43.76	PU	.000	cf. *oswaldi*
TG	3.443	PS	1.477	PS	45.23	PS	.000	
TO	2.637	PS	1.030	PS	8.35	PS	.632	Cercopithecinae
TB	2.767	TB	1.043	MS	10.50	MS	.216	cf. *Theropithecus*
PS	3.144	MS	1.277	TB	12.45	TB	.081	
PH	1.512	PH	.616	PU	30.18	PU	.425	*Papio*
PU	1.698	PU	.693	PS	30.86	PS	.304	species indet.
PC	1.825	PS	.793	PH	31.44	PH	.227	
TB	3.078	TB	1.049	TB	30.15	TB	.690	*Theropithecus*
PH	3.108	PS	1.057	PS	31.81	PS	.302	*brumpti*
PS	3.211	PU	1.207	PE	39.12	PE	.008	
CM	2.153	CA	.894	CM	15.20	CM	.800	Cercopithecinae
M	2.156	CM	.986	MS	18.31	MS	.169	genus indet.
MS	2.267	MS	1.013	PS	22.45	PS	.021	

TABLE 29. Taxonomic assignments of previously unidentified cercopithecoid forelimb elements from the Lower Omo Basin, southern Ethiopia.

Identification Number	*Body Part*	*Present Location*	*Collection Locality*	*Rank Order*	*SCA Similarity Coefficient*	
L16-59p	Left distal humerus	ENM	Shungura Fm. member G4	1	TO	1.203
				2	MS	1.223
				3	PS	1.306
L7-17	Right distal humerus	ENM	Shungura Fm. member G5	1	TB	.885
				2	CA	.914
				3	M	.963
L74-31a	Right distal humerus	ENM	Shungura Fm. member G5	1	TO	.940
				2	TB	.979
				3	PH	.993
L74-32a	Right distal humerus	ENM	Shungura Fm. member G5	1	TB	.637
				2	CM	.697
				3	CA	.729
L7-15	Right distal humerus	ENM	Shungura Fm. member G5	1	CG	1.344
				2	R	1.513
				3	D	1.762
O222-2757	Left distal humerus	ENM	Shungura Fm. member G7	1	R	.932
				2	CG	1.189
				3	CB	1.225
L67-137	Left distal humerus	ENM	Shungura Fm. member G7	1	CB	.591
				2	M	.776
				3	CA	.861
O195-1384	Left distal humerus	ENM	Shungura Fm. member G7	1	MS	.509
				2	TB	.538
				3	CA	.558
L613-48	Right distal humerus	ENM	Shungura Fm. member G7	1	CB	1.007
				2	PE	1.091
				3	CA	1.092
O75N-692	Left distal humerus	ENM	Shungura Fm. member G12	1	MS	.611
				2	TB	.744
				3	CA	.764

TABLE 29. Continued.

PCA Euclidean Distance		*PCA Membership Quotient*		*DFA Mahalanobis Distance*		*DFA Posterior Probability*		*Taxonomic Assignment*
TO	3.947	TO	1.151	TO	47.48	TO	.783	*Theropithecus*
MS	4.024	TB	1.315	PS	51.52	PS	.103	*oswaldi*
TB	4.259	PA	1.656	PA	52.54	PA	.062	
TB	3.165	TB	.691	TB	22.26	TB	.950	*Theropithecus*
CA	3.390	MS	1.101	MS	28.55	MS	.041	*brumpti*
M	3.419	TO	1.116	CA	31.46	CA	.010	
TO	3.553	TB	1.125	TB	21.61	TB	.662	*Theropithecus*
TB	3.722	TO	1.464	PH	24.53	PH	.152	cf. *oswaldi*
PH	3.891	PS	1.505	TO	24.60	TO	.148	
TB	2.001	TB	.649	TB	10.92	TB	.881	*Theropithecus*
MS	2.017	TO	.680	CA	17.25	CA	.037	*brumpti*
CM	2.213	CM	.754	PA	17.84	PA	.028	
CG	4.789	CG	2.128	CG	39.54	CB	1.000	*Colobinae*
R	5.894	PE	3.113	CB	59.04	CB	.000	cf. *Rhinocolobus*
CB	6.514	CG	3.135	PE	81.80	PE	.000	
R	2.749	CG	1.211	CB	35.83	CB	.743	*Rhinocolobus*
CG	4.621	CB	1.432	CG	39.02	CG	.150	*turkanaensis*
CB	4.802	MS	2.089	PE	41.50	PE	.044	
CB	1.617	CB	.758	CB	15.06	CB	.947	Colobinae
M	3.006	PE	1.085	CA	21.64	CA	.035	cf. *Colobus*
CA	3.071	TO	1.371	PE	23.85	PE	.012	
TB	.941	TB	.443	TB	8.86	TB	.377	*Theropithecus*
MS	1.689	TO	.633	PS	9.60	PS	.261	*brumpti*
CA	1.715	PS	.669	MS	10.31	MS	.183	
CB	2.851	PS	1.356	CB	29.34	CB	.609	Colobinae
PE	3.640	CB	1.445	PE	31.94	PE	.166	genus indet.
M	3.953	PE	1.447	TB	32.08	TB	.155	
MS	2.027	TB	.835	TB	13.47	TB	.979	*Theropithecus*
TB	2.534	MS	.843	MS	22.49	MS	.011	cf. *brumpti*
TO	2.972	TO	.877	CA	22.70	CA	.010	

TABLE 30. Taxonomic assignments of previously unidentified cercopithecoid forelimb elements from the Lower Omo Basin, southern Ethiopia.

Identification Number	*Body Part*	*Present Location*	*Collection Locality*	*Rank Order*	*SCA Similarity Coefficient*	
O75N-C200	Left distal humerus	ENM	Shungura Fm. member G12	1 2 3	D CG CB	.651 .661 .701
O75N-C201	Right distal humerus	ENM	Shungura Fm. member G12	1 2 3	TO CA TB	.620 .689 .697
O75N-C203	Left distal humerus	ENM	Shungura Fm. member G12	1 2 3	TO MS TB	.464 .650 .661
O75N-C206	Left distal humerus	ENM	Shungura Fm. member G12	1 2 3	M CB CA	.688 .715 .744
F500-1	Right complete humerus	ENM	Shungura Fm. member G28	1 2 3	R CB M	.962 1.039 1.149
L16-59x	Left scapula fragment	ENM	Shungura Fm. member G4	1 2 3	TO TG CA	.798 .982 .995
O75N-C207	Right scapula fragment	ENM	Shungura Fm. member G12	1 2 3	PS PC MS	1.175 1.340 1.462
F255-8	Left prox ulna	ENM	Shungura Fm. member H3	1 2 3	P D CA	.621 .639 .754

TABLE 30. Continued.

PCA Euclidean Distance		*PCA Membership Quotient*		*DFA Mahalanobis Distance*		*DFA Posterior Probability*		*Taxonomic Assignment*
CG	2.163	CG	.670	CG	18.45	CG	.765	Colobinae
D	2.333	CB	.828	PE	22.41	PE	.106	genus indet.
CB	2.354	PE	1.091	CB	22.63	CB	.095	
CA	2.056	CA	.912	TO	16.69	TO	.474	*Theropithecus*
TO	2.159	TO	.994	CA	16.93	CA	.421	cf. *oswaldi*
MS	2.350	TB	1.014	MS	20.58	MS	.068	
CA	1.345	TO	.518	TB	9.28	TB	.508	*Theropithecus*
TO	1.980	TB	.681	TO	9.88	TO	.376	cf. *oswaldi*
TB	2.097	CA	.812	MS	13.20	MS	.072	
CA	1.921	CB	.837	CB	18.45	CB	.394	Colobinae
CB	2.054	CA	.913	PE	18.77	PE	.335	genus indet.
M	2.275	TO	.997	CA	20.29	CA	.157	
R	2.705	CB	1.283	CB	34.78	CB	.882	*Rhinocolobus*
CB	2.857	CG	1.479	CG	39.67	CG	.076	*turkanaensis*
CG	3.470	TO	1.685	PE	41.34	PE	.033	
TO	2.251	TG	1.160	TG	3.99	TG	.249	*Theropithecus*
TG	2.721	PH	1.172	CA	4.55	CA	.188	*oswaldi*
CA	2.747	CA	1.232	PC	4.75	PC	.171	
MS	3.186	PS	1.384	PS	7.01	PS	.571	Cercopithecinae
PS	3.463	MS	1.419	PC	9.64	PC	.153	cf. *Papio*
PC	4.310	PH	1.870	MS	10.63	MS	.093	
P	1.577	CA	.812	CA	23.17	CA	.645	Colobinae
D	2.090	CG	1.005	CB	24.40	CB	.349	cf. *Paracolobus*
R	2.486	TB	1.063	CM	33.09	CM	.005	

TABLE 31. Summary of total sample of newly identified cercopithecoid postcranial specimens recovered from the Usno and Shungura formations of the Lower Omo Basin, southern Ethiopia, with percentage breakdowns by stratigraphic level and taxon.

Taxon	*Usno Formation*	*Shungura Formation*								TOTAL
	U8	*A*	*B*	*C*	*D*	*E*	*F*	*G*	*H**	
Theropithecus species	3 (25%)	–	1 (11%)	2 (2%)	2 (20%)	1 (10%)	1 (8%)	1 (2%)	–	11 (6%)
Theropithecus brumpti	2 (17%)	–	2 (22%)	82† (89%)	6 (60%)	3 (30%)	6 (46%)	12 (23%)	–	113 (56%)
Theropithecus oswaldi	–	–	–	2 (2%)	–	2 (20%)	2 (15%)	11 (21%)	–	17 (8%)
Papio species	–	–	–	–	–	4 (40%)	–	9 (17%)	–	13 (6%)
Parapapio species	2 (17%)	–	2 (22%)	–	–	–	–	3 (6%)	–	7 (3%)
Cercocebus species	1 (8%)	–	1 (11%)	–	–	–	–	–	–	2 (1%)

Cercopithecus species	1 (8%)	–	–	1 (1%)	–	–	–	–	–	2 (1%)
Cercopithecinae genus indet.	–	–	–	–	–	–	2 (15%)	3 (6%)	–	5 (2.5%)
Colobus species	1 (8%)	1 (100%)	–	1 (1%)	–	–	–	2 (4%)	–	5 (2.5%)
Paracolobus cf. *mutiwa*	–	–	–	1 (1%)	1 (10%)	1 (10%)	–	1 (2%)	1 (100%)	5 (2.5%)
Rhinocolobus turkanaensis	–	–	–	1 (1%)	–	–	–	4 (7%)	–	5 (2.5%)
Colobinae genus indet.	2 (17%)	–	3 (34%)	2 (2%)	1 (10%)	2 (20%)	2 (15%)	6 (12%)	–	18 (9%)
TOTAL	12 (6%)	1 (.5%)	9 (5%)	92 (45%)	10 (5%)	13 (5%)	13 (7%)	52 (26%)	1 (.5%)	203 (100%)

*No primate postcranial remains are known from members J, K, and L in the Shungura Formation.

†Also contains the extinct comparative sample of *T. brumpti*, identified by direct association with cranial remains and by stratigraphic level associations with specimens identified as *T. brumpti* (see Table 46 in Appendix and discussion in Chapter 2.)

The Afar Depression, northeastern Ethiopia

The fossiliferous beds of the Hadar Formation and several nearby sites in the Afar Depression span a time range from somewhat less than 4 Ma to about 2.5 Ma (see Fig. 55). Thus, the Hadar Formation was laid down at about the same time as the Usno Formation and the lower part of the Shungura Formation. However, the mode of deposition of the Hadar Formation was very different from that recorded in the Lower Omo Basin. Lacustrine conditions prevailed in the west-central Afar Sedimentary Basin. Thus, though the total number of cercopithecoids collected from the Hadar area is small (less than 200), individual specimens are relatively more complete, and there is a much higher percentage of identifiable postcranial remains. A total of 27 forelimb postcranial specimens from localities in the Hadar Formation and nearby sites of Ahmado and Leadu are included in this study. Several of these are directly associated with cranial remains and/or other postcranial elements.

All pertinent information regarding each forelimb specimen, together with its taxonomic assignment, appears in Tables 32-35. A summary of all the taxonomic assignments for the Hadar forelimb postcrania, organized by stratigraphic level, is given in Table 36. Complete identification of the Hadar cranial remains of *Theropithecus darti* have recently been published (Eck, 1993). However, since the other cercopithecoid crainal remains have yet to be published in final form, my study will rely on the preliminary identification of *all* Hadar cranial remains made by Delson (1984)—for summary see Fig. 54.

With a sample of only 27 specimens (even though several are represented by more than a single forelimb element), it is difficult to draw too many conclusions. A few observations can be made regarding the relative frequency of cercopithecoid species within the Hadar Formation. For instance, in the earliest horizon (lower SH member), several specimens referable to *Papio* are identified. In the overlying DD member, with its much higher percentage of cercopithecoids, no specimens assignable to *Papio* are identified, yet both *Theropithecus* and *Parapapio* are present. Gray (1980) documents a change in both the fauna and paleoenvironment at Hadar, proceeding from open marshy lake-edge habitats in the lower SH member to more bushy forest-vegetation habitats in the DD member and finally to mixed and then more open habitats in the KH member. Is this paleoenvironmental shift at Hadar reflected in the distribution of cercopithecoid species? From analysis of the Shungura Formation fauna at the Omo, *T. brumpti* is certainly associated with more forested habitats (member C), whereas the frequency of *Papio* dramatically increases in member G where more open lake-shore conditions prevail. Little is known as yet about the paleoenvironmental preferences of *T. darti*, though its type site, the South African cave of Makapansgat, has a fauna indicative of wetter, more closed habitats. So the answer to this paleoenvironmental question appears to be a tentative "yes," but see also comments by Eck (1993).

The presence of *T. brumpti* in the DD member of the Hadar Formation is, of itself, an important range extension. Though it has not been identified cranially at Hadar, it unequivocally exists, based especially on the identification of a distinctive distal humerus, AL 300-1 (see Table 34). This documents the co-existence of *T. brumpti* with the more generalized *T. darti* and suggests that the former species has an independent ancestry extending well into the Pliocene.

Two postcranial specimens have been identified from the KH member. The *Papio* specimen, AL 363-9b, is specifically assignable to *P. hamadryas* (Table 35). From study of the degree of fossilization of the original specimen, together with its morphology, it seems very possible that this humerus was derived from a more recent stratigraphic level (? upper Pleistocene), and its presence in the KH member is therefore intrusive. This opinion, reached during measurement of this specimen, is substantiated by the high scores from all techniques (see Table 35), indicating close affinities with *P. hamadryas*, which is the extant baboon in the Afar region. Subsequent study by Eric Delson (pers. comm.) also suggests that AL 363-9b is intrusive. This demonstrates an additional use for the analytical techniques employed in this study: they can provide an independent check on the possible intrusive nature of a fossil.

Four specimens from the DD member at Hadar as well as four specimens from the nearby site of Ahmado are assignable to *Parapapio* sp. This is the largest number of specimens assigned to a single taxon from the Afar Depression localities. A similar pattern is found at Laetoli in Tanzania, where 71 of 130 identifiable cercopithecoid cranial specimens are assigned to *Parapapio ado* (M. G. Leakey and Delson, 1987). Though the cranial remains of all specimens from the Afar Depression have yet to be published, preliminary reports indicate the presence of a small *Parapapio* species (Delson, 1984). Since Hadar and Laetoli are near-temporal equivalents (F. C. Howell, 1978), this is to be expected. However, whether or not the *Parapapio* from Hadar can be identified as *P. ado* remains to be demonstrated. Leakey and Delson also point out that *P. ado* from Laetoli is nearly identical in dental size and morphology to *Parapapio jonesi*, known from the South African cave localities, and may therefore be synonymous. (If this proves to be the case, *P. ado*, described by Hopwood 1936, has priority.)

With regard to the Colobinae, a large distal humerus from the KH member, AL 363-12 (Table 35), is assigned to *Rhinocolobus*, a large colobine known from the Koobi Fora Formation in Kenya and the Usno and Shungura Formations in southern Ethiopia. *Rhinocolobus* has been identified from cranial remains throughout the Hadar Formation, but not as yet from the KH member (Fig. 54). The AL 363-12 specimen is identical to the humerus of the only associated skeleton of *Rhinocolobus* known from the Koobi Fora Formation, discussed by M. G. Leakey (1982). This helps to establish the species-specific morphology of the forelimb skeleton of this extinct colobine. Furthermore, these analyses further demonstrate the close affinities of *Rhinocolobus* to extant members of the genus *Colobus* (see rank-ordering of AL 363-12 in Table 35).

Another distinctive extinct colobine has been recovered from elsewhere in the Afar Depression in an area called the Leadu Basin. This specimen, AL 2-, consists of a partial skull and most of the postcranial skeleton. Since the entire forelimb

TABLE 32. Taxonomic assignments of previously unidentified cercopithecoid forelimb elements from the Afar Depression, northeastern Ethiopia.

Identification Number	*Body Part*	*Present Location*	*Collection Locality*	*Rank Order*	*SCA Similarity Coefficient*	
AL 196-3d (same individual as AL 196-3c)	Left prox ulna	ENM	Hadar Fm. DD-2	1 2 3	PS TO MS	.709 .726 .790
AL 126-32	Left prox ulna	ENM	Hadar Fm. SH-2s	1 2 3	PA CM CA	.569 .686 .718
AL 2-65 (part of associated skeleton)	Left prox ulna	ENM	Leadu Basin	1 2 3	CB CG PE	.724 .766 .767
AL 133-1	Right prox ulna	ENM	Hadar Fm. DD-2	1 2 3	TB CA TO	.893 .947 1.037
AL 309-4	Right prox ulna	ENM	Hadar Fm. DD-3	1 2 3	MS CA CM	.996 1.007 1.057
AL 332-29	Left prox ulna	ENM	Hadar Fm. DD-1	1 2 3	CA CM MS	.679 .703 .722
AL 133-5	Right prox ulna	ENM	Hadar Fm. DD-3s	1 2 3	PS TB TO	.506 .602 .654
AL 2-37 (part of associated skeleton)	Left distal ulna	ENM	Leadu Basin	1 2 3	CG PE PA	.801 1.003 1.044
AL 29-17	Left distal ulna	ENM	Leadu Basin	1 2 3	CG PA PH	.850 1.080 1.233
AL 187-3	Left prox radius	ENM	Hadar Fm. DD-3s	1 2 3	TG MS TB	.568 .642 .664

TABLE 32. Continued.

PCA Euclidean Distance		*PCA Membership Quotient*		*DFA Mahalanobis Distance*		*DFA Posterior Probability*		*Taxonomic Assignment*
PS	1.886	TO	.771	TO	14.74	TO	.638	*Theropithecus*
TO	2.498	TB	.755	TB	16.88	TB	.219	cf. *darti*
MS	2.805	PS	.758	PS	18.30	PS	.107	
PA	1.603	PH	.742	PA	15.48	PA	.425	Cercopithecinae
PH	2.070	PA	.774	PH	16.03	PH	.324	cf. *Papio*
PG	2.255	TB	.820	CM	17.43	CM	.160	
M	1.774	PE	.641	PE	18.34	PE	.669	Colobinae
CB	1.911	CB	.679	MS	20.74	MS	.202	genus indet.
PE	1.925	TB	.872	CB	22.32	CB	.092	
TB	2.508	TB	.759	TB	25.07	TB	.665	*Theropithecus*
CA	3.019	PE	1.255	CA	27.61	CA	.187	cf. *brumpti*
PE	3.247	TO	1.279	TO	28.36	TO	.129	
CA	2.016	CA	1.055	CM	42.04	CM	.586	Cercopithecinae
CM	2.179	CM	1.367	CA	43.34	CA	.306	cf. *Parapapio*
PE	2.505	MS	1.501	MS	45.53	MS	.102	
CA	1.483	CA	.803	CM	19.31	CM	.727	Cercopithecinae
CM	1.677	MS	.869	MS	22.50	MS	.147	cf. *Parapapio*
MS	1.866	PE	.925	PE	24.48	PE	.055	
TB	1.655	TB	.393	TO	6.34	TO	.913	*Theropithecus*
TO	1.819	PS	.641	TB	11.19	TB	.081	cf. *darti*
PS	2.340	TO	.780	PS	17.31	PS	.004	
CG	2.208	PE	.836	CG	6.78	CG	.640	Colobinae
PE	2.709	CG	.867	PE	8.79	PE	.234	genus indet.
PA	2.906	PA	.959	PA	10.53	PA	.098	
CG	2.550	CG	.887	CG	6.25	CG	.678	Colobinae
PA	2.919	PA	.999	PA	8.53	PA	.216	genus indet.
PH	3.346	PE	1.227	TG	12.12	TG	.036	
TB	1.700	TB	.468	TB	4.18	TB	.323	*Theropithecus*
TG	1.909	PS	.692	CA	5.11	CA	.203	cf. *darti*
MS	1.996	TG	.709	TG	5.65	TG	.155	

TABLE 33. Taxonomic assignments of previously unidentified cercopithecoid forelimb elements from the Afar Depression, northeastern Ethiopia.

Identification Number	*Body Part*	*Present Location*	*Collection Locality*	*Rank Order*	*SCA Similarity Coefficient*	
AL 2-66 (part of associated skeleton)	Left prox radius	ENM	Leadu Basin	1	PE	.728
				2	TG	.783
				3	TO	.795
AL 100-272	Right prox radius	ENM	Ahmado	1	CB	.935
				2	TB	.964
				3	MS	1.103
AL 2-67 (part of associated skeleton)	Left distal radius	ENM	Leadu Basin	1	D	1.168
				2	MS	1.230
				3	TG	1.252
AL 99-49	Right distal radius	ENM	Ahmado	1	TB	.688
				2	MS	.706
				3	PH	.833
AL 116-2a (same individual as AL 116-2b)	Right prox humerus	ENM	Hadar Fm. DD-3s	1	CA	.309
				2	MS	.360
				3	PH	.525
AL 2-63 (part of associated skeleton)	Left prox humerus	ENM	Leadu Basin	1	PH	.448
				2	PE	.544
				3	PU	.576
AL 116-24	Right prox humerus	ENM	Hadar Fm. DD-3s	1	PS	.531
				2	MS	.845
				3	PH	.850
AL 55-16	Right prox humerus	ENM	Hadar Fm. DD-3s	1	TO	.768
				2	P	.805
				3	TG	.839
AL 55-17	Right prox humerus	ENM	Hadar Fm. DD-3s	1	N	.847
				2	PS	.964
				3	MS	.975
AL 285-11b (same individual as AL 285-11a)	Left prox humerus	ENM	Hadar Fm. DD-2	1	TG	.595
				2	PC	.771
				3	PH	.845

TABLE 33. Continued

PCA Euclidean Distance		*PCA Membership Quotient*		*DFA Mahalanobis Distance*		*DFA Posterior Probability*		*Taxonomic Assignment*
PE	2.229	PE	.799	PE	6.74	PE	.483	Colobinae
TG	2.258	TB	.815	CG	8.15	CG	.238	genus indet.
TO	2.412	PA	.854	PH	10.63	PH	.069	
TB	2.615	TB	.756	CB	10.44	CB	.408	Colobinae
CB	2.908	CB	1.170	PE	11.21	PE	.278	genus indet.
MS	3.079	PH	1.171	TB	11.92	TB	.195	
D	3.865	PE	1.246	TG	16.31	TG	.384	Colobinae
PA	4.089	MS	1.434	MS	17.16	MS	.251	genus indet.
MS	4.231	PU	1.497	PE	17.18	PE	.248	
TB	2.062	MS	.616	MS	12.02	MS	.648	Cercopithecinae
MS	2.373	PH	.765	CA	15.85	CA	.095	cf. *Parapapio*
PH	2.421	PC	.770	PH	16.12	PH	.648	
MS	.778	MS	.288	CA	3.68	CA	.551	Cercopithecinae
PH	1.042	CA	.512	MS	4.34	MS	.396	cf. *Cercocebus*
CA	1.109	PH	.582	PH	10.55	PH	.018	
PE	1.132	PH	.567	PE	6.51	PE	.629	Colobinae
PH	1.400	PE	.611	PS	9.33	PS	.154	genus indet.
PU	1.618	PS	.692	PU	10.22	PU	.098	
MS	1.653	MS	.558	PS	8.30	PS	.646	Cercopithecinae
PH	2.621	PS	.955	MS	9.53	MS	.350	cf. *Parapapio*
PS	2.751	PE	.956	PE	19.71	PE	.002	
P	2.595	TB	.979	TG	16.97	TG	.786	*Theropithecus*
TB	2.839	PS	1.263	PH	20.97	PH	.106	cf. *darti*
TO	3.036	PH	1.285	PC	21.21	PC	.094	
PS	2.754	PS	.898	N	9.61	N	.947	Colobinae
N	3.006	N	1.055	PS	16.35	PS	.033	genus indet.
MS	3.197	PU	1.336	MS	18.35	MS	.012	
PU	1.872	TB	.629	TG	11.78	TG	.996	*Theropithecus*
PC	2.034	PC	.867	PC	16.35	PC	.003	cf. *darti*
TB	2.238	PH	.875	PH	18.35	PH	.001	

TABLE 34. Taxonomic assignments of previously unidentified cercopithecoid forelimb elements from the Afar Depression, northeastern Ethiopia.

Identification Number	*Body Part*	*Present Location*	*Collection Locality*	*Rank Order*	*SCA Similarity Coefficient*	
AL 196-3c (same individual as AL 196-3d)	Left distal humerus	ENM	Hadar Fm. DD-2	1	MS	.603
				2	PS	.644
				3	TO	.686
AL 116-2b (same individual as AL 116-2a)	Right distal humerus	ENM	Hadar Fm. DD-3s	1	CA	.802
				2	MS	.859
				3	CB	.878
AL 244-2a	Left distal humerus	ENM	Hadar Fm. DD-3s	1	CM	.636
				2	PA	.639
				3	CA	.679
AL 259-1	Left distal humerus	ENM	Hadar Fm. SH-3s	1	CA	.448
				2	MS	.629
				3	M	.660
AL 100-264	Right distal humerus	ENM	Ahmado	1	PS	.755
				2	MS	.802
				3	TB	.809
AL 187-2a	Left distal humerus	ENM	Hadar Fm. DD-2	1	PA	.753
				2	TB	.930
				3	TO	1.053
AL 145-16	Right distal humerus	ENM	Hadar Fm. SH-2s	1	PC	.735
				2	TG	.738
				3	PU	.762
AL 2-80 (part of associated skeleton)	Left distal humerus	ENM	Leadu Basin	1	D	.521
				2	M	.586
				3	CA	.636
AL 322-10	Right distal humerus	ENM	Hadar Fm. SH-4	1	TG	.609
				2	PS	.683
				3	PH	.691
AL 300-1	Left distal humerus	ENM	Hadar Fm. DD-2	1	CA	1.299
				2	TB	1.351
				3	MS	1.485

TABLE 34. Continued.

PCA Euclidean Distance		*PCA Membership Quotient*		*DFA Mahalanobis Distance*		*DFA Posterior Probability*		*Taxonomic Assignment*
MS	1.803	TO	.699	TB	10.63	TB	.836	*Theropithecus*
TB	1.911	PS	.730	TG	14.98	TG	.095	cf. *darti*
TG	2.136	MS	.836	MS	18.38	MS	.017	
CA	2.128	CA	.812	CA	27.22	CA	.910	Cercopithecinae
CB	2.345	TG	.828	TG	33.45	TG	.040	cf. *Cercocebus*
MS	2.448	CB	1.042	CB	34.98	CB	.019	
CM	1.185	PA	.739	CA	14.82	CA	.548	Cercopithecinae
CA	2.049	CA	.740	MS	17.38	MS	.152	cf. *Parapapio*
PA	2.143	TB	.846	CM	17.73	CM	.128	
CA	1.204	CA	.518	CA	11.28	CA	.775	Cercopithecinae
MS	1.800	MS	.582	MS	14.03	MS	.196	cf. *Cercocebus*
M	1.811	TB	.898	PH	18.13	PH	.025	
MS	2.103	PS	.809	MS	21.46	MS	.564	Cercopithecinae
PS	2.161	MS	.861	PS	23.29	PS	.226	cf. *Parapapio*
PH	2.622	PH	.889	TG	24.27	TG	.138	
TB	2.470	PA	.819	TO	19.38	TO	.946	*Theropithecus*
PA	3.525	TO	.931	PA	25.10	PA	.054	cf. *darti*
TO	3.580	TB	1.138	CM	36.96	CM	.000	
PC	2.306	PC	.952	PS	15.59	PS	.424	Cercopithecinae
TG	2.321	PS	.993	PC	16.81	PC	.230	cf. *Papio*
PU	2.869	PU	1.094	MS	17.92	MS	.132	
M	1.955	CA	.691	PE	14.25	PE	.583	Colobinae
CA	2.014	PE	.897	CA	17.10	CA	.140	genus indet.
D	2.029	MS	.987	CG	17.45	CG	.118	
PS	2.017	TG	.491	PS	15.46	PS	.461	Cercopithecinae
TG	2.145	PS	.503	TG	16.79	TG	.238	cf. *Papio*?
PH	2.376	PH	.950	PH	17.78	PH	.145	
TB	4.754	TB	1.145	TB	40.45	TB	.969	*Theropithecus*
CA	5.336	TO	1.307	CA	52.53	CA	.028	cf. *brumpti*
TO	5.523	MS	1.442	MS	54.21	MS	.003	

TABLE 35. Taxonomic assignments of previously unidentified cercopithecoid forelimb elements from the Afar Depression, northeastern Ethiopia.

Identification Number	*Body Part*	*Present Location*	*Collection Locality*	*Rank Order*	*SCA Similarity Coefficient*	
AL 363-12	Right distal humerus	ENM	Hadar Fm. KH (BKT 1-2)	1	R	1.025
				2	CB	1.039
				3	CG	1.064
AL 285- 11a (same individual as AL 285-11b)	Right distal humerus	ENM	Hadar Fm. DD-2	1	MS	.570
				2	TG	.654
				3	TB	.655
AL 363-9b	Right distal humerus	ENM	Hadar Fm. KH (BKT 1-2) (intrusive?)	1	PH	.556
				2	TG	.633
				3	PS	.709
AL 99-26	Right distal humerus	ENM	Ahmado	1	CM	.845
				2	MS	.998
				3	TB	1.011
AL 99-50	Right distal humerus	ENM	Ahmado	1	MS	.675
				2	CA	.870
				3	TB	.877
AL 2-62 (part of associated skeleton)	Left scapula fragment	ENM	Leadu Basin	1	PE	1.116
				2	CB	1.203
				3	MS	1.420

skeleton is represented, it is possible to measure and analyze all of the elements, resulting in 7 separate assessments of the affinities of this unknown colobine (one for each proximal and distal portion and a final one for the scapula). The results are presented in Tables 32-35, beginning first with the ulna elements and ending with the scapula. In 22 of the 28 separate analyses performed (78% of cases), the AL 2-skeleton was assigned to *Presbytis entellus* based on rank order 1 or 2 scores. It was next most commonly assigned to *Dolichopithecus* or *Mesopithecus*, which is still significant, since the sample sizes of these two extinct species are too small to be included in either PCA membership or DFA. Finally, in a small percentage of the analyses it was assigned to *Colobus*. The AL 2- colobine has yet to be formally described or named. The results of my analysis show that it is not closely related to *Rhinocolobus* or *Paracolobus*. From analysis of the colobine material in South Africa assigned to *Cercopithecoides* (see next section), the Leadu AL 2- skeleton also does not appear to be closely related to that extinct colobine. Perhaps the Leadu skeleton will prove to be the first evidence of a colobine of Eurasian affinities in sub-Saharan Africa.

TABLE 35 Continued.

PCA Euclidean Distance		*PCA Membership Quotient*		*DFA Mahalanobis Distance*		*DFA Posterior Probability*		*Taxonomic Assignment*
R	3.076	CG	.908	CB	31.72	CB	.652	*Rhinocolobus*
CB	3.670	CB	1.494	CG	33.41	CG	.280	*turkanaensis*
CG	4.019	TO	1.504	TO	36.50	TO	.060	
TB	1.288	MS	.514	TB	9.72	TB	.569	*Theropithecus*
MS	1.535	TB	.684	MS	12.52	MS	.140	cf. *darti*
PS	1.803	PS	.762	CA	13.07	CA	.107	
PH	1.641	PH	.692	PH	10.43	PH	.818	*Papio*
PS	1.682	PS	.837	TG	14.48	TG	.108	*hamadryas*
TG	1.848	TG	.908	PS	17.92	PS	.019	
CM	2.804	MS	1.296	CM	37.50	CM	.976	Cercopithecinae
MS	3.204	CM	1.299	MS	46.12	MS	.013	cf. *Parapapio*?
TB	3.512	TB	1.326	TB	46.52	TB	.011	
MS	1.646	MS	.665	MS	17.16	MS	.940	Cercopithecinae
CA	2.021	TB	.850	TB	23.96	TB	.031	cf. *Parapapio*
TB	2.085	PS	.993	CA	24.83	CA	.020	
CB	2.337	CB	1.052	PE	9.93	PE	.804	Colobinae
PE	2.686	PE	1.149	CB	14.68	CB	.075	genus indet.
MS	3.001	MS	1.294	MS	16.03	MS	.038	

The Transvaal Caves, Republic of South Africa

The limestone cave deposits of the Transvaal region of South Africa are best known for their important accumulation of australopithecine fossils. In addition to these early hominids, the dolomite caverns of the Transvaal contain a rich and varied cercopithecoid fauna with at least 6 genera and more than 20 species recognized. The first systematic treatment of the South African fossil Cercopithecoidea was undertaken by Freedman (1957), followed by numerous other reviews (e.g., see Freedman, 1960, 1965, 1976; Freedman and Brain, 1972; Maier, 1970, 1971, 1972b; Eisenhart, 1974; Szalay and Delson, 1979; Dechow and Singer, 1984; Delson, 1984, 1988; Cooke, 1990).

A persistent problem complicating nearly all analyses of the South African cave faunas is the lack of relative and absolute dating for the various cave localities. Not only is the absolute age of the cave accumulations difficult to establish, even their relative chronology is often disputed. F. C. Howell (1955) was the first to utilize a broad program of fauna correlation to determine the relative chronology of the

TABLE 36. Summary of total sample of newly identified cercopithecoid postcranial specimens recovered from the Hadar Formation and other sites in the Afar Depression, northeastern Ethiopia, with percentage breakdowns by stratigraphic level and taxon.

Taxon	*Hadar Formation*			*Ahmado*	*Leadu*	*TOTAL*
	KH Member	*DD Member*	*SH Member*			
Theropithecus cf. *darti*	–	6 (43%)	–	–	–	6 (22%)
Theropithecus brumpti	–	2 (14%)	–	–	–	2 (7%)
Papio species	3 (75%)	–	1* (50%)	–	–	4 (15%)
Parapapio species	–	4 (29%)	–	4 (80%)	–	8 (30%)
cf. *Cercocebus* species	1 (25%)	1 (7%)	–	–	–	2 (7%)
Rhinocolobus turkanaensis	–	–	1 (50%)	–	–	1 (4%)
Colobinae genus indet.	–	1 (7%)	–	1 (20%)	2 (100%)	4 (15%)
TOTAL	4 (15%)	14 (52%)	2 (7%)	5 (19%)	2 (7%)	27 (100%)

*Specimen assigned specifically to *Papio hamadryas* (see Table 35); very possibly represents an intrusive element in the KH member. For further discussion see text.

Transvaal caves and their probable absolute ages. More recent studies have attempted to resolve this issue through the use of geomorphological dating (Partridge, 1973, 1982), paleomagnetism (McFadden et al., 1979) and a chronological scheme based on fossil Bovidae (Vrba, 1975, 1982). Delson (1984) provides the best synthesis of all pre-existing data, together with his own biozonation scheme based on the co-occurrence of cercopithecoid taxa from well-dated sites in East Africa. Fig. 55 depicts this chronological arrangement, which will be followed in this study. In summary, the major South African limestone-cave faunas span a time range from about 3.2 Ma to about 1.5 Ma. The oldest cave locality is Makapansgat and the youngest is Kromdraai (see Fig. 55). There is substantial overlap in the ages of Swartkrans, Kromdraai, and Bolt's Farm, on the one hand, and between Taung and Sterkfontein on the other.

A total of 58 cercopithecoid forelimb postcranial specimens were measured from the six Transvaal cave sites of Makapansgat, Taung, Sterkfontein, Swartkrans, Kromdraai, and Bolt's Farm. Though the total number of cercopithecoid fossils from these sites is in excess of 1500, the overwhelming majority are dental remains. The paucity of postcranial specimens is a direct result of the mode of deposition of fossil remains and the existence of bone-crushing predators at the sites (Brain, 1981). All pertinent information regarding each of the 58 forelimb specimens, along with its taxonomic assignment, appears in Tables 37-42. Table 43 is a summary of all these assignments, organized by site.

Even though the 58 identified postcranial remains amount to less than 10% of the total for identified cranial remains from the South African caves (see Freedman, 1976; Szalay and Delson, 1979; Delson, 1984), there is a remarkable similarity when the percentage breakdown of taxa is compared. At Makapansgat only 2 postcranial specimens are identified, yet they represent the two most common taxa at the site—*Theropithecus darti* and *Parapapio* (Table 43). It is worth noting that the *T. darti* specimen, a proximal radius, is the only known occurrence of a postcranial element in association with identified cranial remains. Quite fortuitously, Maier (1973:85, fig. 4) photographed this radius in a block of breccia juxtaposed to the complete skull of *T. darti*. The fact that the analytical techniques used in this study independently assigned the proximal radius to *Theropithecus* is a reaffirmation of the method's reliability.

At Sterkfontein there are 27 forelimb elements (Table 42). Analysis reveals a distribution of taxa somewhat different from the identifications based on cranial remains. My analytical techniques indicate the presence of *T. darti* and *Papio* cf. *robinsoni* at Sterkfontein. Since *T. darti* has been identified from the similar-aged site of Makapansgat, its occurrence at Sterkfontein is not unexpected. However, the identification of early *Papio* from Sterkfontein may be a bit more problematic. These specimens could possibly be variants of misidentified elements actually attributable to *Parapapio*. Finally, one distal humerus, STS-1087, is allocated to *Cercocebus* (see Table 39). The existence of this genus has not yet been confirmed in Transvaal cave deposits, yet Eisenhart (1974), in an unpublished thesis, did refer 5 specimens from Makapansgat to *Cercocebus*.

TABLE 37. Taxonomic assignments of previously unidentified cercopithecoid forelimb elements from the limestone caves of the Transvaal, Republic of South Africa.

Identification Number	*Body Part*	*Present Location*	*Collection Locality*	*Rank Order*	*SCA Similarity Coefficient*	
M3120	Left	BPIP	Makapansgat	1	PS	.691
(same individual	prox		member 4-5	2	MS	.712
as M3118)	ulna		upper phase I	3	PH	.753
M3119	Left	BPIP	Makapansgat	1	TG	.766
	prox		member 4-5	2	PC	.773
	radius		upper phase I	3	PA	.779
M3118	Left	BPIP	Makapansgat	1	PC	.744
(same individual	distal		member 4-5	2	CA	.884
as M3120)	humerus		upper phase I	3	PU	1.023
UCMP 56658	Right	UCMP	Taung	1	PS	.915
	distal		Australopith.	2	TG	1.047
	humerus		breccia, V4738	3	PH	1.141
SWP-525	Right	UWMS-A	Sterkfontein	1	TG	.840
	prox		Dump 18	2	TO	.900
	ulna		= member 4	3	TB	.930
SWP-515	Right	UWMS-A	Sterkfontein	1	PC	.773
	prox radius		Dump 15	2	PA	.862
	w/shaft		= member 4	3	PU	.916
SWP-517	Right	UWMS-A	Sterkfontein	1	MS	.966
	prox		Dump 13	2	CA	1.084
	radius		= member 4	3	PS	1.117
SWP-516	Right	UWMS-A	Sterkfontein	1	TO	.572
	prox		Dump 13	2	TG	.581
	radius		= member 4	3	PA	.602
SWP-518	Right	UWMS-A	Sterkfontein	1	CA	.260
	prox		Dump 13	2	MS	.410
	radius		= member 4	3	PA	.417
SWP-522	Left	UWMS-A	Sterkfontein	1	TG	.512
	prox		Dump 13	2	PS	.551
	radius		=member 4	3	CA	.566

TABLE 37. Continued.

PCA Euclidean Distance		*PCA Membership Quotient*		*DFA Mahalanobis Distance*		*DFA Posterior Probability*		*Taxonomic Assignment*
PS	1.402	MS	.698	MS	13.52	MS	.443	Cercopithecinae
M	2.361	PS	.712	PC	14.72	PC	.244	cf. *Parapapio*
TO	2.547	TO	.884	PH	15.00	PH	.212	
TG	2.389	PA	.721	TG	6.54	TG	.597	*Theropithecus*
PA	2.441	TG	.840	PA	8.76	PA	.197	cf. *darti*
PU	2.654	PA	.856	PU	9.53	PU	.133	
PC	2.688	MS	.933	PC	27.60	PC	.827	Cercopithecinae
CA	3.085	PA	1.042	MS	30.83	MS	.165	cf. *Parapapio*
PU	3.745	PU	1.237	PA	37.16	PA	.007	
PS	3.372	PS	1.053	PS	25.72	PS	.993	Cercopithecinae
TG	3.715	PH	1.500	MS	35.85	MS	.006	cf. *Parapapio*
PH	3.933	MS	1.527	TG	42.34	TG	.001	
TG	2.473	TB	.828	TG	19.48	TG	.658	*Theropithecus*
TO	3.278	TO	.888	PS	20.84	PS	.333	cf. *darti*
PS	3.464	TG	1.046	TO	29.28	TO	.005	
PC	1.978	PC	.747	PC	12.10	PC	.507	*Papio*
PA	2.579	PA	.760	PA	14.95	PA	.122	cf. *robinsoni*
PS	2.737	PU	.820	PS	15.07	PS	.115	
MS	3.128	PS	1.180	PS	10.25	PS	.760	*Parapapio*
CA	3.378	MS	1.250	MS	14.43	MS	.094	species indet.
PS	3.387	TG	1.265	CA	14.67	CA	.083	
TG	1.832	PA	.569	TG	4.08	TG	.466	*Theropithecus*
TO	1.843	TO	.602	TO	5.93	TO	.185	cf. *darti*
PA	2.021	TG	.716	PH	6.70	PH	.126	
CA	.624	MS	.220	MS	10.79	MS	.988	*Parapapio*
PH	1.597	PH	.487	CA	23.37	CA	.002	species indet.
MS	1.732	PA	.520	PU	23.65	PU	.002	
TG	1.622	CA	.468	CA	4.08	CA	.201	*Parapapio*
CA	1.774	TG	.484	PS	4.38	PS	.173	species indet.
MS	1.883	MS	.546	TG	4.62	TG	.154	

TABLE 38. Taxonomic assignments of previously unidentified cercopithecoid forelimb elements from the limestone caves of the Transvaal, Republic of South Africa

Identification Number	*Body Part*	*Present Location*	*Collection Locality*	*Rank Order*	*SCA Similarity Coefficient*	
SWP-514	Right prox radius	UWMS-A	Sterkfontein Dump 13 = member 4	1	PC	.691
				2	PA	.832
				3	PU	.926
SWP-521	Left prox radius	UWMS-A	Sterkfontein Dump 13 = member 4	1	MS	.502
				2	TG	.606
				3	CA	.714
SWP-519	Left prox radius	UWMS-A	Sterkfontein Dump 18 = member 4	1	TG	.642
				2	TB	.891
				3	TO	.973
SWP-520	Left prox radius	UWMS-A	Sterkfontein Dump 13 = member 4	1	TG	.346
				2	TO	.554
				3	MS	.800
STS-554	Left prox radius	TMP	Sterkfontein member 4 type site	1	MS	.615
				2	TG	.628
				3	PC	.712
STS-550	Left distal radius	TMP	Sterkfontein member 4 type site	1	MS	.867
				2	CA	1.006
				3	TG	1.070
SWP-513	Left distal radius w/shaft	UWMS-A	Sterkfontein Dump 13 = member 4	1	MS	.948
				2	CM	.986
				3	CA	.988
STS-1106	Left prox humerus	TMP	Sterkfontein member 4 type site	1	N	.468
				2	MS	.517
				3	CB	.556
SWP-505	Left prox humerus	TMP	Sterkfontein Dump 13 = member 4	1	TB	.977
				2	TG	.983
				3	N	1.115
STS-377c	Right distal humerus	TMP	Sterkfontein member 4 type site	1	MS	.648
				2	PS	.736
				3	PH	.855

TABLE 38. Continued

PCA Euclidean Distance		*PCA Membership Quotient*		*DFA Mahalanobis Distance*		*DFA Posterior Probability*		*Taxonomic Assignment*
PC PA PU	2.203 2.820 3.160	PC PA PU	.730 .803 .904	PC PA PS	7.24 10.21 12.11	PC PA PS	.713 .161 .062	*Papio* cf. *robinsoni*
MS TG CM	1.378 2.201 2.236	TG PA MS	.583 .788 .815	MS TG CM	10.59 19.22 19.26	MS TG CM	.959 .014 .013	*Parapapio* species indet.
TG PC TO	1.516 2.792 2.909	TG TB PC	.659 .988 .990	TG TO CM	7.68 13.02 14.63	TG TO CM	.865 .060 .027	*Theropithecus* cf. *darti*
TG TO MS	1.155 1.740 2.710	TG TO PA	.522 .689 .903	TG TO PH	1.43 6.07 8.61	TG TO PH	.838 .082 .023	*Theropithecus* cf. *darti*
TG MS PC	1.990 2.006 2.142	CA MS PC	.510 .642 .733	MS CA TG	5.70 6.52 6.71	MS CA TG	.226 .150 .136	*Parapapio* species indet.
TG MS PS	2.960 3.421 3.768	MS PS CA	1.080 1.138 1.244	PS MS TO	19.10 19.37 22.74	PS MS TO	.482 .422 .078	*Parapapio* species indet.
MS CA TO	3.075 3.412 3.512	MS CA TO	.837 1.019 1.120	MS PS CA	11.78 12.09 13.83	MS PS CA	.424 .363 .152	*Parapapio* species indet.
N CB MS	1.518 1.885 2.038	N CB PS	.698 .727 .777	N CB MS	6.36 6.53 10.13	N CB MS	.403 .370 .061	Colobinae cf. *Cercopithecoides*
TB TG MS	2.866 3.190 3.477	TB PS TG	.978 1.247 1.428	TB TG N	28.43 28.62 35.24	TB TG N	.434 .394 .014	*Theropithecus* cf. *darti*
CA MS PS	2.290 2.822 2.950	PC CA MS	.945 .978 1.177	MS PH CA	102.52 103.33 109.31	MS PH CA	.579 .387 .019	*Parapapio* species indet.

TABLE 39. Taxonomic assignments of previously unidentified cercopithecoid forelimb elements from the limestone caves of the Transvaal, Republic of South Africa.

Identification Number	*Body Part*	*Present Location*	*Collection Locality*	*Rank Order*	*SCA Similarity Coefficient*	
SWP-504	Right prox humerus w/shaft	TMP	Sterkfontein Dump 13 member 4	1	MS	.582
				2	M	.731
				3	CA	.841
STS-1087	Left distal humerus	TMP	Sterkfontein level unknown	1	CA	1.050
				2	CM	1.059
				3	TB	1.150
STS-1151	Left distal humerus	TMP	Sterkfontein level unknown	1	PA	.967
				2	PH	1.094
				3	PU	1.239
STS-1741	Left distal humerus	TMP	Sterkfontein level unknown	1	CA	.714
				2	MS	.806
				3	CM	.966
SWP-509	Right distal humerus	UWMS-A	Sterkfontein Dump 14 member 4	1	PC	.745
				2	PH	.774
				3	PU	.928
SWP-507	Right distal humerus	UWMS-A	Sterkfontein Dump 13 member 4	1	CA	.850
				2	MS	.933
				3	TG	.934
SWP-506	Right distal humerus w/shaft	UWMS-A	Sterkfontein Dump 18 member 4	1	PC	.682
				2	PH	.710
				3	PA	.759
SWP-508	Right distal humerus	UWMS-A	Sterkfontein Dump 13 member 4	1	CM	.739
				2	TG	.793
				3	PS	.844
SWP-512	Left distal humerus	UWMS-A	Sterkfontein Dump 13 member 4	1	TG	.815
				2	PU	.871
				3	CA	.879
SWP-511	Left distal humerus	UWMS-A	Sterkfontein Dump 13 member 4	1	MS	.541
				2	PH	.550
				3	PS	.614

TABLE 39. Continued.

PCA Euclidean Distance		*PCA Membership Quotient*		*DFA Mahalanobis Distance*		*DFA Posterior Probability*		*Taxonomic Assignment*
MS	1.776	MS	.837	MS	12.33	MS	.480	*Parapapio*
M	1.970	PS	1.016	CA	12.80	CA	.379	species indet.
CA	2.655	CA	1.124	CM	15.52	CM	.098	
CA	3.203	CA	1.247	CA	31.67	CA	.553	Cercocebus
CM	3.831	MS	1.322	CM	32.48	CM	.368	species indet.
TB	4.299	CM	1.391	TB	35.58	TB	.078	
PA	3.261	PA	1.262	PA	17.68	PA	.999	*Papio*
PU	3.759	PU	1.408	PH	32.60	PH	.001	cf. *robinsoni*
PH	4.249	PH	1.409	PU	36.26	PU	.000	
MS	2.074	MS	.996	CM	15.96	CM	.681	*Parapapio*
PA	2.677	CM	1.103	MS	17.98	MS	.247	species indet.
CA	3.617	PA	1.134	PS	20.93	PS	.057	
PC	2.567	PC	1.123	PC	14.52	PC	.827	*Papio*
PH	2.619	PH	1.178	PH	18.31	PH	.124	cf. *robinsoni*
PA	2.794	PA	1.188	PA	20.48	PA	.042	
MS	2.452	CA	1.073	MS	24.60	MS	.621	*Parapapio*
PS	2.855	MS	1.099	TG	27.31	TG	.160	species indet.
CA	2.903	PS	1.117	PS	29.18	PS	.063	
PC	1.859	PC	.797	PA	17.48	PA	.824	*Papio*
PA	2.210	PH	.889	PC	21.22	PC	.127	cf. *robinsoni*
PH	2.239	PU	.891	PH	23.70	PH	.037	
MS	1.874	CA	.772	CM	22.62	CM	.242	*Parapapio*
TG	2.144	MS	.807	MS	22.84	MS	.216	species indet.
CM	2.715	CM	.880	CA	23.34	CA	.168	
TG	1.077	TG	.817	TG	30.34	TG	.268	*Theropithecus*
PC	1.768	TO	.865	TO	30.77	TO	.216	cf. *darti*
PH	1.851	PC	.873	CM	31.09	CM	.184	
PH	1.720	PH	.657	MS	9.48	MS	.366	*Parapapio*
MS	1.804	PS	.693	CM	10.37	CM	.235	species indet.
PS	2.039	TB	.743	TG	11.06	TG	.166	

TABLE 40. Taxonomic assignments of previously unidentified cercopithecoid forelimb elements from the limestone caves of the Transvaal, Republic of South Africa.

Identification Number	*Body Part*	*Present Location*	*Collection Locality*	*Rank Order*	*SCA Similarity Coefficient*	
SWP-510	Left	UWMS-A	Sterkfontein	1	M	.740
	distal		Dump 13	2	PE	.826
	humerus		= member 4	3	CA	.829
SK-591d	Left	TMP	Swartkrans	1	PH	.569
	prox		member 1	2	PC	.627
	ulna		pink breccia	3	PA	.634
SK-1870	Left	TMP	Swartkrans	1	N	.602
	prox		member 1	2	CB	.621
	radius		pink breccia	3	D	.735
SK-1867	Left	TMP	Swartkrans	1	PH	.694
	prox		member 1	2	PC	.708
	radius		pink breccia	3	PA	.769
SK-1814	Right	TMP	Swartkrans	1	PU	.660
	prox radius		level unknown	2	PH	.684
	w/shaft		channel infill?	3	PA	.692
SK-591c	Left	TMP	Swartkrans	1	PH	.666
	complete		member 1	2	TG	.837
	humerus (prox)		pink breccia	3	PC	.893
SK-1506	Right	TMP	Swartkrans	1	PS	.674
	distal		member 1	2	PH	.782
	humerus		pink breccia	3	PU	.848
SK-#?	Left	TMP	Swartkrans	1	PC	.594
	distal		member 1	2	PA	.679
	humerus (distal)		pink breccia	3	PH	.707
SK-591c	Left	TMP	Swartkrans	1	PC	.319
	complete		member 1	2	PH	.653
	humerus		pink breccia	3	PU	.654
SK-860	Left	TMP	Swartkrans	1	TO	.904
	distal		member 1	2	CA	.951
	humerus		pink breccia	3	TB	.990

TABLE 40. Continued.

PCA Euclidean Distance		*PCA Membership Quotient*		*DFA Mahalanobis Distance*		*DFA Posterior Probability*		*Taxonomic Assignment*
M	2.657	PE	.814	PE	25.94	PE	.325	Colobinae
CA	2.682	PU	.846	CA	25.97	CA	.320	cf.
PE	2.774	CA	.850	PC	27.39	PC	.158	*Cercopithecoides*
PH	2.058	PH	.558	PH	10.93	PH	.835	*Papio*
PC	2.101	PC	.658	PA	14.98	PA	.110	cf. *robinsoni*
TG	2.131	PA	.693	TG	18.31	TG	.021	
D	2.132	N	.630	N	30.53	N	.418	Colobinae
CB	2.192	CB	.674	CB	31.26	CB	.290	cf.
N	2.377	TO	.686	PS	32.23	PS	.178	*Cercopithecoides*
PH	2.124	PC	.744	PC	5.13	PC	.369	*Papio*
PC	2.161	PA	.748	PA	5.28	PA	.342	cf. *robinsoni*
PA	2.328	PS	.882	PH	7.21	PH	.130	
PA	1.503	PU	.444	PU	14.03	PU	.573	*Papio*
PU	1.803	PA	.506	PA	15.99	PA	.215	cf. *ursinus*
PC	1.843	PH	.562	PH	17.88	PH	.083	
PH	1.739	PH	.677	PH	11.75	PH	.973	*Papio*
PC	1.910	PU	.887	PC	20.56	PC	.012	cf. *robinsoni*
TG	2.262	TG	.988	TG	20.62	TG	.012	
PS	2.155	PH	.647	PH	15.55	PH	.576	*Papio*
PH	2.315	PS	1.055	PS	17.22	PS	.250	cf. *robinsoni*
PU	2.963	PU	1.129	PU	17.96	PU	.173	
PC	1.241	PC	.521	PC	16.29	PC	.596	*Papio*
PA	1.924	PA	.704	PA	18.58	PA	.190	cf. *robinsoni*
PU	2.220	PU	.720	PH	21.67	PH	.041	
PH	1.562	PH	.571	PC	3.46	PC	.998	*Papio*
PC	1.978	PC	.764	PU	16.63	PU	.001	cf. *robinsoni*
PU	2.523	PU	.812	PH	18.18	PH	.001	
TB	1.518	TO	.737	TO	27.33	TO	.821	*Theropithecus*
TO	1.737	TB	.763	PU	30.70	PU	.153	*oswaldi*
MS	1.844	MS	.976	CA	34.81	CA	.019	

TABLE 41. Taxonomic assignments of previously unidentified cercopithecoid forelimb elements from the limestone caves of the Transvaal, Republic of South Africa.

Identification Number	*Body Part*	*Present Location*	*Collection Locality*	*Rank Order*	*SCA Similarity Coefficient*	
KB 3128	Left	TMP	Kromdraai B	1	CA	1.192
	prox			2	MS	1.208
	ulna			3	CM	1.223
KB 420	Left	TMP	Kromdraai B	1	TO	1.175
	prox			2	TB	1.247
	ulna			3	PS	1.327
KB 2902	Left	TMP	Kromdraai B	1	TO	1.038
	prox			2	TB	1.178
	ulna			3	PS	1.200
KB 412	Right	TMP	Kromdraai B	1	PC	.564
	prox			2	MS	.576
	radius			3	TG	.614
KB 2925	Right	TMP	Kromdraai B	1	PA	.892
	prox			2	PC	.984
	radius			3	PH	1.109
KB 405	Left	TMP	Kromdraai B	1	PA	.706
	prox			2	MS	.732
	radius			3	PC	.757
KB 411	Left	TMP	Kromdraai B	1	MS	.519
	prox			2	TO	.671
	radius			3	CM	.719
KB 407	Right	TMP	Kromdraai B	1	CG	.809
	distal			2	MS	.922
	radius			3	N	.995
KB 3130	Right	TMP	Kromdraai B	1	TB	.897
	distal			2	CA	.913
	humerus			3	TO	1.045
KB 408	Right	TMP	Kromdraai B	1	PA	.737
	distal			2	MS	.787
	humerus			3	CM	.806

TABLE 41. Continued.

PCA Euclidean Distance		*PCA Membership Quotient*		*DFA Mahalanobis Distance*		*DFA Posterior Probability*		*Taxonomic Assignment*
CA	4.051	MS	1.242	CA	42.33	CA	.660	Cercopithecinae
CM	4.233	CA	1.249	MS	43.75	MS	.325	cf. *Cercocebus*
PG	4.452	TB	1.811	CM	50.62	CM	.010	
TO	4.515	TO	1.045	TO	38.41	TO	.984	*Theropithecus*
TB	4.790	TB	1.280	TB	47.18	TB	.012	cf. *oswaldi*
PG	4.885	PS	1.787	PS	50.38	PS	.002	
TO	3.495	TO	1.074	TO	29.43	TO	.757	*Theropithecus*
PG	4.074	TB	1.236	TB	32.67	TB	.150	cf. *oswaldi*
PS	4.250	PS	1.832	PH	36.03	PH	.028	
PC	1.641	PS	.594	PC	3.40	PC	.285	*Papio*
MS	1.742	MS	.603	MS	4.11	MS	.200	cf. *robinsoni*
TG	2.069	PC	.619	TG	4.51	TG	.164	
PA	2.522	PA	.697	PA	9.64	PA	.444	*Papio*
PC	2.685	PC	.854	PH	10.48	PH	.291	cf. *robinsoni*
PH	3.120	PU	.904	PU	13.16	PU	.076	
CA	1.946	PA	.721	PC	5.24	PC	.302	*Papio*
CM	2.276	MS	.747	MS	6.19	MS	.187	cf. *robinsoni*
PA	2.279	PC	.772	PA	6.45	PA	.165	
CM	1.931	MS	.663	MS	6.65	MS	.316	Cercopithecinae
MS	2.167	TB	.696	TG	7.39	TG	.217	cf. *Parapapio*?
TG	2.212	CM	.743	CM	8.02	CM	.159	
CG	2.533	CG	.791	CG	11.83	CG	.435	Colobinae
MS	3.487	N	1.111	N	11.75	N	.418	cf.
PC	3.952	MS	1.308	PC	15.60	PC	.064	*Cercopithecoides*
TB	2.196	TO	.850	TB	25.11	TB	.905	*Theropithecus*
MS	2.807	TB	1.061	TO	30.67	TO	.056	cf. *oswaldi*
CA	2.865	CA	1.571	CA	31.71	CA	.033	
PA	1.699	PA	.748	CM	19.12	CM	.347	Cercopithecinae
MS	2.471	CA	.854	PA	19.47	PA	.291	cf. *Papio*
PC	2.535	PC	1.099	CA	20.07	CA	.216	*robinsoni*

TABLE 42. Taxonomic assignments of previously unidentified cercopithecoid forelimb elements from the limestone caves of the Transvaal, Republic of South Africa.

Identification Number	*Body Part*	*Present Location*	*Collection Locality*	*Rank Order*	*SCA Similarity Coefficient*	
KA 166	Right	TMP	Kromdraai A	1	PU	.720
	distal		Faunal Site	2	PH	.853
	humerus			3	PA	.863
BF 65	Left	UWMS-A	Bolts Farm	1	PC	.580
UCMP 56829	distal ulna		V4888	2	PH	.746
	w/shaft			3	TG	.786
BF 61	Right	UWMS-A	Bolts Farm	1	CM	.694
UCMP 56825	prox radius		V4888	2	MS	.707
	w/shaft			3	CA	.724
BF 62	Right	UWMS-A	Bolts Farm	1	CB	.969
UCMP 56826	distal		V4888	2	PE	1.009
	radius			3	TG	1.046
BF 55	Left	UWMS-A	Bolts Farm	1	MS	1.044
UCMP 56819	prox		V4888	2	TG	1.104
	humerus			3	CA	1.179
BF 48	Right	UWMS-A	Bolts Farm	1	TO	.458
UCMP 56812	prox humerus		V4888	2	TB	.551
	w/shaft			3	TG	1.074
BF 56	Right	UWMS-A	Bolts Farm	1	PA	.525
UCMP 56820	distal		V4888	2	PC	.607
	humerus			3	PU	.679
BF 49	Right	UWMS-A	Bolts Farm	1	PH	.915
UCMP 56813	distal		V4888	2	PS	.960
	humerus			3	PC	.988
BF 47	Right	UWMS-A	Bolts Farm	1	PE	.602
UCMP 56811	distal		V4888	2	D	.998
	humerus			3	TG	1.118

TABLE 42. Continued.

PCA Euclidean Distance		*PCA Membership Quotient*		*DFA Mahalanobis Distance*		*DFA Posterior Probability*		*Taxonomic Assignment*
PH	2.128	PH	.973	PU	10.73	PU	.996	*Papio*
PU	2.681	PU	1.032	PH	22.31	PH	.003	*angusticeps*
PC	3.066	PS	1.266	PA	25.75	PA	.001	
PC	1.572	PC	.614	PC	3.33	PC	.361	*Papio*
PH	2.025	PH	.771	PH	4.60	PH	.191	cf. *robinsoni*
TG	2.114	TG	.865	TG	5.56	TG	.118	
CM	2.210	MS	.836	MS	5.95	MS	.229	Cercopithecinae
CA	2.334	PA	.870	CM	6.11	CM	.210	cf. *Parapapio*
TG	2.533	CA	.877	TG	6.95	TG	.139	
PE	2.876	CB	.809	PE	27.16	PE	.476	Colobinae
CB	3.233	TG	.839	CB	28.33	CB	.265	cf.
TG	3.337	PE	.951	TG	29.60	TG	.141	*Cercopithecoides*
MS	3.128	PS	1.284	CA	26.13	CA	.712	Cercopithecinae
TG	3.318	CA	1.302	PH	29.28	PH	.148	cf. *Parapapio*
PE	3.335	MS	1.422	MS	30.25	MS	.091	
TO	1.828	TB	.634	TB	24.96	TB	.491	*Theropithecus*
TB	2.671	PH	1.336	TG	25.06	TG	.467	cf. *oswaldi*
TG	2.886	PU	1.455	PH	30.58	PH	.030	
PA	.927	PA	.525	PA	11.10	PA	.901	*Papio*
PC	1.555	PC	.684	PC	16.04	PC	.076	cf. *robinsoni*
PU	1.938	PU	.812	PU	19.65	PU	.013	
PA	3.046	PS	1.143	PA	34.71	PA	.319	*Papio*
PH	3.145	PA	1.309	PH	35.36	PH	.230	cf. *robinsoni*
PS	3.181	PH	1.363	PS	35.69	PS	.196	
PE	1.886	PE	.772	PE	9.72	PE	.999	Colobinae
D	3.020	TG	1.444	TG	26.03	TG	.000	cf.
TG	3.324	CA	1.777	CA	26.56	CA	.000	*Cercopithecoides*

TABLE 43. Summary of total sample of newly identified cercopithecoid postcranial specimens recovered from the limestone caves in the Transvaal region of the Republic of South Africa, with percentage breakdowns by stratigraphic level and taxon.

Taxon	*Makapansgat*	*Taung*	*Sterkfontein*		*Swartkrans*	*Kromdraai*		*Bolt's Farm*	*TOTAL*
	members 4–5	*faunal level*	*member 4*	*member ?*	*member 1*	*level A*	*level B*	*Pit 23*	
Theropithecus darti	1 (50%)	–	6 (25%)	–	–	–	–	–	7 (12%)
Theropithecus oswaldi	–	–	–	–	1 (11%)	–	3 (30%)	1 (12%)	5 (9%)
Papio angusticeps	–	–	–	–	–	1 (100%)	–	–	1 (2%)
Papio cf. *robinsoni*	–	–	4 (17%)	1 (33%)	6 (67%)	–	4 (40%)	3 (38%)	18 (31%)
Papio cf. *ursinus*	–	–	–	–	1* (11%)	–	–	–	1 (2%)
Parapapio species	1 (50%)	1 (100%)	12 (50%)	1 (33%)	–	–	1? (10%)	2 (25%)	18 (31%)
Cercocebus species	–	–	–	1 (33%)	–	–	1 (10%)	–	2 (3%)
Cercopithecoides williamsi	–	–	2 (8%)	–	1 (11%)	–	1 (10%)	2 (25%)	6 (10%)
TOTAL	2 (3%)	1 (2%)	24 (41%)	3 (5%)	9 (16%)	1 (2%)	10 (17%)	8 (14%)	58 (100%)

*Specimen assigned specifically to *Papio ursinus* (see Table 40), possibly derived from a cave in filling formed subsequent to the formation of member 1. For further discussion see text.

At Swartkrans the distribution of taxa identified on the basis of postcranial remains is very similar to that based on cranial remains. However, it should be noted that the multivariate results could not be evaluated in a manner that would allow the sorting of *Papio robinsoni* from *P. (Dinopithecus) ingens*. There was simply no baseline sample for comparison. It is therefore possible that some specimens identified as *P. robinsoni* could be *P. ingens*. It is also worth mentioning that the *Theropithecus* distal humerus SK-860 from Swartkrans is most certainly *T. oswaldi* rather than *T. darti* (see Table 40). The fact that Makapansgat and Swartkrans are separated by about 1.5 m.y., yet have sister species of the same genus, argues for the existence of an evolving lineage or chronospecies (also see Jablonski, 1993b). My postcranial identification of *T. oswaldi* at Swartkrans is further substantiated by the recent recovery of cranial remains of this taxon from the Swartkrans member 1, "Hanging Remnant" by C.K. Brain (reported by Delson, 1988). Finally, the proximal radius SK-1814 identified as *Papio* cf. *ursinus,* is probably *not* contemporary with the rest of the Swartkrans postcrania. Its degree of fossilization and essentially modern nature (see rank-orderings in Table 40) indicate that its occurrence in the deposits is possibly the result of a channel infilling.

At Kromdraai A only one forelimb element was available for study, and it is assigned to *Papio angusticeps* (Table 42). Kromdraai B yielded 10 postcranial specimens. Three are identified as *T. oswaldi*, though no cranial specimens have yet been allocated to that species. However, since Swartkrans and Kromdraai are temporal equivalents, there is every reason to expect that the ubiquitous *T. oswaldi* should be found at both sites. Kromdraai B also yielded further evidence for the existence of *Cercocebus* (see KB 3128 in Table 41) in South Africa, though there is no cranial evidence from this site confirming the identification. Finally, the forelimb specimens from Bolt's Farm (Table 42) reveal a very close match to the distribution of cranially identified specimens. The only exception is the single occurrence of a specimen of *T. oswaldi.*

An overview of Table 43 reveals several interesting trends in the distribution of fossil cercopithecoids in the Transvaal caves of South Africa. For instance, the earlier sites (Makapansgat, Taung, and Sterkfontein) have a predominance of *Parapapio*, yet are nearly devoid of *Papio*. The reverse is true when the later sites (Swartkrans, Kromdraai, and Bolt's Farm) are considered. Delson (1984, 1988) found this change in frequency of *Papio* quite useful in his biozonation scheme. Another trend is the presence of the colobine *Cercopithecoides williamsi* throughout the sequence (based on cranial identifications). Six specimens of this colobine genus have also been identified postcranially in this study. They are all distinctively colobine, yet there is no common pattern of affinity. For instance, the proximal radius and proximal humerus are like those of *Nasalis*, the distal radius like *Colobus guereza,* and the distal humerus like *Presbytis entellus* (see Tables 38-42). M. G. Leakey (1982) has described a skull and associated partial skeleton of *Cercopithecoides* from East Africa which I have not yet had the chance to study. An investigation of this associated skeleton could do much to shed light on the adaptations and evolutionary relationships of this aberrant, predominantly South African, colobine.

Other Circum-Mediterranean and Eastern African Sites

During the course of data collection for the comparative sample, a number of unassigned cercopithecoid forelimb postcranial specimens which did not fit neatly into one of the above-mentioned site-groupings were located. Since two of these were from European sites and the others from a variety of African sites, I have decided to present them here (see Table 44) to further demonstrate the effectiveness of the assignment techniques, even when applied outside the main study area.

The first specimen in Table 44 is V.J. 130, a proximal ulna from the site of Villafranca d'Asti in northwestern Italy. This historically important site is now dated to a time interval about 3.5 Ma (Savage and Russell, 1983). Two species of cercopithecoids, *Macaca sylvanus* and cf. *Mesopithecus monspessulanus,* have been identified at Villafranca d'Asti from dental remains (Szalay and Delson, 1979). On the concurrence of three of the four multivariate techniques, V.J. 130 is assigned to *Macaca* cf. *sylvanus*. The posterior probability score is especially indicative of this assignment. After this analysis was completed, Delson (pers. comm.) confirmed that there was actually a close association at the site between this proximal ulna and the dental remains identified as *M. sylvanus.*

The next specimen is a left distal humerus from the late Pleistocene site of Tamar Hat in Algeria. On the basis of faunal correlation, this site is dated to less than 50,000 years B.P. (Szalay and Delson, 1979). The only primate known in the vertebrate fauna at Tamar Hat is *M. sylvanus*, identified on the basis of dental remains. The uncataloged left distal humerus that I studied is also assigned to *M. sylvanus*, based on the concurrence of all four analytical techniques. This assignment is in full agreement with both the fact that extant populations of *M. sylvanus* occur today in Algeria and the identification of fossil primate dental remains at Tamar Hat.

The third specimen in Table 44 is a distal humerus from the site of Sables de Montpellier in France. This lower Pliocene site is placed in Neogene mammal zone MN 14 by Mein (1979), which dates it about 4.5 Ma. The primates known from Sables de Montpellier are *Macaca sylvanus, Dolichopithecus rusinensis,* and *Mesopithecus monspessulanus*. Surprisingly, this unassigned distal humerus, MM09.139, ranks very strongly with *Colobus badius*, followed by *C. guereza*. Both of these species have wholly African distributions. On the basis of its small size alone, MM09.139 cannot be *Dolichopithecus*. The rank-ordering of all four multivariate techniques assures that the specimen's affinities lie with the Colobinae, ruling out an assignment to *M. sylvanus*. This leaves two options: (1) that MM09.139 represents an unidentified European colobine of African origin, or (2) that MM09.139 actually represents part of the previously poorly known forelimb of *Mesopithecus monspessulanus*. The latter option is adopted here, with the proviso that these results could indicate that the forelimb skeleton of *M. monspessulanus* exhibited a much more pronounced arboreal adaptation than its sister species *M. pentelici*, a view also adopted by Delson (1973) and Ardito and Mottura (1987).

The next two entries in Table 44 represent the proximal and distal ends of a complete radius from Olduvai Gorge in Tanzania. The precise stratigraphic level of this unassigned specimen, 2-0037, is not known. The overwhelming majority of primate postcrania from Olduvai Gorge are assigned to *Theropithecus oswaldi* and are often associated with identified cranial remains; these postcranial elements form part of the extinct comparative sample used in this study (see Table 48 in Appendix). The 2-0037 radius differs in morphology from the *T. oswaldi* sample, however, and is therefore analyzed here as an unknown. Separate analyses of the proximal and the distal radius yield the same conclusion, that this specimen is referable to *Papio*, specifically *P.* cf. *cynocephalus*. This accords well with Jolly's (1965) brief notes regarding this specimen. *P.* cf. *cynocephalus* is the species of *Papio* that currently inhabits Tanzania. The 2-0037 radius might therefore be derived from one of the later stratigraphic intervals at Olduvai at a time when *P. cynocephalus* was already distinct.

The sixth entry is KNM-ER 2023, a right distal radius from the sub-Burgi level of the Koobi Fora Formation, east of Lake Turkana in Kenya. This level predates 2.6 Ma, based on biostratigraphic and geochronologic correlations (Delson, 1984; Feibel et al., 1989). KNM-ER 2023 is associated with a proximal and distal femur, a tibia, and a partial skull. The skull has been referred to *Theropithecus brumpti* by M. G. Leakey (1976), but Delson (1984; pers. comm.) considers it to be *T. darti*. My statistical analyses of the KNM-ER 2023 distal radius yield an assignment to *T. brumpti* based on the first two techniques; the latter two techniques cannot be considered here, since there were not large enough samples of *T. brumpti* to be applicable. Furthermore, from the analyses of *T. darti* specimens at Makapansgat, Sterkfontein, and Hadar presented previously, it is notable that this species bears a morphological affinity to *T. gelada*. The distal radius from East Turkana analyzed here shows no such affinity. My analysis therefore supports M. G. Leakey's interpretation, suggesting that KNM-ER 2023 is assignable to *T. brumpti*. This assignment adds further credibility to the view that *T. brumpti* represents a distinct, independent lineage stretching back into the early Pliocene (e.g., Eck and Jablonski, 1987; Jablonski, 1993b).

The next entries come from two sites in Kenya, the lower member of Lothagam, on the west side of Lake Turkana, and Lukeino in Baringo Basin to the south of Lake Turkana. The lower member of Lothagam is dated to approximately 5.0 Ma by faunal correlation bracketed by radiometric dates; Lukeino is dated to about 6.5 Ma by the same means (Delson, 1984; F. C. Howell, 1978; see also Fig. 55). At Lothagam there are three main stratigraphic levels, numbered 1-3. An isolated molar of *Theropithecus* sp. is known from the uppermost level, Lothagam-3; no cercopithecoid dental remains have been recovered from the two lower members. In 1967 Brian Patterson recovered a cercopithecoid distal humerus, Lothagam 73-67K, from high up in the lower member, probably Lothagam-1. This specimen is analyzed here for the first time. The rank-ordering of taxa in Table 44 shows affinities with *Macaca, Cercopithecus,* and *Papio,* yet the specimen exhibits no strong attachment for any specific taxon. This pattern of affinity is similar to that seen at Sterkfontein and Hadar—two Pliocene sites with a significant number of *Parapapio* specimens.

TABLE 44. Taxonomic assignments of cercopithecoid forelimb elements from sites in Europe and Africa not previously considered in this study.

Identification Number	*Body Part*	*Present Location*	*Collection Locality*	*Rank Order*	*SCA Similarity Coefficient*	
V.J. 130	Left prox ulna	NMB	Villafranca d'Asti (Italy)	1	MS	.425
				2	PS	.692
				3	CM	.744
uncataloged	Left distal humerus	MNHN-P	Tamar Hat (Algeria)	1	MS	.511
				2	PC	.588
				3	CA	.596
MM09.139	Right distal humerus	FSUL	Sables de Montpellier (France)	1	CB	.469
				2	CG	.921
				3	N	1.004
2-0037 FU 4/01	Complete right radius (prox)	KNM-P	Olduvai Gorge bed unknown	1	PC	1.024
				2	TG	1.056
				3	PS	1.176
2-0037 FU 4/01	Complete right radius (distal)	KNM-P	Olduvai Gorge bed unknown	1	PC	.723
				2	PA	.746
				3	MS	.755
KNM-ER 2023	Right distal radius	KNM-P	Koobi Fora Fm. sub-Burgi level	1	TB	.791
				2	PH	.831
				3	PU	.843
KNM-LU 860	Left prox humerus w/shaft	KNM-P	Baringo Basin Lukeino	1	CA	.823
				2	PS	.974
				3	PU	1.006
73-67K	Right distal humerus	KNM-P	Lothagam lower member	1	PA	.761
				2	MS	.804
				3	TO	.805
KNM-CH 117	Right scapula fragment	KNM-P	Baringo Basin Chesowanja	1	TO	.660
				2	PA	1.353
				3	PH	1.577

TABLE 44. Continued.

PCA Euclidean Distance		*PCA Membership Quotient*		*DFA Mahalanobis Distance*		*DFA Posterior Probability*		*Taxonomic Assignment*
MS PS CM	1.373 2.036 2.630	PS MS TB	.606 .650 .711	MS CM TB	6.54 13.06 16.78	MS CM TB	.956 .037 .006	*Macaca* cf. *sylvanus*
MS CA PU	1.430 1.707 1.801	MS CA PH	.637 .760 .837	MS CA PC	5.52 9.47 9.78	MS CA PC	.726 .101 .087	*Macaca* cf. *sylvanus*
CB CG PE	1.318 3.178 3.445	CB CG PE	.524 1.198 1.332	CB CG N	8.53 30.87 35.05	CB CG N	1.000 .000 .000	aff. *Colobus* (*Mesopithecus monspessulanus*)
PC TG PS	3.401 3.460 3.736	PC PS TB	1.088 1.122 1.198	PC PS TG	72.11 74.73 75.80	PC PS TG	.630 .170 .100	*Papio* cf. *cynocephalus*
PC PS MS	1.753 1.865 1.990	PA MS PS	.861 .903 .928	PC MS PH	7.23 10.75 11.31	PC MS PH	.686 .118 .089	*Papio* cf. *cynocephalus*
TB PU PH	1.778 2.346 2.591	PU PA PH	.921 .954 1.033	PH PU MS	6.97 10.61 12.04	PH PU MS	.782 .126 .062	*Theropithecus brumpti*
PS MS PU	2.479 3.232 3.239	PS CA PU	.669 .868 1.159	PS MS PU	32.26 32.64 38.37	PS MS PU	.529 .439 .025	Cercopithecinae cf. *Parapapio*
MS TO CM	1.719 1.756 1.864	CM TO MS	.495 .758 .820	MS PA TO	24.41 26.61 31.40	MS PA TO	.719 .239 .022	Cercopithecinae cf. *Parapapio*
TO PA PH	2.533 2.945 3.491	PA PH PE	1.381 1.472 1.503	PA PU PE	13.36 15.53 15.87	PA PU PE	.582 .196 .165	*Theropithecus* cf. *oswaldi*

Additional morphological comparisons of 73-67K coupled with these multivariate patterns yield an assignment to cf. *Parapapio*.

At Lukeino no cercopithecoid dental remains have been recovered to date. However, in 1973 Martin Pickford collected a proximal humerus of a small cercopithecoid, KNM-LU 860, which is analyzed here for the first time. The rank-orderings of taxa in Table 44 indicate resemblances to *Papio sphinx* (the most primitive *Papio* species—see Chapter 6), *Cercocebus,* and *Macaca*, yet as was the case with the Lothagam specimen, no affinity with any specific taxon is indicated. Based on further morphological comparisons and the rank-ordering of these analytical techniques, the Lukeino proximal humerus is also assigned to cf. *Parapapio*. The implications of two of the oldest cercopithecoid specimens in East Africa both being assigned to *Parapapio* will be discussed further in Chapter 8.

The final specimen in Table 44 is KNM-CH 117, a partial scapula from the site of Chesowanja in the Baringo Basin of Kenya. Chesowanja can be broadly considered as early Pleistocene on the basis of faunal correlation (F. C. Howell, 1978). The only cercopithecoid dental remains recovered from this site are attributed to *?Theropithecus* sp. indet. (Delson, 1984; see also Fig. 54, this chapter). In the early 1970s, a field team headed by Bill Bishop recovered the cercopithecoid partial scapula which is analyzed here. The rank-orderings of the first two analytical techniques indicate affinities with *T. oswaldi*. The other two techniques cannot be considered, since there were not large enough samples of *T. oswaldi* to be applicable. Therefore, KNM-CH 117 can be assigned to *T. oswaldi* with some degree of certainty. The existence of *T. oswaldi* at the early Pleistocene site of Chesowanja is fully consistent with the occurrence of this species at a variety of other similar-aged sites in East Africa, most notably Olduvai Gorge.

8

DISCUSSION AND CONCLUSIONS

In the preceding chapters of this monograph I have presented the forelimb morphology of extant and fossil cercopithecoids step-by-step in a series of analyses. Now I will attempt to draw together the results of these analyses and present an overview of what this study can tell us about the evolution of the Cercopithecoidea. I will first consider the morphological differences in the forelimb that separate colobines and cercopithecines, and the origin and evolution of these differences through comparisons with Miocene and Oligocene forms. Second, I will discuss the evidence that my studies of the forelimb provide for interpreting the locomotion and ecology of the extinct taxa considered in previous chapters. Finally, I will draw some conclusions about the phyletic relationships of various Plio-Pleistocene cercopithecoids, specifically focusing on the new range-extensions for some taxa that are indicated by this study.

The Cercopithecine/Colobine Dichotomy

Despite the oft-touted uniformity of Old World monkeys, there seems to be little doubt that all living members of the family can be placed in one of two subfamilies. Extant cercopithecids are universally divided into two subfamilies—cercopithecines and colobines—with no intermediate or controversial forms. Such a sharp division is most unusual among families of extant primates. Consider, for example, the Lemuridae (*Hapalemur*), Cheirogaleidae (*Allocebus*), Indriidae (*Lepilemur*), Pongidae (*Hylobates*), Cebidae (*Callicebus*), and Callitrichidae (*Callimico*). In all of these, there are genera (listed in parentheses) that seem to bridge the gaps between subfamilial or familial divisions, falling uncomfortably between recognized groupings.

Indeed, the most obvious pattern that emerged from the multivariate analyses was the consistent separation of colobines and cercopithecines, reinforcing the view that this is indeed a real, and trenchant, dichotomy among extant Old World monkeys. To what degree is it possible to reconstruct the primitive cercopithecoid morphology in light of this great difference between cercopithecines and colobines? Four sources of evidence seem germane: (1) outgroup comparisons with *Aegyptopithecus* and other early anthropoids such as *Apidium*, (2) outgroup comparisons with early fossil cercopithecoids such as *Victoriapithecus*, (3) interpretations of previous cladistic analyses of early catarrhine and cercopithecoid evolution (e.g., Harrison, 1987; Strasser and Delson, 1987; Jablonski, 1993b), and (4) interpretations of quasicladistic behavioral/ecological analyses of cercopithecoid evolution (e.g., Andrews, 1982; Andrews and Aiello, 1984; Pickford, 1987b).

Table 45 summarizes the major differences in forelimb morphology between extant colobines and cercopithecines, derived from the analyses in Chapter 6. Many of these differences can be easily noted in the photographs on Plates 1-3. I have also added to Table 45 my own assessment of these forelimb features in three relevant fossil primate outgroups, where data were available. From these comparisons it should be possible to answer the following: (1) Are colobines persistently primitive? Or (2) do cercopithecines display the least derived morphology? Or (3) are both groups a mixture of primitive and derived character states?

It can be seen from Table 45 and Plates 1-3 that extant colobines and cercopithecines each exhibit distinctive and uniform forelimb morphologies, with virtually no overlap between them. This was apparent from the PCA multivariate plots in Chapter 6 and can be visually documented here. In Plate 1, the shoulder joints of cercopithecines *Theropithecus* [a] and *Papio* [b] are compared with those of colobines *Colobus* [c] and *Nasalis* [d]. Using the anatomical features listed in Table 45 as a reference, it can be seen that the acromion is more laterally placed and the humeral head more elongate in the two cercopithecines [a, b]. The humeral head of colobines [c, d] is more rounded, and the greater tuberosity lies below the level of the humeral head, not above it as in many cercopithecines—compare Plate 1 [c, d] with [a, b]. It can also be seen that the degree of coraco-acromial projection is greater in the two colobines than in the cercopithecines.

In Plate 2, the elbow joints of the same four cercopithecoid genera are compared. Referring to the features in Table 45, it can be seen that the medial trochlear ridge (trochlear flange) of the humerus projects quite far distally in cercopithecines, yet it is nearly absent in colobines—compare Plate 2 [a, b] with [c, d]. The humeral trochlear articular surface is wide in colobines and the capitulum is more rounded, sometimes almost ball-like. The cercopithecine ulna has a narrow trochlear notch articular surface and a wide radial notch surface, while the colobine ulna exhibits the opposite polarities, with a wide trochlear notch surface and a narrow radial notch. Finally, it can be seen that the radial head of colobines is relatively larger and more nearly spherical than that of cercopithecines. The bicipital tuberosity is also significantly wider in colobines.

In Plate 3, the wrist joints (minus the carpals, which were not included in this study) of *Theropithecus* [a], *Papio* [b], *Colobus* [c], and *Nasalis* [d] are shown. Of the features listed in Table 45, it is possible to see how the styloid process of the radius is more distinct and projects much further distally in cercopithecines [a, b] than in colobines [c, d]. The scapholunate articular surface is also expanded in cercopithecines. The distal ulna of colobines is generally more elongate in the antero-

TABLE 45. Summary of the major anatomical features of the forelimb characterizing colobines and cercopithecines*, with outgroup comparisons to the extinct taxa *Victoriapithecus*, *Aegyptopithecus*, and *Apidium* for the 15 features highlighted in boldface.

Forelimb Features	*Colobinae*	*Cercopithecinae*	Victoriapithecus	Aegyptopithecus	Apidium
SCAPULA					
Coracoid process lengthened	Yes	No	–†	–	–
Laterally placed acromion	No	Yes	–	–	–
Coraco-acromial projection	Yes	No	–	–	–
Wide and deep glenoid cavity	No	Yes	–	–	
HUMERUS					
Ant.-Post. elongation of humeral head	No	Yes	Yes	No	No
Broad bicipital groove	No	Yes	Yes	Yes	Yes
Projecting greater tuberosity	No	Yes	No	No	No
Projecting medial trochlear ridge	No	Yes	Yes	Yes	Yes
Wide and rounded capitulum	Yes	No	No	No	No
Med. epicondyle expanded posteriorly	No	Yes	Yes	No	No
Wide and deep olecranon fossa	Yes	No	No	No	No
Projecting olecranon fossa lat. margin	No	Yes	Yes	Yes	Yes
Post. trochlea articular surface wide	Yes	No	No	No	No
Trochlea gutter distance wide	Yes	No	Yes	Yes	Yes
ULNA					
Olecranon length and width expanded	No	Yes	Yes	Yes	Yes
Wide radial notch articular surface	No	Yes	Yes	No	No
Radial notch articular height expanded	Yes	No	No	No	No
Radial notch positioned anteriorly	Yes	No	No	Yes	Yes
Wide trochlear notch articular surface	Yes	No	No	Yes	Yes
Ant.-Post. elongation of distal ulna	Yes	No	–	–	–
Wide ulnar head articular surface	Yes	No	–	–	–
RADIUS					
Med.-Lat. head diameter expanded	Yes	No	–	–	–
Ant.-Post. head diameter expanded	Yes	No	–	–	–
Head articular surface expanded	Yes	No	–	–	–
Med.-Lat. neck width expanded	No	Yes	–	–	–
Wide bicipital tuberosity	Yes	No	–	–	–
Wide scapho-lunate articular surface	No	Yes	–	–	–
Projecting styloid process	No	Yes	–	–	–
Wide ulnar notch articular facet	Yes	No	–	–	–
Wide extensor digitorum groove	No	Yes	–	–	–

*The polarities in this table often impose a dichotomy on an anatomical continuum in order to facilitate discussion and make possible the outgroup comparisons. Very few of the features listed can be classified as wholly discrete traits; they are more accurately described as morphocline polarities derived from multivariate analyses. The polarities of the features in boldface are summarized in Table 46 and are also mapped on the cladogram in Fig. 56 for all five groups listed above.

†Denotes a lack of fossil evidence for this forelimb feature.

posterior dimension, and the ulnar head articular surface is considerably expanded in colobines when compared with cercopithecines.

I have demonstrated that a complete sorting of colobines and cercopithecines on the basis of forelimb morphology can be accomplished through the use of multivariate morphometrics, and can be further documented using more traditional comparative anatomical techniques. The next step is to determine what can be inferred about the primitive vs. derived nature of the forelimb in these two groups. For this determination, several outgroups that were not part of the initial analysis will be used as references for comparison.

Recent cranio-dental studies have shown that *Victoriapithecus* from the early to middle Miocene of East Africa is a very primitive cercopithecoid and actually predates the colobine/cercopithecine divergence (M. G. Leakey, 1985; Benefit and Pickford, 1986; Benefit, 1987; Benefit and McCrossin, 1989; 1991). This conclusion is based, in part, on in-depth analyses of the dentition of *Victoriapithecus* along with its sister taxon *Prohylobates*. In particular, *Victoriapithecus* exhibits: (1) incomplete molar bilophodonty, (2) frequent development of a crista obliqua on the upper molars, (3) most lower molars retaining a small but distinct hypoconulid, (4) lack of elongation of the M_1, and (5) no reduction in the buccalward flare of the mesio-buccal P_4 flange (Von Koenigswald, 1969; M. G. Leakey, 1985; Benefit and Pickford, 1986; Harrison, 1987). In contrast, all later cercopithecoids (both colobines and cercopithecines) share the derived character-states for all of these features. *Victoriapithecus* is thus most easily interpreted as the sister group of both extant subfamilies of cercopithecoids. If one does not wish to consider *Victoriapithecus* as a pre-colobine/pre-cercopithecine cercopithecoid, the only likely alternative is that the derived states of the above features evolved in parallel in the two modern subfamilies. This is certainly not the most parsimonious interpretation (Benefit and Pickford, 1986; Benefit, 1987; Harrison, 1987, 1989). Since forelimb elements of *Victoriapithecus* have also been recovered along with the dental remains, this taxon thus provides the ideal outgroup for interpreting the polarities of the colobine/cercopithecine forelimb.

Another useful taxon for outgroup comparisons is *Aegyptopithecus*, from the early to middle Oligocene of the Fayum region of Egypt (Simons, 1965). *Aegyptopithecus* is now considered the earliest known common ancestor of the Catarrhini (Fleagle and Kay, 1983; Fleagle, 1988). It is therefore the sister group of all living and extinct Old World monkeys, apes, and humans. *Aegyptopithecus* is represented by four nearly complete crania, dozens of lower jaws, and parts of the postcranial skeleton, including some forelimb elements. The inclusion of this taxon in the analysis should make it possible to determine which forelimb features are apomorphic for the clade Cercopithecoidea and which features may be plesiomorphic for all of the Catarrhini.

A final outgroup useful for interpreting the polarity of character states within the Catarrhini is the early anthropoid *Apidium*. This non-catarrhine may actually be most closely related to the South American Platyrrhini (Fleagle and Kay, 1987), though it has not as yet been formally assigned to that infraorder. One fact is clear: *Apidium* shares few, if any, derived features with the Catarrhini (Harrison, 1987; Fleagle,

1994). It thus provides an ideal outgroup for interpreting character-states within the Catarrhini.

Fig. 56 is a cladogram depicting the relationships presented above. The Colobinae and Cercopithecinae are represented as sister groups; they in turn share a sister-group relationship with the basal cercopithecoid *Victoriapithecus*, which is placed in its own subfamily, the Victoriapithecinae (Von Koenigswald, 1969; M. G. Leakey, 1985; Strasser and Delson, 1987). The basal catarrhine *Aegyptopithecus* then joins as the sister group of all Cercopithecoidea, and *Apidium* occupies the position of sister group of all the Catarrhini.

Within the context of this cladogram (Fig. 56), which is based on cranio-dental characters, it is now possible to interpret some forelimb features of Table 45 in a cladistic framework. I am not performing a cladistic analysis of these features, but rather mapping the features onto this pre-determined cladogram in order to chart the changes in their character-states (see Table 46). Such information will have a direct bearing on deciphering the primitive vs. derived nature of the cercopithecine/colobine forelimb.

As can be seen in the center of Table 45, forelimb remains of *Victoriapithecus* (proximal ulna, KNM-MB 32, distal humerus, KMN-MB 19, and proximal humerus KNM-MB 12044) show near identity with the Cercopithecinae in all comparable features. In 13 of the 15 features compared, (underscored for emphasis) they show an identical morphology. In contrast, only two features of the *Victoriapithecus* forelimb exhibit the colobine polarity (3 and 10 in Table 56). Since *Victoriapithecus* is now considered a basal pre-colobine/pre-cercopithecine cercopithecoid, there is only one way to interpret this information: the forelimb morphology of cercopithecines is closer to the primitive morphotype for the cercopithecoid clade, and the colobine forelimb morphology represents the derived condition.

This interpretation is supported by further outgroup comparisons with *Aegyptopithecus* and *Apidium*. In 8 of the 13 features shared by *Victoriapithecus* and the Cercopithecinae, *Aegyptopithecus* and *Apidium* exhibit the same polarity. Thus these 8 features can be thought of as plesiomorphic for the Catarrhini (see Fig. 56 and Table 46). Finally, the Colobinae is characterized by 8 apomorphic features of the forelimb which are not found in any other catarrhines. Five additional features of the colobine forelimb (1, 6, 12, 14, 15) are derived vis-à-vis all other cercopithecoids, yet have the same character-states as seen in early fossil catarrhines. The most parsimonious interpretation of these 5 features is that they are reversals. In a recent study, Harrison (1989) describes newly-found postcranial elements of *Victoriapithecus* which include a proximal humerus, proximal radius, and distal ulna. Of these, it was possible to include the proximal humerus, KNM-MB 12044, in my study. Harrison's (1989) analysis of the other two forelimb elements of *Victoriapithecus* serves to augment and reinforce the interpretation presented here, as do the conclusions of McCrossin and Benefit (1992) in their study of a newly recovered ischium of *Victoriapithecus*.

Evidence that the colobine forelimb is derived fits the pattern of an accumulating body of data indicating that the Colobinae are the more specialized (and derived)

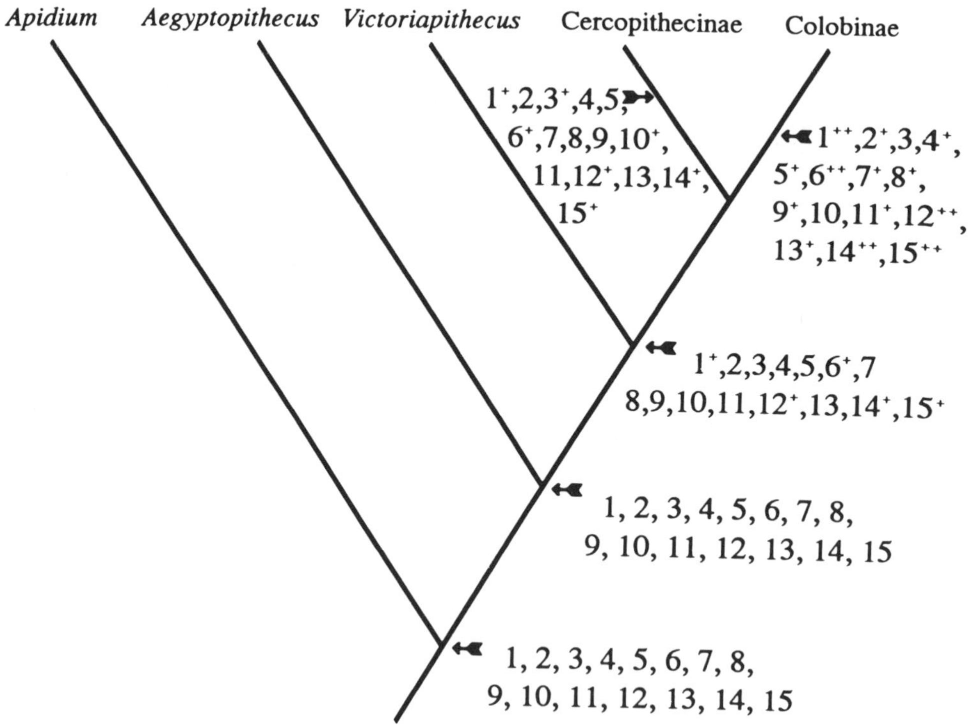

Figure 56. Cladogram depicting relationships of the non-hominoid Catarrhini and *Apidium*, based on cranio-dental analyses following the work of Fleagle and Kay (1983, 1987), Andrews (1985), M. G. Leakey (1985), Fleagle (1986), Benefit (1987), Strasser and Delson (1987), and Harrison (1987). This cladogram serves as baseline for analysis of the anatomical features of the forelimb listed in Table 45. No cladistic analysis of these forelimb features is attempted here; rather, the features are mapped onto a pre-determined cladogram, based on cranio-dental evidence, in an effort to better understand the sequence of development of forelimb features in the Catarrhini, and especially within the Cercopithecoidea. The numbers 1-15 refer to the 15 numbered forelimb features highlighted in bold in Table 45 and listed separately in Table 46. Each character has only two states: X = primitive, and X^{+} = derived. Within the Colobinae, the most parsimonious interpretation for the five features here signified as X^{++} is that they have been reversed (see Table 46).

subfamily of the Cercopithecoidea (see also Strasser and Delson, 1987; Strasser, 1988). Their possession of a sacculated (ruminant-like) stomach, coupled with an ability to produce the enzymes necessary for the digestion of cellulose, is a unique adaptation not found elsewhere in the primate order (Hill, 1972). Given this unique adaptation, it is not surprising that colobines are almost exclusively folivorous and primarily arboreal. Their high-crested molars with shortened trigonids reflect this dietary preference; and their mandibles with expanded gonial regions can also be considered synamorphies (Delson, 1975; Strasser and Delson, 1987). Colobines

TABLE 46. Character-states of the 15 forelimb features highlighted in boldface in Table 45.

Primitive Features	*Derived Features*
1, 1^{++}: Rounded humeral head† →	1^{+}: Antero-posterior elongation of humeral head
2: Broad bicipital groove →	2^{+}: Narrowing biciptial groove
3: Lack of projection of greater tuberosity →	3^{+}: Greater tuberosity projects above humeral head
4: Projecting medial trochlear ridge →	4^{+}: Lack of a projecting medial trochlear ridge
5: Narrow and flattened capitulum →	5^{+}: Wide and rounded captulum
6, 6^{++}: Lack of posterior expansion of medial epicondyle† →	6^{+}: Medial epicondyle expanded posteriorly
7: Narrow and shallow olecranon fossa →	7^{+}: Wide and deep olecranon fossa
8: Lateral margin of olecranon fossa projects posteriorly →	8^{+}: Lack of a projecting lateral margin of the olecranon fossa
9: Narrow posterior trochlear articular surface →	9^{+}: Widened posterior trochlear surface
10: Trochlear gutter distance widened →	10^{+}: Trochlear gutter distance narrow
11: Long and widened olecranon process →	11^{+}: Shortened and narrower olecranon process
12, 12^{++}: Narrow radial notch articular surface →	12^{+}: Wide radial notch articular surface
13: Radial notch positioned anteriorly →	13^{+}: Radial notch articular height expanded
14, 14^{++}: Radial notch positioned anteriorly →	14^{+}: Radial notch positioned more posteriorly
15, 15^{++}: Wide trochlear articular surface →	15^{+}: Narrow trochlear articular surface

†Only two character-states (X = primitive and X^{+} = derived) have been defined for each feature in this study. During the evolution of the Colobinae, five features (1, 6, 12, 14, 15) appear to have become reversed and are thus designated X^{++} in the above table and in Fig. 56. The primitive character-state (X) and the reversed character-state (X^{++}) are morphologically indistinguishable.

also share uniquely derived postcranial features such as reduction of the pollex and a remodeling of the joints of the foot, resulting in a novel grasping pattern during leaping (Strasser, 1988). All of these apomorphic features can be related to an arboreal adaptation which was also secondarily derived (Benefit, 1987; Strasser, 1988). Thus, when the polarities have been interpreted correctly (see Table 46), this essentially phenetic study of the cercopithecoid forelimb can provide an independent source of evidence which impacts on the phylogenetic position of the Colobinae within the Catarrhini.

Another issue should be explored: Given an understanding of the polarities presented in Table 45, how does this study of the forelimb relate to current theories regarding the origin of the Cercopithecoidea? Based on an analysis of the habitat usage, ecological polarity, dietary preference, and positional behavior of extant catarrhines, together with a reinterpretation of the fossil record, Andrews and Aiello (1984:451) have proposed the following evolutionary scenario: "the cercopithecoids emerged from the Anthropoidea by about the beginning of the Miocene as a result of both an ecological shift towards more open savanna conditions (Andrews, 1982) and a dietary shift from frugivory toward a degree of folivory." According to this scenario, their immediate ancestors were forest-dwelling, arboreal frugivores. Temerin and Cant (1983), propose that the development in cercopithecoids of a more eclectic diet and increasing amounts of terrestriality could have been brought about by a shift toward cooler, drier, more seasonal climates that began in the early Miocene. This increased "patchiness" in the environment would have brought about a reduction in the available fruit sources. The earliest cercopithecoids may have adapted to the situation by exploiting lower-quality but more prevalent food sources such as leaves, unripe fruit, etc. In doing so, they developed their specialized bilophodont dentition and a digestive system that could tolerate plant secondary compounds (Andrews and Aiello, 1984).

An interesting corollary of the above scenario is that the ancestral habitat type of cercopithecoids was savanna-woodland (Andrews, 1982), *not* a tropical forest biome. Therefore, the locomotor adaptations of the earliest cercopithecoids must have tended toward cursoriality. Indeed, many of the earliest fossil cercopithecoids show evidence of terrestrial adaptations in the postcranial skeleton and have often been recovered from drier, more open paleoenvironments (ibid.). Certainly, the cercopithecine-like morphology of the earliest cercopithecoid, *Victoriapithecus* (see Table 45), is in full agreement with at least a semi-terrestrial mode of life for this species (Harrison, 1989). Looking at fossil colobines from this perspective, it is interesting to note that some of the most cercopithecine-like fossil forms, such as *Mesopithecus* and *Dolichopithecus* (see plots in Chapter 6), are also among the earliest known colobines, and furthermore are considered by many authorities to be terrestrially adapted (see comments in following section). The logical outgrowth of this scenario is that the wholly arboreal adaptations of most living colobines and some arboreal cercopithecines may in fact represent a *secondary* arboreal adaptation or reinvasion of tropical forest biome (e.g., Strasser, 1988). Ripley (1979) and Andrews (1982) emphasize this point by demonstrating that arboreal Old World monkeys locomote through the trees in an "above-branch," often cursorial fashion which arose from an initial terrestrial adaptation. This contrasts with the below-branch arm-swinging, hanging, and suspensory behaviors of most other arboreal primates. Strasser (1988:242) also observes that as a result of competition with already arboreally adapted apes, formative colobines could have developed their specialized arboreal locomotor abilities when they "were forced to the periphery of crowns where leaping between trees was a reasonable alternative to descent along the trunk." Thus the terrestrial legacy of the ancestral cercopithecoid and proto-colobine may be still recognizable today through careful studies of the positional behavior of living species.

Locomotor Adaptations of Fossil Cercopithecoids

In previous chapters I have included the skeletal remains of various identified and unknown extinct cercopithecoids in analyses of different anatomical regions of the forelimb. These univariate and multivariate analyses provided a wealth of information regarding the similarities and differences of these extinct species. In this section, I will attempt to summarize what we can say about the likely locomotor habits of these extinct taxa, usually based on several independent analyses of their forelimb skeletons.

Up to this point the major focus of this study has been on taxonomic/phylogenetic interpretations of this data set. As discussed in Chapter 1 (see Fig. 1), there is also functional/locomotor information that can be derived from each analysis. Many of the PCA plots in Chapter 6 present a primary taxonomic/phylogenetic sort of the taxa. However, *within* each taxonomic grouping there is also a functional/locomotor sorting of the taxa. This information, coupled with comments from previous studies, will be used to describe the locomotor adaptations of the extinct species included in this study.

Mesopithecus

Mesopithecus is a small to medium-sized colobine from the late Miocene to Pliocene of southern and central Europe, extending into Iran. There are two known species, *M. pentelici* and *M. monspessulanus*. Gaudry (1862) originally pointed out that *M. pentelici* had the skull and teeth of a colobine but the limb structure of a macaque. The analyses in Chapter 6 both confirm and amplify Gaudry's observations. *M. pentelici* appears in six of the Chapter 6 plots: the proximal humerus, distal humerus, proximal ulna, distal ulna, proximal radius, and distal radius. In four of these analyses it lies very near the center of the plot, twice occupying the position of the most cercopithecine-like colobine. In plots of the proximal radius, distal radius, and proximal ulna it lies nearest its sister taxon *Dolichopithecus*. When compared with extant colobines, *Mesopithecus* is quite similar to *Presbytis entellus*, the most terrestrially adapted of the colobine species. Compared with cercopithecines it usually falls nearest *Macaca sylvanus*, a terrestrial and arboreal quadruped, and *Cercocebus albigena*, an arboreal quadruped. Only in the analysis of the proximal humerus does it fall near the arboreal species *Colobus*.

Delson (1973:480), in his extensive analysis of the circum-Mediterranean fossil Colobinae, concludes that "it may be supposed that *Mesopithecus pentelici* lived in troops in gallery forest and savannah fringe, spending a good part of the day feeding terrestrially or moving between arboreal feeding sites." The present study confirms Delson's scenario and furthermore underscores the terrestrial adaptations of *M. pentelici*.

M. monspessulanus, the apparent descendant of *M. pentelici* (Delson, 1973), is rather poorly known. In this study only one specimen of *M. monspessulanus*, a distal humerus, was analyzed (see Table 44 in Chapter 7). In all of the analyses this specimen ranked very near *Colobus badius*, an arboreal quadruped and leaper. It is interesting to note that on the basis of other postcranial specimens and analyses of the paleoenvironment, Delson (1973, 1975) and Szalay and Delson (1979) conclude that *M. monspessulanus* was very likely arboreally adapted, inhabiting a well-watered and more wooded environment.

Dolichopithecus

Dolichopithecus is a colobine of moderately large size from the latest Miocene and Pliocene of central, eastern, and southern Europe. Only one species, *D. ruscinensis*, is now recognized (Szalay and Delson, 1979). Gabis (1961) considered *Dolichopithecus* a fully terrestrial quadruped much like living savanna baboons whereas Deperet (1890) and Jolly (1967) viewed its basic adaptation as more macaque-like. Delson (1973:586) concludes that "*Dolichopithecus* shows a locomotor adaptive level beyond that in the most terrestrial colobines, but not equal to any fully terrestrial cercopithecines." In my analyses in Chapter 6, *Dolichopithecus* appears in four of the plots: the distal humerus, proximal ulna, proximal radius, and distal radius. In all four cases it lies nearest its sister taxon *Mesopithecus*. Its next-closest approach is to *Presbytis entellus*, a predominantly terrestrially adapted colobine. In none of the four analyses does it lie particularly close to any of the species of *Papio* or to *Macaca sylvanus*. Thus the observations of Gabis, Deperet, and Jolly are not confirmed by this study. In fact, based on my analysis, I can see no reason to infer a locomotor adaptation for *Dolichopithecus* any different from that hypothesized for *Mesopithecus*. In closing, it is interesting to note the observation of Szalay and Delson (1979:417) that "the habitus of *Dolichopithecus* is best seen as a convergence to mandrills, foraging on the forest floor, and at times ascending into the trees." This is an intriguing scenario, but it cannot be supported by any of the analyses in my study.

Paradolichopithecus

Paradolichopithecus is a large cercopithecine from the Pliocene of Spain, France, Italy, Romania, and Russian Central Asia. Two species are currently recognised: *P. geitcus*—European in distribution (Necrasov et al., 1961), and *P. sushkini*—Asian in distribution (Trofimov, 1977). Szalay and Delson (1979) question whether *P. geticus* might be better considered synonymous with "*P. arvernensis*," also European in distribution (see also discussion by Ardito and Mottura, 1987). Whatever the proper species nomen, my comments pertain to the European form. In the analyses in Chapter 6, *Paradolichopithecus* only appears in two of the plots: the distal humerus and proximal ulna. Based on the distal humerus, it is nearest the cluster of *Papio* species, though it also shows some affinity with *Theropithecus brumpti* and *Macaca* on component 1. Analysis of the distal ulna reveals similarities to *Papio sphinx, Theropithecus brumpti,* and *Macaca*. All of these species are large-bodied, primarily terrestrially adapted, cursorial forms inhabiting either primary forest or savanna-woodland. According to Aguirre and Soto (1978), the dispersal of *Paradolichopithecus* in Europe appears to coincide with a period of increasing humidity and forestation, so perhaps an ecological model based on the mandrill *Papio sphinx* might be an appropriate analogue for *Paradolichopithecus*.

Theropithecus darti

Theropithecus darti is a medium-sized cercopithecine from the middle to late Pliocene of southern and eastern Africa. Its identification and analysis in this study are based primarily on specimens from the sites of Makapansgat, Sterkfontein, and Hadar. In most cases the postcranial remains tend to resemble *T. gelada*. From this it is possible to infer a locomotor adaptation of habitual terrestriality as seen in *T. gelada*. However, I agree with Jolly (1972), Eck and Jablonski (1984), and Dunbar (1993) that the locomotor behavior and habitat preference of *T. gelada* represents the adaptations of a specialized relict population that only developed in the Ethiopian highlands. Eck and Jablonski (1984:129) conclude that *T. darti* "appears to have preferred lowland, woodland/grassland communities." In an effort to more precisely determine the habitat preferences of *T. darti*, Eck (1993) compares the frequency of specimens of *T. darti* to those of the family Bovidae throughout the Hadar Formation. Eck (1993:72-73) concludes: "The pattern of associations suggests, perhaps, that *T. darti*, like *Aepyceros* and the alcelaphines, preferred drier woodlands and grasslands rather than the moist swamps and woodlands preferred by the reduncines." This conclusion accords well with the view that since *T. darti* and *T. oswaldi* form a continuous evolving lineage and are likely a chronospecies (Jablonski, 1993b), they would share similar habitat preferences.

Theropithecus brumpti

Theropithecus brumpti is a very large cercopithecine from the middle to late Pliocene of eastern Africa. Its distribution is limited to the Lower Omo River Basin of southern Ethiopia, the Turkana Basin of northern Kenya, and possibly the Awash River Basin of northeastern Ethiopia. In the analyses in Chapter 6, *T. brumpti* appears in all seven of the plots where extinct species were analyzed. Therefore, all parts of the forelimb of *T. brumpti* were available for study. In plots of the scapula and proximal humerus (the shoulder joint), *T. brumpti* lies nearest *T. oswaldi* among all taxa in the analysis. Based on the distal humerus, it lies between the large-bodied papionins and small-bodied papionins on component 1 and near *T. oswaldi* on component 2. On the proximal ulna, *T. brumpti* lies very near *Papio sphinx* and not far from *Paradolichopithecus* and *T. oswaldi*. On the proximal radius, it also lies very near *P. sphinx* and equally near *T. gelada; T. oswaldi* is also not far removed. On the distal ulna, *T. brumpti* lies once again very near *P. sphinx* and rather close to the small-bodied papionins. Finally, on the distal radius, *T. brumpti* lies nearest *P. sphinx* and *T. gelada* on component 1 but falls closest to *T. oswaldi* on component 2.

Some conclusions are apparent in these analyses that can shed some light on the potential locomotor adaptations of *T. brumpti*. First, it is not "savanna baboon-like" in its forelimb anatomy. My analyses also show that *T. brumpti* is not "gelada baboon-like" in its forelimb. Both of these represent the most extreme terrestrial adaptations among the cercopithecines. Of all the extant species it is most similar to

Papio sphinx, a forest-dwelling terrestrial and partly arboreal species. Since *T. brumpti* was a very large primate, it probably did not spend all of its time in an arboreal habitat—but whatever differences separate the extant mandrill from the savanna baboon also separate *T. brumpti* from other wholly terrestrial and cursorial species. Eck and Jablonski (1984) suggest that *T. brumpti* preferred "lowland, riverine forest and/or back-swamp communities." Certainly much of the fauna recovered from the Lower Omo Basin Shungura Formation, members A-E, accords well with such an interpretation, as do dietary reconstructions of increased folivory and frugivory in the *T. brumpti* lineage (Benefit and McCrossin, 1990; Teaford, 1993). Furthermore, lowland riverine forest is one of the preferred habitats of the mandrill *P. sphinx*. Jablonski (1986) has analyzed the hand morphology of *T. brumpti*, showing that though it was more powerfully built than the hand of *T. gelada*, it had the same high degree of precise opposability of thumb and index finger, indicating a similar feeding adaptation. Jablonski concludes that this specialized feeding adaptation is a heritage trait that was likely primitive for the genus as a whole. However, Jablonski (1986) notes that since the phalanges of *T. brumpti* are curved, as is characteristic of many climbing, arboreal mammals, *T. brumpti* was better adapted to a forested, arboreal environment than were other species of the *Theropithecus* group. This view is also supported by Krentz (1993) who recently analyzed the entire postcranial anatomy of *Theropithecus*. He concludes that *T. brumpti* is best characterized as an arboreal quadruped.

Theropithecus oswaldi

Theropithecus oswaldi is another very large cercopithecine which is known from the late Pliocene to middle Pleistocene of northern, eastern, and southern Africa. Like *T. brumpti, T. oswaldi* appears in all seven analyses of the extinct species (see Chapter 6). Thus all elements of the forelimb are available for study. Though *T. oswaldi* often appears quite near *T. brumpti* in many of the plots discussed in the preceding section, if this information is re-evaluated it can be seen that, in those ways that *T. oswaldi* and *T. brumpti* are separated, *T. oswaldi* almost always moves in the direction of the *Papio* cluster. Thus, based upon the proximal humerus, distal humerus, proximal ulna, distal ulna, and distal radius, *T. oswaldi* diverges from *T. brumpti* on either component 1 or component 2 *in the direction* of the *Papio* (savanna baboon) cluster. Thus the close relationship of these two *Theropithecus* species reflects their common heritage, while the subtle differences that separate them appear to reflect habitus preferences.

T. oswaldi, especially the middle Pleistocene representatives, has long been considered among the most terrestrially adapted of all the extinct Plio-Pleistocene cercopithecoids (Jolly, 1965, 1967, 1970, 1972). Because of its large size (sometimes reaching that of a female gorilla), unique postcranial adaptations, and open-country-adapted faunal associations, it has often been the subject of unique and provocative speculation (Shipman et al., 1981). Jolly (1972:115) views *T. oswaldi* as quite closely related to the living gelada baboon. He concludes: "Like their living relative,

[*T. oswaldi*] were inhabitants of open country, rarely if ever climbing trees, and preferred the margins of shallow lakes where seasonal flooding inhibited the growth of trees but enriched the ground-flora, especially of grasses. Blades, seeds, and rhizomes of grass, picked by hand, probably supplied the bulk of the diet, but would have been supplemented by other small food-objects."

Recent work on the adaptations and ecological energenics of *T. oswaldi* by Lee and Foley (1993) has led to some interesting speculations concerning the dietary preferences and locomotor habits of this species just prior to its extinction. Lee and Foley (1993:494) suggest: "*Theropithecus oswaldi* could have become more selective in its diet and concentrated on foods that gave a higher energetic return per pick such as fruits. However, both its morphological adaptations to a predominately grass diet and the presence of the generalist frugivore *Papio* would limit the foods available to *T. oswaldi*, especially if the *Papio* species had greater competitive abilities in terms of locating and defending fruit patches." Another adaptive strategy is also possible—the large, open-grassland-adapted *oswaldi* could have become *less* selective in its diet and "...maximized intake at the expense of quality as do many grazing species when confronted with declining food biomass and a correlated decrease in food quality (ibid., p. 495)."

As *T. oswaldi* increased in body size throughout the Pleistocene, it would have have converged on the ruminants, the most specialized grazers. Lee and Foley (1993:495) conclude: "The adaptive radiation of many medium sized modern bovids during the early and middle Pleistocene (Vrba, 1985) may have exerted strong competitive pressure on the relatively less efficient grazing *Theropithecus* and limited the range of foods available for exploitation." Whatever scenario one supports regarding the extinction of *T. oswaldi*, it is clear that in terms of locomotor adaptations and diet, this species of *Theropithecus* was very specialized.

Paracolobus

Paracolobus is a very large colobine known from the middle Pliocene to the early Pleistocene of eastern Africa. Three species have been identified, but only two have been named as yet: *P. chemeroni* and *P. mutiwa* (M. G. Leakey, 1982). The former species has a nearly complete postcranial skeleton and is thus the focus of this analysis of locomotor adaptations. *Paracolobus* appears in five of the seven analyses of the extinct taxa in Chapter 6. It is not found in the analyses of the distal ulna and distal radius because the hands and lower ends of the extremities became detached from this nearly complete skeleton prior to its deposition (Birchette, 1982).

Paracolobus presents an intriguing mosaic of locomotor adaptations on the basis of its locations in the plots of Chapter 6. Based on the scapula, it is most like *Nasalis* and *Colobus*, both arboreal quadrupeds and leapers, on component 1, yet is far removed from them on component 2. In analyses of the proximal and distal humerus, it falls very near *Presbytis entellus*, a terrestrial and arboreal quadruped, on both component 1 and 2. In the analysis of the proximal ulna, *Paracolobus* falls very near *Colobus guereza* on both components. Finally, based on analysis of the proximal radius, *Paracolobus* falls at the extreme colobine pole, equidistant from the two *Colobus* species on the one hand and from *Dolichopithecus* and *Mesopithecus* on the other.

These analyses of *Paracolobus* present an interesting combination of resemblances to both arboreal and terrestrial quadrupeds. Depending on which limb element is emphasized, one could argue that this large extinct colobine was a tree-dweller or a cursorial ground-dweller. The total weight of the evidence does, however, slightly favor an arboreal adaptation. Birchette (1982:426-427), in his detailed review of the postcranium of *Paracolobus*, also struggled with this interpretation, finally concluding: "on the basis of a detailed examination of several skeletal elements, and with the modified view of the functional significance of various morphological traits with respect to locomotor behavior, the clearest image that emerges of *Paracolobus* is one of a predominantly arboreal quadruped that was capable of reasonably efficient movement on the ground and that probably spent a small but regular percentage of its time on a daily basis engaged in terrestrial locomotor behavior." I find my conclusions based on multivariate analyses of the *Paracolobus* forelimb very similar to Birchette's (1982) conclusions, derived from more traditional comparative anatomical analyses.

Rhinocolobus

Rhinocolobus is a large colobine known from the middle Pliocene to early Pleistocene of eastern Africa. Only one species, *R. turkanaensis*, is currently recognized (M. G. Leakey, 1982). One associated partial skeleton is known from the Koobi Fora Formation, East Turkana (ibid.) and is the subject of this analysis. *Rhinocolobus* appears in analyses of the proximal humerus, distal humerus, proximal ulna, and distal ulna in Chapter 6. In all four of these analyses the results are remarkably consistent for this extinct colobine. Based on the proximal humerus, *Rhinocolobus* is close to the *Colobus* species on component 1 and near *Nasalis* on component 2. Analysis of the distal humerus shows *Rhinocolobus* nearest *Colobus badius* and *C. guereza* on both components. On the proximal ulna, *Rhinocolobus* is at the extreme colobine pole, falling nearest *Nasalis* on component 1. Based on component 2, it is identical with *C. badius*. Finally, on the distal ulna, *Rhinocolobus* once again falls right at the mean for *C. badius* on component 1 and lies between *C. guereza* and *Presbytis entellus* on component 2.

From these analyses, the most straightforward conclusion is that among living colobines *Rhinocolobus* is most like the arboreal quadrupeds and leapers of the *Colobus* species-group. It certainly had a predominantly arboreal adaptation based on my analysis of the forelimb elements. This accords well with the paleoenvironmental evidence suggesting that *Rhinocolobus* is always associated with a riverine forest or woodland habitat (ibid.). This is further supported by its ubiquitous occurrence in the Shungura Formation of the Lower Omo Basin—a predominantly riverine forest habitat up through Member G when a change in paleoclimate to dryer conditions causes the disappearance of *Rhinocolobus* (M.G. Leakey, 1987). My analysis of the *Rhinocolobus* forelimb is in full agreement with M. G. Leakey's (1982, p. 170) final conclusion that "*Rhinocolobus* was thc most similar to extant *Colobus* in dietary and locomotor adaptation." It was certainly similar to *Colobus* in habitat preference as well.

Cercopithecoides

Cercopithecoides is another large colobine known from the middle Pliocene to early Pleistocene of eastern and southern Africa. Two species are recognized: *C. williamsi* from eastern and southern Africa and *C. kimeui*, known only from eastern Africa (M. G. Leakey, 1982). A new associated partial skeleton of *Cercopithecoides* was recently recovered from the Koobi Fora Formation of East Turkana (ibid.). After analysis of these postcranial remains, Birchette (1982:442) concludes: "the humerus, scapula, and ulna of *C. williamsi* display numerous features indicative of a 'hyperterrestrial' adaptation and radically different from (most) comparable features of *Paracolobus chemeroni*." Leakey adds that *Cercopithecoides* is often found associated with savanna and open-country paleohabitats.

Unfortunately, the discovery of this associated skeleton occurred after my visit to the Kenya National Museums and therefore was not included in this study. However, I did identify a number of colobine specimens from the South African Transvaal cave sites where *C. williamsi* is the only recognized colobine, on the basis of cranio-dental remains. The results of these analyses (see Tables 37-42 in Chapter 6) do not seem to indicate the extreme terrestrial adaptations noted for the East Turkana species. Perhaps the absence of competition from other fossil colobines inhabiting South Africa at this time made it possible for these southern populations of *Cercopithecoides* to maintain a more eclectic locomotor adaptation (however, also see comments on phyletic relationships below).

Phyletic Relationships of Plio-Pleistocene Cercopithecoids

Because proper outgroups for comparison are absent, it is difficult to draw significant conclusions about the mega-phyletic relationships of Plio-Pleistocene cercopithecoids on the basis of the multivariate-generated phenetic plots of the forelimb presented in Chapter 6. However, the identification of 234 previously unidentified forelimb elements from a variety of African Plio-Pleistocene localities (see Chapter 7) does provide a wealth of new information about the temporal ranges of many fossil cercopithecoid species. When these range-extensions are viewed along with the forelimb plots of Chapter 6, several interesting patterns emerge.

Fig. 57 presents the temporal ranges of African fossil cercopithecoids based solely on the identification of forelimb elements in this study. With the exception of

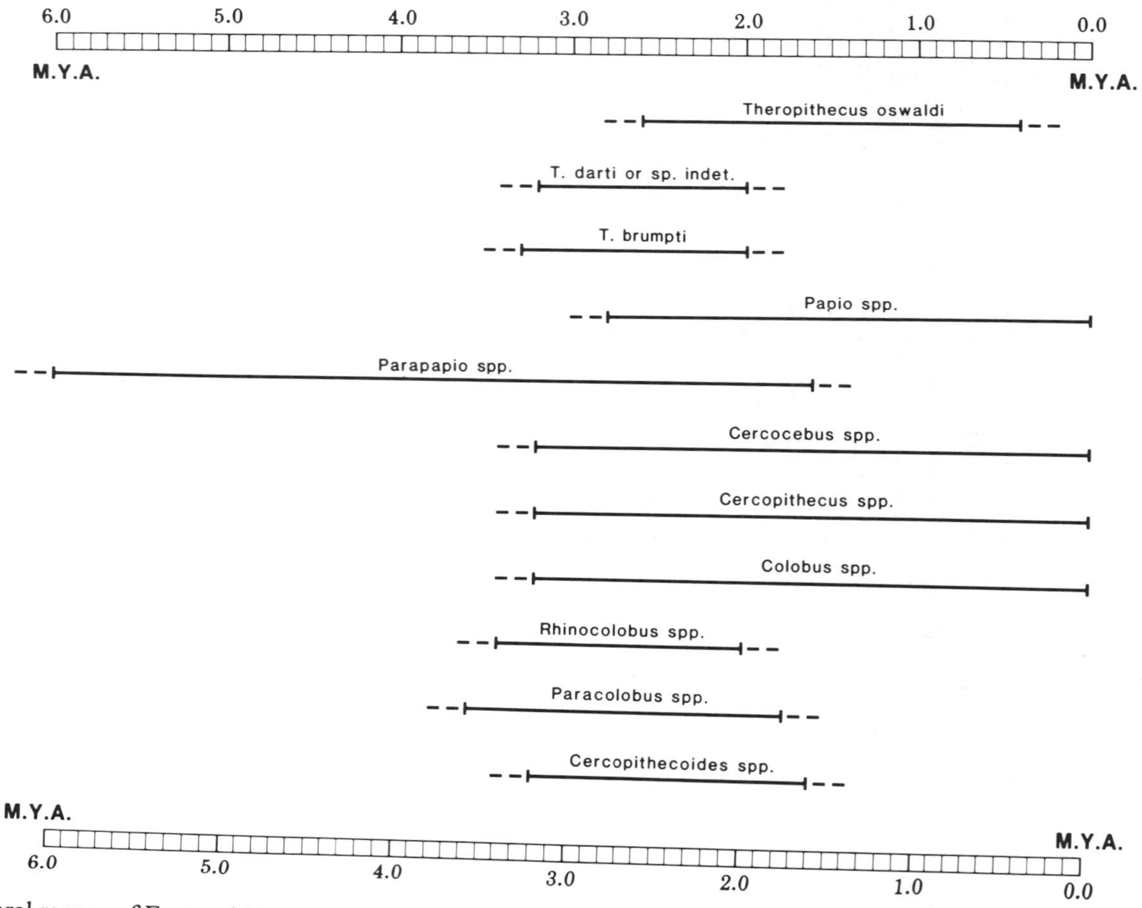

Figure 57. Temporal ranges of East and South African cercopithecoid taxa, based solely on identifications of the forelimb skeleton provided by this study.

Parapapio, all of the fossil lineages appear to begin rather suddenly between 3 and 4 million years ago. This is *not* evidence for a step-like punctuated origin of these lineages, but rather represents the lack of an African fossil record between the latest Miocene through early Pliocene (see Fig. 55 in Chapter 7). There are simply very few early Pliocene sites that preserve mammalian remains, and those few usually preferentially preserve dental remains rather than postcrania.

The longest enduring lineage in Fig. 57 is *Parapapio*. It is now first documented at the latest Miocene East African site of Lukeino (see Table 44 in Chapter 7) and survives until the early Pleistocene. Actually *Parapapio* probably has an even longer temporal range, perhaps stretching back to the middle Miocene. Delson (1975, 1983) views it as the ancestral cercopithecine stock which gave rise to the Papionini, including both the Asian macaque radiation and the African *Cercocebus* and *Papio* radiations. Given its temporal range and primitive morphology, it certainly is the most likely candidate. *Parapapio* was not directly included in any of the analyses of Chapter 6 because associated cranial and postcranial remains are lacking. However, it is worth noting that one of its proposed direct modern descendants, *Macaca sylvanus*, was included. As can be seen from a survey of the plots in Chapter 6, *M. sylvanus* nearly always occupies a position at the center of each plot, indicating that it is a possible plesiomorph for the common ancestors of the papionins. It is also worth noting that *Victoriapithecus* also shares a number of similarities with *M. sylvanus* and *Cercocebus*, yet all of these are symplesiomorphic features.

Another interesting problem in the evolution of the cercopithecoids is the origin of the *Theropithecus* species-group. The conventional wisdom is that this group was derived from a primitive African stock of Papionini (Jolly, 1972). Current studies on the phylogenetic history of *Theropithecus* (Delson, 1993; Jablonski, 1993b) suggest that the genus *Papio* is *the* sister taxon of the *Theropithecus* species-group. This view accords well with the cranio-dental evidence, since the three extinct species of *Theropithecus*, *T. darti*, *T. oswaldi*, and *T. brumpti*, are all characterized by a very specialized (and derived) dentition and unique cranial anatomy (especially *T. brumpti*). My study of the forelimb skeleton produces evidence that all three lineages extend back to the middle Pliocene (see Fig. 57). The longest temporal range is that of *T. brumpti*. This species also appears to be the sister group of the other three, *T. darti, oswaldi,* and *gelada*. Though my method sometimes resulted in confusion in the identification of late *T. darti* and early *T. oswaldi* (indicative of the ancestral-descendant relationship hypothesized for these species—see Jablonski, 1993b), and whereas *T. darti* often was identified as *T. gelada*, there was never any problem with the identification of *T. brumpti*. It was even possible to identify *T. brumpti* at Hadar (see Tables 32 and 34 in Chapter 7), where no previous cranial remains have yet been identified. Based on my analysis, one can propose the *Theropithecus* species-group as having a much longer ancestry than previously considered, likely extending back to the beginning of the Pliocene (>5 Ma).

The radiation of the genus *Papio* is in some way tied to the decline and extinction of all the species of *Theropithecus* except for the relict form *T. gelada*. Recently, Eck and Jablonski (1984, 1987) have reviewed the East African fossil record of *Papio* and have transferred a number of specimens from *Papio* to *Theropithecus*. Eck and Jablonski (1984: 130) state: "the East African sample of *Papio* (now) consists of about 21 specimens other than isolated teeth, and most of these are so poorly preserved that their proper identification is difficult." My analysis of the forelimb of previously unidentified East African cercopithecoids adds about 17 specimens of *Papio* to that total but with no new range extensions. This was a very small percentage of the total number of East African specimens analyzed. Thus, the evolutionary history of the presently ubiquitous African savanna baboon still remains a mystery.

The final group I will discuss is the large extinct colobines of East and South Africa. These were identified on the basis of associated cranial and postcranial material derived entirely from East Turkana and the Baringo Basin. The sample of the identified material was small, so the question arises as to whether some terrestrially adapted colobines were actually identified as cercopithecines. This remains a possibility, but visual inspection of the multivariately identified fossil colobine material always revealed at least some heritage characters of the Colobinae, even in the most terrestrially adapted specimens, such as those belonging to *Cercopithecoides*. With regard to determining the phyletic relationships of *Rhinocolobus, Paracolobus,* and *Cercopithecoides*, all that can be inferred from Fig. 57 and the multivariate forelimb plots in Chapter 6 is a ranking of these three species in descending order vis-à-vis their relationship to extant *Colobus* species. My analysis shows that *Rhinocolobus* is the most closely related, followed by *Paracolobus* and finally *Cercopithecoides* as the least related to extant *Colobus*. These determinations accord well with those derived by M. G. Leakey (1982, 1987), based primarily on studies of the cranio-dental remains of these extinct colobines. It should be further added that none of these extinct African forms showed any affinities to the European fossil colobines *Mesopithecus* and *Dolichopithecus* on the basis of my studies of the forelimb.

A final point should be addressed regarding the affinities of the South African species of *Cercopithecoides, C. williamsi*. M. G. Leakey (1982) reports the discovery of a partial skeleton of this species designated KNM-ER 4420, from the late Pliocene of East Turkana. As mentioned previously, Birchette (1982) considers this skeleton to have "hyperterrestrial" adaptations while my analysis of other specimens of *C. williamsi* from South Africa does not indicate anywhere near the same degree of terrestrial adaptation. My studies also do not support M. G. Leakey's (ibid., p. 171) claim that *Cercopithecoides* is "an early offshoot from the main colobine lineage [that] may share a common ancestry with the rather terrestrial *Mesopithecus* and *Dolichopithecus*." Until I have a chance to study directly the newfound postcranial skeleton of KNM-ER 4420 from East Africa and compare it with the specimens of *C. williamsi* that I have identified in South Africa, I question the identification of KNM-ER 4420 as *C. williamsi*.

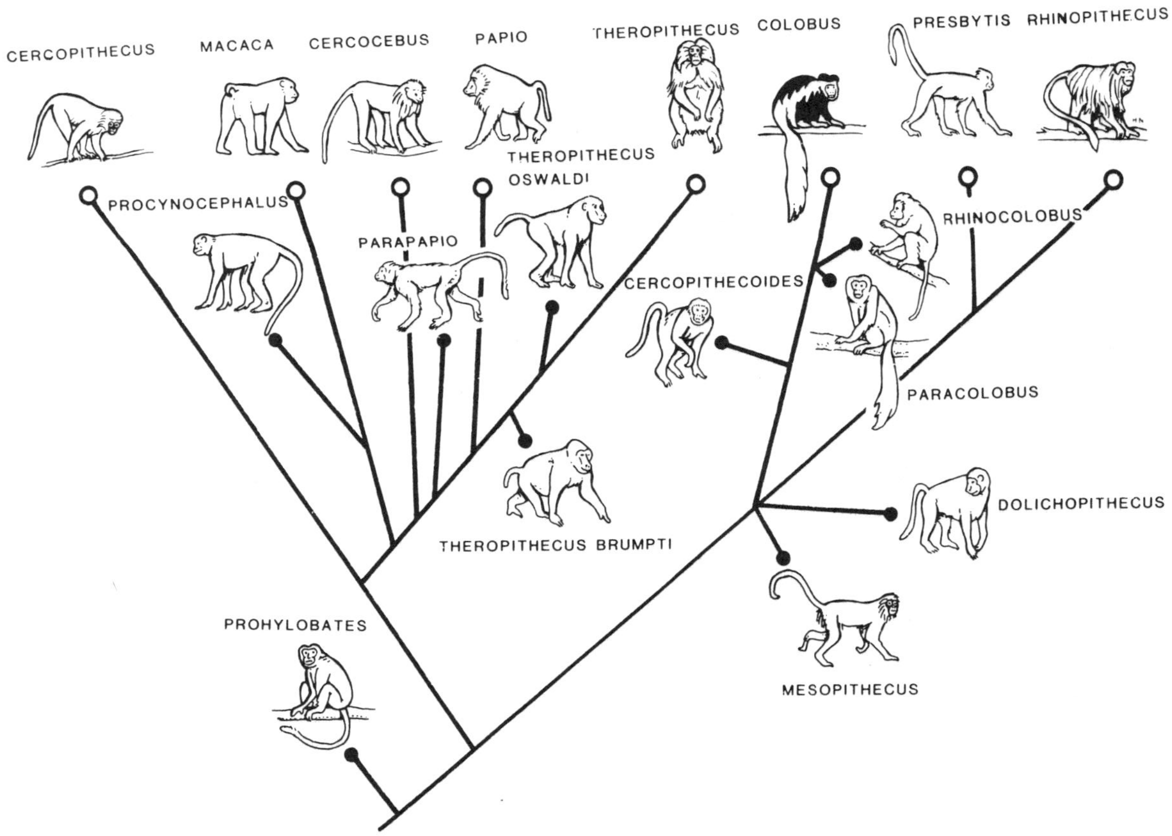

Figure 58. Cladogram illustrating the phylogenetic and locomotor diversity of the Plio-Pleistocene radiation of the Cercopithecoidea. During this adaptive radiation, the Old World monkeys became the world's most diverse and successful group of primates. This cladogram is from Fleagle (1988), and is reproduced here with the permission of the author and the publisher, Academic Press.

Final Comments

In summary, the results of this study serve to reinforce and amplify current views of the history of cercopithecoid monkey evolution. The origin of the group can be traced to the late Oligocene, probably to a primitive catarrhine not unlike *Aegyptopithecus* from the Fayum. The earliest Old World monkey, *Victoriapithecus*, seems to have a forelimb that suggests a semi-terrestrial, wholly quadrupedal adaptation that most resembles the morphology (and adaptations) of extant cercopithecines. Some of the forelimb features of *Victoriapithecus* are also best interpreted as primitive catarrhine retentions.

Sometime during the middle Miocene there occurred a major phyletic split of the early cercopithecids into the Cercopithecinae and Colobinae, with the former retaining the primitive (*Victoriapithecus*-like) forelimb morphology and the latter acquiring the derived (colobine) forelimb morphology after its reinvasion of the tropical forest biome. By the late Miocene and early Pliocene, there is evidence of numerous large cercopithecids from Eurasia. Colobines appear to be the first Old World monkeys to invade Eurasia, about 10 million years ago. Interestingly, many of the colobines of this Eurasian radiation seem more terrestrial than their living counterparts, lending credence to the hypothesis that the cercopithecine forelimb morphology and terrestrial habitat preference are primitive for early Cercopithecinae and Colobinae alike.

In the Pliocene of Africa, there is evidence of an extensive radiation of relatively large Old World monkeys representing both cercopithecines (i.e., *Parapapio, Theropithecus darti, T. brumpti*) and colobines (i.e., *Cercopithecoides, Paracolobus,* and *Rhinocolobus*). The evidence from the forelimb structure of the giant "gelada," *T. brumpti*, indicates it was quite divergent in adaptation from its closest living relative, *T. gelada*, as well as from its sister taxon, the *T. darti-T. oswaldi* lineage. In comparison with the earlier Eurasian colobines, the large extinct colobines of Africa appear to be more arboreal, like their smaller living relatives. However, there is one form, *Cercopithecoides*, which is thought to be more terrestrially adapted than any living colobine.

In conclusion, this monograph demonstrates that the fossil record of cercopithecoid monkeys contains an even more interesting and diverse array of Old World higher primates than previously thought. This diversity of living and extinct cercopithecoids is well illustrated in Fig. 58, reproduced here from Fleagle (1988). Excluding humankind, it easy to see why the Old World monkeys can today be considered the most successful group of primates to inhabit the Earth.

Appendix

THE EXTINCT COMPARATIVE SAMPLE

The four tables that follow contain all the vital information on the individual specimens making up the extinct comparative sample. Since some of the postcranial specimens composing the extinct comparative sample are identified to species here for the first time, I thought it important to present them in tabular form as a supplement to the identification of the unknown fossil forelimb specimens presented in Chapter 7. The key to the institutional abbreviations used in this appendix appears in the front matter of this monograph. A complete listing of the individual specimens making up the extant comparative sample can be found in Tables 51-63 in Ciochon (1986). For a detailed discussion of both the extant and extinct comparative samples, see Chapter 2 of this monograph.

TABLE 47. Comparative sample of *Theropithecus brumpti* from the Omo Shungura Formation.

Specimen Number	*Skeletal Element*	*Collection Locality*	*Present Location*
L78-37c	left prox ulna	member C8	ENM
L32-144	right prox ulna	member C7	ENM
L32-38	right prox ulna	member C7	ENM
L78-60	left prox ulna w/shaft	member C8	ENM
L32-1591	right prox ulna	member C7	ENM
L345-8	left prox ulna	member C9	ENM
L42-40	right prox ulna	member C5	ENM
L193-42	right prox ulna w/shaft	member C8	ENM
L23-66	right prox ulna w/shaft	member C6	ENM
L345-19	left prox ulna w/shaft	member C9	ENM
L193-43	left prox ulna	member C8	ENM
L885-1	right prox ulna	member C4	ENM
L881-1	right prox ulna	member C8	ENM
L858-1	right prox ulna	member C7	ENM
L47-41	left prox ulna	member C8	ENM
L78-95	right prox ulna	member C8	ENM
L199-6	left prox ulna	member C7	ENM
L546-7	right prox ulna	member C8	ENM
L62-12	right distal ulna	member C5	ENM
Ll7-45c	left prox radius	member C8	ENM
L78-37b	right prox radius	member C8	ENM
L32-l55h	right complete radius	member C7	ENM
L345-18a	left prox radius w/entire shaft	member C9	ENM
L345-236	left prox radius w/entire shaft	member C9	ENM
O18-701	left prox radius	member C6	ENM
L886-14	right prox radius	member C8	ENM

(continued on next page)

TABLE 47 Continued.

Specimen Number	*Skeletal Element*	*Collection Locality*	*Present Location*
L312-4	left prox radius w/shaft	member C8	ENM
L303-4	right prox radius	member C8	ENM
018-72-2	right prox radius	member C6	ENM
L70-6	left prox radius	member C6	ENM
L18-24	right prox radius	member C8	ENM
L327-18	right prox radius	member C5	ENM
0132-724	left prox radius	member C6	ENM
L818-5	right prox radius	member C8	ENM
L54-2c	left distal radius	member C8	ENM
L17-45b	left distal radius	member C8	ENM
L319-1	left distal radius w/shaft	member C8	ENM
L806-2	left distal radius	member C8	ENM
L818-2	right distal radius	member C8	ENM
018-703	left distal radius	member C6	ENM
L32-159i	right prox humerus	member C7	ENM
L34-2	left prox humerus w/shaft	member C9	ENM
L32-100a	left prox humerus	member C7	ENM
L32-159h	right prox humerus	member C7	ENM
L32-159g	right prox humerus	member C7	ENM
L70-7	left complete humerus	member C8	ENM
0158-410a	left prox humerus	member C8	ENM
L32-155g	left complete humerus	member C7	ENM
L32-159k	left prox humerus	member C7	ENM
L32-159j	left prox humerus	member C7	ENM
L304-9	right prox humerus	member C4	ENM
O18-2281	left prox humerus w/shaft	member C6	ENM

(continued on next page)

TABLE 47 Continued.

Specimen Number	*Skeletal Element*	*Collection Locality*	*Present Location*
O18-28	left prox humerus w/shaft	member C7	ENM
O18-40	left prox humerus	member C7	ENM
L344-15	left prox humerus	member C8	ENM
L345-56	left prox humerus	member C9	ENM
L443-1	right prox humerus	member C9	ENM
L55-48	left prox humerus	member C6	ENM
L69-3	right prox humerus	member C8	ENM
L193-82	right prox humerus	member C8	ENM
L32-267	left prox humerus	member C7	ENM
L47-42	left prox humerus	member C8	ENM
L55-6	left distal humerus	member C6	ENM
L47-87	left distal humerus	member C8	ENM
L23-4	left distal humerus	member C6	ENM
L15-5	right distal humerus	member C6	ENM
L110-1	right distal humerus w/shaft	member C	ENM
L32-10la	left distal humerus	member C7	ENM
L18-20	left distal humerus	member C8	ENM
L193-30	right distal humerus	member C8	ENM
L292-8	left distal humerus	member C7	ENM
0158-410b	left distal humerus	member C8	ENM
L886-15	left distal humerus	member C8	ENM
L143-45	right distal hunerus	member C9	ENM
0165-608	left distal humerus	member C6	ENM
0217-2620	left distal humerus	member C8	ENM
L373-2	left distal humerus	member C1	ENM
L345-42	scapular fragment	member C9	ENM

TABLE 48. Comparative sample of *Theropithecus oswaldi* from sites in East Africa.

Specimen Number	*Skeletal Element*	*Collection Locality*	*Present Location*
OLD 1953/3 BK II	left prox ulna	Olduvai Upper Bed II	KNM-P
OLD/69 S.188 Long K	left prox ulna	Olduvai Upper Bed IV	KNM-P
KNM-OG 1310	left prox ulna	Olorgesailie	KNM-P
KNM-ER 286	left prox ulna	East Turkana KF L. member	KNM-P
KNM-OG 1467	left prox ulna	Olorgesailie	KNM-P
OLD/63 DK I 3060 L/2	left prox ulna	Olduvai Lower Bed I	KNM-P
M 18723	right prox ulna w/shaft	Kanjera	BMNH-P
M 18724	left prox ulna w/shaft	Kanjera	BMNH-P
M 18803	left prox ulna	Kanjera	BMNH-P
M 11546	left prox ulna w/shaft	Kanjera	BMNH-P
OLD/62 MCK II067/5604	left prox ulna w/shaft	Olduvai Upper Bed II	KNM-P
M 18726	right prox ulna w/shaft	Kanjera	BMNH-P
KNM-OG	right distal ulna	Olorgesailie	KNM-P
OLD/69 S.45 MNK	left prox radius	Olduvai Lower Bed II	KNM-P
OLD 1953/4 BK II	left prox radius	Olduvai Upper Bed II	KNM-P
OLD 1941 IS KAR T '30	left prox radius	Olduvai Bed ??	KNM-P
KNM-OG 1109	right prox radius	Olorgesailie	KNM-P
KNM-ER 28c	left prox radius	East Turkana KF L. member	KNM-P
M 18801	right prox radius w/shaft	Kanjera	BMNH-P
M 18802	left prox radius	Kanjera	BMNH-P
M 18717	left prox radius	Kanjera	BMNH-P

(continued on next page)

TABLE 48 Continued.

Specimen Number	*Skeletal Element*	*Collection Locality*	*Present Location*
M 11545	right prox radius	Kanjera	BMNH-P
M 11544	left complete radius	Kanjera	BMNH-P
OLD/62 MCK II 067/5609	left prox radius	Olduvai Upper Bed II	KNM-P
KNM-OG 1462	left distal radius w/shaft	Olorgesailie	KNM-P
KNM-ER 2131	left complete humerus	East Turkana KF U. member	KNM-P
OLD/62 MCK II 067/560	left complete humerus	Olduvai Upper Bed II	KNM-P
OLD/63 DK I 103 L/2	right prox humerus	Olduvai Lower Bed I	KNM-P
SWK II F 2953	right distal humerus	Olduvai Bed II	KNM-P
OLD/62 DK I A 2967	right distal humerus	Olduvai Lower Bed I	KNM-P
OLD S.I-1941-F921	left distal humerus	Olduvai Bed ??	KNM-P
OLD 957 BK II E 1384	left distal humerus w/shaft	Olduvai Upper Bed II	KNM-P
KNM-OG 1056	left distal humerus w/shaft	Olorgesailie	KNM-P
KNM-OG 1062	right distal humerus	Olorgesailie	KNM-P
KNM-OG 1460	right distal humerus	Olorgesailie	KNM-P
KNM-OG 1064	right distal humerus	Olorgesailie	KNM-P
KNM-OG 1318	right distal humerus	Olorgesailie	KNM-P
OLD/63 DK I II 18 3103	left distal humerus	Olduvai Lower Bed I	KNM-P
M 11542	left distal humerus w/shaft	Kanjera	BMNH-P
M 18789	right distal humerus	Kanjera	BMNH-P
M 18721	left distal humerus w/shaft	Kanjera	BMNH-P
M 11543	left distal humerus w/shaft	Kanjera	BMNH-P

(continued on next page)

TABLE 48 Continued.

Specimen Number	*Skeletal Element*	*Collection Locality*	*Present Location*
KNM-OG 1454	right scapular fragment	Olorgesailie	KNM-P
M 32071	left scapular fragment	Kanjera	BMNH-P
M 32072	left scapular fragment	Kanjera	BMNH-P
M 32070	right scapular fragment	Kanjera	BMNH-P
OLD/62 MCK II 067/6606	left scapular fragment	Olduvai Upper Bed II	KNM-P

TABLE 49. Comparative sample of *Dolichopithecus* and *Mesopithecus*.

Specimen Number	*Skeletal Element*	*Collection Locality*	*Present Location*
Dolichopithecus ruscinensis			
Pp 22	right prox ulna	Serrat d'en Vacquer	MHNL
MP 1	right prox ulna	Serrat d'en Vacquer	MHNP
40.986	left prox ulna	Serrat d'en Vacquer	FSUL
Pp 21	right complete radius	Serrat d'en Vacquer	MHNL
Pp 21"	right prox radius	Serrat d'en Vacquer	MHNL
Per 013	left prox radius w/shaft	Serrat d'en Vacquer	MNHN-P
Per 015	right prox radius w/shaft	Serrat d'en Vacquer	MNHN-P
MP 2	left prox radius w/shaft	Serrat d'en Vacquer	MHNP
Pp 23	right distal humerus w/shaft	Serrat d'en Vacquer	MHNL
Per 011	left distal humerus w/shaft	Serrat d'en Vacquer	MNHN-P
Per 012	left distal humerus	Serrat d'en Vacquer	MNHN-P
Per 010	left distal humerus w/shaft	Serrat d'en Vacquer	MNHN-P
MP 3	right distal humerus	Serrat d'en Vacquer	MHNP
MP 4	right distal humerus w/shaft	Serrat d'en Vacquer	MHNP
Mesopithecus pentelici			
Pik 248	left prox ulna	Pikermi	MNHN-P
Pik 249	left prox ulna	Pikermi	MNHN-P
Pik 032	left prox ulna	Pikermi	MNHN-P
Pik 080	right prox ulna w/shaft	Pikermi	MNHN-P
Pik 250	left distal ulna	Pikermi	MNHN-P
Pik 1723	left prox radius	Pikermi	MNHN-P
Pik 1724	left prox radius w/shaft	Pikermi	MNHN-P
Pik 1725	right prox radius	Pikermi	MNHN-P
Pik 080	right complete radius	Pikermi	MNHN-P
Pik 1720	left distal radius	Pikermi	MNHN-P
Pik 1721	left distal radius	Pikermi	MNHN-P
Pik 1722	right distal radius	Pikermi	MNHN-P
Pik 1726	right distal radius w/shaft	Pikermi	MNHN-P
Pik 244	right complete humerus	Pikermi	MNHN-P
Pik 1727	right complete humerus	Pikermi	MNHN-P
Pik 1728	left distal humerus w/shaft	Pikermi	MNHN-P
Pik 1729	left distal humerus w/shaft	Pikermi	MNHN-P
Pik 245	left distal humerus w/shaft	Pikermi	MNHN-P
Pik 1730	left distal humerus w/shaft	Pikermi	MNHN-P

TABLE 50. Comparative sample of *Paradolichopithecus, Paracolobus,* and *Rhinocolobus.*

Specimen Number	*Skeletal Element*	*Collection Locality*	*Present Disposition*
Paradolichopithecus geticus			
Bug 303	right distal humerus	Graunceanu, Romania	CARB
Paracolobus chemeroni			
KNM-BC 3 D	left prox ulna w/shaft	Chemeron Fm., JM90	KNM-P
KNM-BC 3 V	left prox radius w/shaft	Chemeron Fm., JM90	KNM-P
KNM-BC 3 F	right complete humerus	Chemeron Fm., JM90	KNM-P
KNM-BC 3 H	right scapula	Chemeron Fm., JM90	KNM-P
Rhinocolobus turkanaensis			
KNM-ER 1542 T	left prox ulna	Koobi Fora Fm., sub-KBS	KNM-P
KNM-ER 1542 S	left distal ulna	Koobi Fora Fm., sub-KBS	KNM-P
KNM-ER 1542 O	left prox humerus	Koobi Fora Fm., sub-KBS	KNM-P
KNM-ER 1542 P	left distal humerus	Koobi Fora Fm., sub-KBS	KNM-P
KNM-ER 1542 X	right scapular frag	Koobi Fora Fm., sub-KBS	KNM-P

Literature Cited

Aguirre, E., and E. Soto
1978 *Paradolichopithecus* in La Puebla de Valverde, Spain: Cercopithecoidea in European Neogene stratigraphy. Jour. Human Evol. 7:559-565.

Aiello, L. C.
1981 The allometry of primate body proportions, pp. 331-358 *in* Vertebrate Locomotion, M. H. Day, ed. Symp. Zool. Soc. London, no. 48.

Albrecht, G. H.
1978 The craniofacial morphology of the Sulawesi macaques: Multivariate approaches to biological problems. Contr. to Primatol. 13:1-151.
1980 Multivariate analysis and the study of form, with special reference to canonical variate analysis. Am. Zool. 20:679-693.

Amtmann, E.
1979 Biomechanical interpretation of form and structure of bones: Role of genetics and function in growth and remodeling, pp. 347-366 *in* Environment, Behavior, and Morphology: Dynamic Interactions in Primates, M. E. Morbeck, H. Preuschoft, and N. Gomberg, eds. New York: Gustav Fisher.

Anderson, T. W.
1958 An Introduction to Multivariate Statistical Analysis. New York: John Wiley & Sons. 347 pp.

Andreasen, J. K.
1976 A computer routine to draw Hubbs and Hubbs graphs using an automatic plotter. Syst. Zool. 25:82.

Andrews, P.
1982 Ecological polarity in primate evolution. Zool. Jour. Linn. Soc. 74:233-244.
1985 Family group systematics and evolution among catarrhine primates, pp. 14-22 *in* Ancestors: The Hard Evidence, E. Delson, ed. New York: Alan R. Liss.

Andrews, P., and L. C. Aiello
1984 An evolutionary model for feeding and positional behavior, pp. 429-466 *in* Food Acquisition and Processing in Primates, D. J. Chivers, B. A. Wood, and A. Bilsborough, eds. New York: Plenum Press.

Andrews, P., and D. B. Williams
1973 The use of principle components analysis in physical anthropology. Am. Jour. Phys. Anthropol. 39:291-304.

Ardito, G. and A. Mottura
1987 An overview of the geographic and chronologic distribution of West European cercopithecoids. Human Evol. 2:29-45

Ashton, E. H., R. M. Flinn, W. J. Moore, C. E. Oxnard, and T. F. Spence
1981 Further quantitative studies of form and function in the primate pelvis, with special reference to *Australopithecus*. Trans. Zool. Soc. London 360:1-98.

Ashton, E. H., and C. E. Oxnard
1963 The musculature of the primate shoulder. Trans. Zool. Soc. London 29:553-650.

Barry, J.C.
1987 The history and chronology of Siwalik cercopithecids. Human Evol. 2:47-58.

Benefit, B. R.
1987 The Molar Morphology, Natural History, and Phylogenetic Position of the Middle Miocene Monkey *Victoriapithecus*, and Their Implications for Understanding the Evolution of Old World Monkeys. Ph.D. dissertation, New York University.

Benefit, B. R. and M. L. McCrossin
1989 The facial morphology of *Victoriapithecus*. Am. Jour. Phys. Anthropol. 78:191.
1990 Diet, species diversity and distribution of African fossil baboons. Kroeber Anthropol. Soc. Papers 71/72:77-93.
1991 Ancestral facial morphology of Old World higher primates. Proc. Natl. Acad. Sci. USA 88:5267-5271.

Benefit, B. R. and M. Pickford
1986 Miocene fossil cercopithecoids from Kenya. Am. Jour. Phys. Anthropol. 69:441-464.

Birchette, M. G.
1982 The Postcranial Skeleton of *Paracolobus chemeroni*. Ph.D. dissertation, Harvard University. 494 pp.

Blackith, R. E., and R. A. Reyment
1971 Multivariate Morphometrics. London: Academic Press. 412 pp.

Bookstein, F. L.
1989 "Size and shape": A comment on semantics. Syst. Zool. 38:173-180.

Bookstein, F. L,. B. Chernoff, R. L. Elder, J. M. Humphries, G. R. Smith, and R. E. Strauss
1985 Morphometrics in Evolutionary Biology. Philadelphia: Academy of Natural Sciences, Spec. Publ. no. 15.

Booth, C. P.
1962 Some observations on behavior of *Cercopithecus* monkeys. Annals N. Y. Acad. Sci. 102:477-487.

Booth, S. N., and L. Freedman
1970 Multivariate discriminant analysis applied to cranial features of *Papio ursinus* and *P. cynocephalus*. Folia Primatol. 12:296-304.

Brain, C. K.
1981 The Hunters or the Hunted? An Introduction to Cave Taphonomy. Chicago: University of Chicago Press. 395 pp.

Bronowski, J., and W. M. Long
1952 Statistics of discrimination in anthropology. Am. Jour. Phys. Anthropol. 10:385-394.

Brown, F. H., I. McDougall, T. Davies, and R. Maurer
1985 An integrated Plio-Pleistocene chronology for the Turkana Basin, pp. 82-90 *in* Ancestors: The Hard Evidence, E. Delson, ed. New York: A. R. Liss.

Bromage, T.G., and F. Schrenk
1986 A cercopithecoid tooth from the Pliocene of Malawi. Jour. Human Evol. 15:497-500.

Campbell, N. A.
1978 Multivariate analysis in biological anthropology: Some further considerations. Jour. Human Evol. 7:197-203.

Chalmers, N. R.
1968 Group composition, ecology, and daily activities of free-living mangabeys in Uganda. Folia Primatol. 8:247-262.

Ciochon, R. L.
1977 A methodological approach to the study of the Omo Group cercopithecoid postcranial remains. Am. Jour. Phys. Anthropol. 47:123.
1986 The Cercopithecoid Forelimb: Anatomical Implications for the Evolution of African Plio-Pleistocene Species. Ph.D. dissertation, University of California, Berkeley. 489 pp.
1993 Utility of the size-shape correction in multivariate analysis: Example from the cercopithecoid forelimb. Am. Jour. Phys. Anthropol. (to be submitted).

Ciochon, R. L., and R. S. Corruccini
1975 Morphometric analysis of platyrrhine femora, with taxonomic implications and notes on two fossil forms. Jour. Human Evol. 4:193-217.
1976 Shoulder joint of Sterkfontein *Australopithecus*. South Afr. Jour. Sci. 72:80-82.
1977a The coraco-acromical ligament and projection index in man and other anthropoid primates. Jour. Anat. 124:627-632.
1977b The phenetic position of *Pliopithecus* and its phylogenetic relationship to the Hominoidea. Syst. Zool. 26:290-299.

Ciochon, R.L., and R. Nisbett
1993 Functional and phyletic significance of ulnar olecranon length and retroflexion in cercopithecoid primates. Jour. Human Evol. (to be submitted).

Clutton-Brock, T. H., and P. H. Harvey
1977 Species differences in feeding and ranging behavior in primates. pp. 557-584 *in* Primate Ecology, T. H. Clutton-Brock, ed. London: Academic Press.

Conroy, G. C.
1976 Primate postcranial remains from the Oligocene of Egypt. Contr. to Primatol. 8:1-123.

Constandse-Westermann, T. S.
1972 Coefficients of Biological Distance. Oosterhout, Netherlands: Anthropological Publications.

Cooke, H.B.S.
1990 Taung fossils in the University of California collections, pp. 119-134 *in* From Apes to Angels: Essays in Anthropology in Honor of Phillip V. Tobias. New York: Wiley-Liss.

Cooley, W. W., and P. R. Lohnes
1971 Multivariate Data Analysis. New York: John Wiley & Sons.

Corruccini, R. S.
1972 Allometry correction in taximetrics. Syst. Zool. 21:375-383.
1973 Size and shape in similarity coefficients based on metric characters. Am. Jour. Phys. Anthropol. 38:743-754.
1975 Multivariate analysis in biological anthropology: Some considerations. Jour. Human Evol. 4:1-19.
1978a Morphometric analysis: Uses and abuses. Yearbook Phys. Anthropol. 21:134-150.
1978b Relative growth and shape analysis. Homo 28:222-226.
1983 Principal components for allometric analysis. Am. Jour. Phys. Anthropol. 60:451-453.
1987 Shape in morphometrics: Comparative analysis. Am. Jour. Phys. Anthropol. 73:289-303.

Corruccini, R. S., and R. L. Ciochon
1976 Morphometric affinities of the human shoulder. Am. Jour. Phys. Anthropol. 45:19-38.
1978 Morphoclinal variation in the anthropoid shoulder. Am. Jour. Phys. Anthropol. 48:539-542.
1979 Primate facial allometry and interpretations of australopithecine variation. Nature (London) 281:62-64.

Corruccini, R. S., R. L. Ciochon, and H. M. McHenry
1975 Osteometric shape relationships in the wrist joint of some anthropoids. Folia Primatol. 24:250-274.
1976 The postcranium of Miocene hominoids: Were dryopithecines merely "dental apes"? Primates 17:205-223.

Crook, J. H., and F. P. G. Aldrich-Blake
1968 Ecological and behavioral contrasts between sympatric ground-dwelling primates in Ethiopia. Folia Primatol. 8:192-227.

Croxton, F. E.
1953 Elementary Statistics with Application in Medicine and the Biological Sciences. New York: Dover. 376 pp.

Day, M. H.
1967 Olduvai hominid 10: A multivariate analysis. Nature (London) 215:323-324.

1979 The locomotor interpretation of fossil primate postcranial bones, pp. 245-258 *in* Environment, Behavior, and Morphology: Dynamic Interactions in Primates, M. E. Morbeck, H. Preuschoft, and N. Gomberg, eds. New York: Gustav Fisher.

Day, M. H., and J. R. Napier
1966 A hominid toe bone from bed 1, Olduvai Gorge, Tanzania. Nature (London) 211:929-930.

Dechow, P. C.
1983 Estimates of body weights from craniometric variables in baboons. Am. Jour. Phys. Anthropol. 60:113-123.

Dechow, P. C., and R. Singer
1984 Additional fossil *Theropithecus* from Hopefield, South Africa: A comparison with other African sites and a re-evaluation of its taxonomic status. Am. Jour. Phys. Anthropol. 63:405-435.

de Heinzelin, J.
1983 The Omo Group: Archives of the International Omo Research Expedition, 2 vols. Annales Musee Royale de L'Afrique Centrale, Tervuren, Belgium, ser. 8, no. 85.

Deino, A., and R. Potts
1990 Single-crystal $^{40}Ar/^{39}Ar$ dating of the Olorgesailie Formation, southern Kenya rift. Jour. Geophys. Res. 95:8453-8470.

Delson, E.
1973 Fossil Colobine Monkeys of the Circum-Mediterranean Region and the Evolutionary History of the Cercopithecidae (Primates, Mammalia). Ph.D. dissertation, Columbia University. 856 pp.
1975 Evolutionary history of the Cercopithecidae, pp. 167-217 *in* Approaches to Primate Paleobiology, F. S. Szalay, ed. (Contr. to Primatol., vol. 5.)
1983 Evolutionary tempos in catarrhine primates, pp. 101-106 *in* Modalites, Rythmes, et Mecanismes de L'Evolution Biologique, J. Chaline, ed. Colloques Internationaux du C.N.R.S., no. 330.
1984 Cercopithecoid biochronology of the African Plio-Pleistocene: Correlation among eastern and southern hominid-bearing localities, pp. 199-218 *in* The Early Evolution of Man, P. Andrews and J. L. Franzen, eds. Cour. Forsch.-Inst. Senckenberg 69:1-277.
1988 Chronology of South African australopith site units, pp. 317-324 *in* Evolutionary History of the "Robust" Australopithecines, F. E. Grine, ed. New York: Aldine de Gruyter.
1993 *Theropithecus* fossils from Africa and India and the taxonomy of the genus, pp. 157-189 *in Theropithecus*: The Rise and Fall of a Primate Genus, N.G. Jablonski, ed. Cambridge: Cambridge University Press.

Delson, E. and D. Dean
1993 Are *Papio baringensis* R. Leakey, 1969, and *P. quadratirostris* Iwamoto, 1982, species of *Papio* or *Theropithecus*?, pp. 125-156 *in Theropithecus*: The Rise and Fall of a Primate Genus, N.G. Jablonski, ed. Cambridge: Cambridge University Press.

Delson, E. and R. Hoffstetter
1993 *Theropithecus* from Ternifine, Algeria, pp. 191-208 *in Theropithecus*: The Rise

and Fall of a Primate Genus, N.G. Jablonski, ed. Cambridge: Cambridge University Press.

Delson, E., and D. Plopsor
1975 *Paradolichopithecus*, a large terrestrial monkey (Cercopithecidae, Primates) from the Plio-Pleistocene of southern Europe, and its importance for mammalian biochronology, pp. 91-96 *in* Proc. Sixth Session, Reg. Comm. Mediterranean Neogene, Bratislava.

Delson, E., and A. Rosenberger
1984 Are there any anthropoid primate living fossils?, pp. 50-61 *in* Living Fossils, N. Eldredge and S. M. Stanley, eds. New York: Springer Verlag.

Dempster, W. T.
1955 The anthropometry of body action. Annals N. Y. Acad. Sci. 63:559-585.
1961 Free body dynamics as an approach to the mechanics of human posture and motion, pp. 81-93 *in* Biomechanical Studies of the Musculo-skeletal System, F. G. Evans, ed. Springfield, Ill.: Charles C. Thomas.

Deperet, C.
1890 Les animaux pliocènes du Roussillon. Mem. Soc. Geol. de France, Paleontol. 3:1-126.

Dixon, W. J.
1977 BMD Biomedical Computer Programs, 3rd ed. Berkeley and Los Angeles: University of California Press.

Dunbar, R.M.
1993 Socioecology of the extinct theropiths: A modelling approach, pp. 465-486 *in* *Theropithecus*: The Rise and Fall of a Primate Genus, N.G. Jablonski, ed. Cambridge: Cambridge University Press.

Eck, G. G.
1977 Diversity and frequency distribution of Omo Group Cercopithecoidea. Jour. Human Evol. 6:55-63.
1987 *Theropithecus oswaldi* from the Shungura Formation, Lower Omo Basin, southwestern Ethiopia, pp. 123-139 *in* Les Faunes Plio-Pléistocènes de la Basse Vallée de l'Omo (Éthiopie), Tome 3, Cercopithecidae de la Formation de Shungura. Paris: Éditions du Centre National de la Recherche Scientifique.
1993 *Theropithecus darti* from the Hadar Formation, Ethiopia, pp. 15-83 *in Theropithecus*: The Rise and Fall of Primate Genus, N.G. Jablonski, ed. Cambridge: Cambridge University Press.

Eck, G. G., and N. G. Jablonski
1984 A reassessment of the taxonomic status and phyletic relationships of *Papio baringensis* and *Papio quadratirostris* (Primates, Cercopithecidae). Am. Jour. Phys. Anthropol. 65:109-134.
1987 The skull of *Theropithecus brumpti* compared with those of other species of the genus *Theropithecus*, pp. 11-122 *in* Les Faunes Plio-Pléistocènes de la Basse Vallée de l'Omo (Éthiopie), Tome 3, Cercopithedae de la Formation de Shungura. Paris: Éditions du Centre National de la Recherche Scientifique.

Eisenhart, W.
1974 The Fossil Cercopithecoids of Makapansgat and Sterkfontein. B. A. thesis, Harvard University.

Epps, M. G.
1968 The Differences in the Forelimb Bones of Some East African Monkeys: The Clavicle, Scapula, and Humerus. Ph.D. dissertation, University College of North Wales. 86 pp.

Feibel, C. S., F. H. Brown, and I. McDougall
1989 Stratigraphic context of fossil hominids from the Omo Group deposits: Northern Turkana Basin, Kenya, and Ethiopia. Am. Jour. Phys. Anthropol. 78:595-622.

Feldesman, M. R.
1976 The primate forelimb: A morphometric study of locomotor diversity. Univ. of Oregon Anthropol. Papers, no. 10:1-154.

Fisher, R. A.
1936 The use of multiple measurements in taxonomic problems. Annals of Eugenics 7:179-188.

Fleagle, J. G.
1977 Locomotor behavior and skeletal anatomy of sympatric Malaysian leaf monkeys (*Presbytis obscura* and *Presbytis melalophus*). Yrbk. Phys. Anthropol. 20:440-453.
1983 Locomotor adaptations of Oligocene and Miocene hominoids and their phyletic implications, pp. 301-324 *in* New Interpretations of Ape and Human Ancestry, R. L. Ciochon and R. S. Corruccini, eds. New York: Plenum Press.
1986 The fossil record of early catarrhine evolution, pp. 130-149 *in* Major Topics in Primate and Human Evolution, B. Wood, L. Martin, and P. Andrews, eds. Cambridge, England: Cambridge University Press.
1988 Primate Adaptation and Evolution. San Diego: Academic Press. 486 pp.
1994 Anthropoid origins, pp.17-35 *in* Integrative Paths to the Past: Paleoanthropological Advances in Honor of F. Clark Howell, R.S. Corrucini and R.L. Ciochon, eds. Englewood Cliffs, NJ: Prentice Hall.

Fleagle, J. G., and R. F. Kay
1983 New interpretations of the phyletic position of Oligocene hominoids, pp. 181-210 *in* New Interpretations of Ape and Human Ancestry, R. L. Ciochon and R. S. Corruccini, eds. New York: Plenum Press.
1987 The phyletic position of the Parapithecidae. Jour. Human Evol. 16:483-532.

Fleagle, J. G., R. F. Kay, and E. L. Simons
1982 The humerus of *Aegyptopithecus zeuxis*, a primitive anthropoid. Am. Jour. Phys. Anthropol. 59:175-193.

Fleagle, J. G., E. L. Simons, and G. C. Conroy
1975 Ape limb bones from the Oligocene of Egypt. Science 189:135-137.

Fooden, J.
1969 Taxonomy and evolution of the monkeys of the Celebes (Primates: Cercopithecidae). Biblio. Primatol. 10:1-148.

Freedman, L.
1957 The fossil Cercopithecoidea of South Africa. Annals Transvaal Mus. 23:121-162.
1960 Some new cercopithecoid specimens from Makapansgat, South Africa. Paleontol. Africana 7:7-45.
1965 Fossil and subfossil primates from the limestone deposits at Taung, Bolt's Farm and Witkrans, South Africa. Paleontol. Africana 9:19-48.
1976 South African fossil Cercopithecoidea: A reassessment, including a description of new material from Makapansgat, Sterkfontein and Taung. Jour. Human Evol. 5:297-315.

Freedman, L., and C. K. Brain
1972 Fossil cercopithecoid remains from the Kromdraai australopithecine site (Mammalia: Primates). Annals Transvaal Mus. 28(1):1-16.

Gabis, R.V.
1961 Os des membres de *Dolichopithecus*, singe fossile du Pliocene du Roussillon. C.R. Acad. Sci., Paris 252:1368-1370.

Gartlan, J. S., and C. K. Brain
1968 Ecology and social variability in *Cercopithecus aethiops* and *C. mitis*, pp. 253-293 *in* Primates: Studies in Adaptation and Variability, P. C. Jay, ed. New York: Holt, Rinehart, and Winston.

Gaudry, A.
1862 Animaux Fossiles et Géologie de l'Attique. Paris: F. Savy.

Geraads, D.
1987 Dating the northern African cercopithecoid fossil record. Human Evol. 2:19-27.

Gerven, D. P. van
1972 The contribution of size and shape variation to patterns of sexual dimorphism of the human femur. Am. Jour. Phys. Anthropol. 37:49-60.

Gould, S. J.
1966 Allometry and size in ontogeny and phylogeny. Biol. Rev. 41:587-640.

Grand, T.
1968 Functional anatomy of the upper limb, pp. 104-124 *in* Biology of the Howler Monkey (*Alouatta caraya*), M. R. Malinow, ed. (Biblio. Primatol. 7.)

Gray, B. T.
1980 Environmental Reconstruction of the Hadar Formation (Afar, Ethiopia). Ph.D. dissertation, Case Western Reserve University. 431 pp.

Hall, K. R. L.
1966 Distribution and adaptation of baboons. Symp. Zool. Soc. London 17:49-73.

Harrison, T.
1987 The phylogenetic relationships of the early catarrhine primates: A review of the current evidence. Jour. Human Evol. 16:41-80.
1989 New postcranial remains of *Victoriapithecus* from the middle Miocene of Kenya. Jour. Human Evol. 18:3-54.

Hill, W. C. O.
1972 Evolutionary Biology of the Primates. New York: Academic Press.

Hopkins, J. W.
1966 Some considerations in multivariate allometry. Biometrics 22:747-760.

Hopwood, A. T.
1936 New and little-known fossil mammals from the Pleistocene of Kenya colony and Tanganyika territory. Ann. Mag. Nat. Hist., ser. 10, 17:636-638.

Howell, A. Brazier
1944 Speed in Animals: Their Specialization for Running and Leaping. Chicago: University of Chicago Press. 270 pp.

Howell, F. C.
1955 The age of the australopithecines of southern Africa. Am. Jour. Phys. Anthropol. 13:635-662.
1978 Hominidae, pp. 154-248 *in* Evolution of African Mammals, V. J. Maglio and H. B. S. Cooke, eds. Cambridge, Mass.: Harvard University Press. 641 pp.

Howells, W. W.
1973 Cranial variation in man: A study by multivariate analysis of patterns of difference among recent human populations. Papers Peabody Mus. Archaeol. and Ethnol. 67:1-269.

Hulzinga, J.
1962 From DD to D^2 and back: The quantitative expression of resemblance. Proc. Koninkl. Nederlandse Akad. van Wetenschappen, ser. C, 65:380-391.

Huxley, J. S., J. Needham, and I. M. Lerner
1941 Terminology of relative growth-rates. Nature (London) 148:225.

Isaac, G. L.
1977 Olorgesailie: Archeological Studies of a Middle Pleistocene Lake Basin in Kenya. Prehistoric Archeology and Ecology Series, K. W. Butzer and L. G. Freeman, eds. Chicago: University of Chicago Press. 227 pp.

Jablonski, N. G.
1986 The hand of *Theropithecus brumpti*, pp. 173-182 *in* Primate Evolution, vol. 1, Selected Proceedings of the Tenth Congress of the International Primatological Society, Nairobi, J. G. Else and P. C. Lee, eds. Cambridge, England: Cambridge University Press.
1993a Introduction, pp. 1-11, *in Theropithecus*: The Rise and Fall of a Primate Genus, N.G. Jablonski, ed. Cambridge: Cambridge University Press.
1993b The phylogeny of *Theropithecus*, pp. 209-224 *in Theropithecus*: The Rise and Fall of a Primate Genus, N.G. Jablonski, ed. Cambridge: Cambridge University Press.

Jablonski, N.G., ed.
1993 *Theropithecus*: The Rise and Fall of a Primate Genus. Cambridge: Cambridge University Press. 536 pp.

Jay, P. C.
1965 The common langur of north India, pp. 197-249 *in* Primate Behavior, I. De Vore, ed. New York: Holt, Rinehart, and Winston.

Jenkins, F. A.
1973 The functional anatomy of the mammalian humero-ulnar articulation. Am. Jour. Anat. 137:281-298.

Jolicoeur, P.
1959 Multivariate geographical variation in the wolf, *Canis lupus L.* Evolution 13:283-299.

Jolly, C. J.
1965 Origins and Specializations of the Long-Faced Cercopithecoidea. Ph.D. dissertation, University of London.
1967 The fossil history of the baboons, pp. 23-50 *in* The Baboon in Medical Research, vol. 2, H. Vagtborg, ed. Austin: University of Texas Press.
1970 The seed-eaters, a new model of hominid differentiation based on a baboon analogy. Man 5:1-26.
1972 The classification and natural history of *Theropithecus (Simopithecus)* (Andrews, 1916), baboons of the African Plio-Pleistocene. Bull. Brit. Mus. Nat. Hist., Geol. 22:1-123.

Jones, C., and J. Sabater Pi
1968 Comparative ecology of *Cercocebus albigena* (Gray) and *Cercocebus torquatus* (Kerr) in Rio Muni, West Africa. Folia Primatol. 9:99-113.

Jouventin, P.
1975 Observations sur la socio-ecology du mandrill. Terre Vie 29:493-532.

Jungers, W. L.
1984 Scaling of the hominoid locomotor skeleton, with special reference to lesser apes, pp. 146-169 *in* The Lesser Apes: Evolutionary and Behavioural Biology, H. Preuschoft, D. J. Chivers, W. Y. Brockelman, and N. Creel, eds. Edinburgh: University of Edinburgh Press.
1985 Body size and scaling of limb proportions in primates, pp. 345-381 *in* Size and Scaling in Primate Biology, W. L. Jungers, ed. New York: Plenum Press.

Kawabe, M., and T. Mano
1972 Ecology and behavior of the wild proboscis monkey, *Nasalis larvatus* (Wurmb) in Sabah, Malaysia. Primates 13:213-227.

Kingdon, J.
1971 East African Mammals: An Atlas of Evolution in Africa, vol. 1. London: Academic Press. 446 pp.

Krentz, H.B.
1992 The Forelimb Anatomy of *Theropithecus brumpti* and *Theropithecus oswaldi* from the Shungura Formation, Ethiopia. Ph.D. dissertation, University of Washington.
1993 Postcranial anatomy of extant and extinct species of *Theropithecus,* pp. 383-422 *in Theropithecus*: The Rise and Fall of a Primate Genus, N.G. Jablonski, ed. Cambridge: Cambridge University Press.

Kummer, H.
1968 Social organization of Hamadryas baboons. Biblio. Primatol. 6:1-189.

Larson, S., and J. T. Stern
1986 EMG of scapulohumeral muscles in the chimpanzee during reaching and "arboreal" locomotion. Am. Jour. Anat. 176:171-190.
1989 Role of supraspinatus in the quadrupedal locomotion of vervets (*Cercopithecus aethiops*): Implications for interpretation of humeral morphology. Am. Jour. Phys. Anthropol. 79:369-377.

Leakey, M. G.
1976 Cercopithecoidea of the East Rudolf succession, pp. 345-350 *in* Earliest Man and Environments in the Lake Rudolf Basin, Y. Coppens, F. C. Howell, G. L. Isaac, and R. E. F. Leakey, eds. Chicago: University of Chicago Press.
1982 Extinct large colobines from the Plio-Pleistocene of Africa. Am. Jour. Phys. Anthropol. 58:153-172.
1985 Early Miocene cercopithecids from Buluk, northern Kenya. Folia Primatol. 44:1-14.
1987 Colobinae (Mammalia, Primates) from the Omo valley, Ethiopia, pp. 149-169, *in* Les Faunes Plio-Pléistocènes de la Basse Vallée de l'Omo (Éthiopie), Tome 3, Cercopithecidae de la Formation de Shungura. Paris: Éditions du Centre National de la Recherche Scientifique.
1993 Evolution of *Theropithecus* in the Turkana Basin, pp. 85-123 *in Theropithecus*: The Rise and Fall of Primate Genus, N.G. Jablonski, ed. Cambridge: Cambridge University Press.

Leakey, M. G., and E. Delson
1987 Fossil Cercopithecidae from the Laetolil beds, pp. 91-107 *in* Laetoli: A Pliocene Site in Northern Tanzania, M. D. Leakey and J. M. Harris, eds. Oxford: Clarendon Press.

Leakey, M. G., and R. E. F. Leakey
1973 Further evidence of *Simopithecus* (Mammalia, Primates) from Olduvai and Olorgesailie, pp. 101-120 *in* Fossil Vertebrates of Africa, vol. 3, L. S. B. Leakey, R. J. G. Savage, and S. C. Coryndon, eds. London: Academic Press.

Leakey, R. E. F.
1969 New Cercopithecidae from the Chemeron beds, Kenya, pp. 53-69 *in* Fossil Vertebrates of Africa, vol. 1, L. S. B. Leakey, ed. London: Academic Press.

Lee, P.C. and R.A. Foley
1993 Ecological energenics and extinction of giant gelada baboons, pp. 487-498 *in Theropithecus*: The Rise and Fall of a Primate Genus, N.G. Jablonski, ed. Cambridge: Cambridge University Press.

Lewis, O. J.
1971 The contrasting morphology found in the wrist joints of semibrachiating monkeys and brachiating apes. Folia Primatol. 16:248-256.
1972 Osteological features characterizing the wrists of monkeys and apes, with a reconstruction of this region in *Dryopithecus (Proconsul) africanus*. Am. Jour. Phys. Anthropol. 36:45-58.

McCrossin, M.L. and B.R. Benefit
1992 Comparative assessment of the ischial morphology of *Victoriapithecus macinnesi*. Am. Jour. Phys. Anthropol. 87:277-290.

McFadden, P. L., A. Brock, and T. C. Partridge
1979 Paleomagnetism and the age of the Makapansgat hominid site. Earth Plan. Sci. Lett. 44:373-382.

McHenry, H. M., and R. S. Corruccini
1975 Distal humerus in hominoid evolution. Folia Primatol. 23:227-244.
1976 Affinities of Tertiary hominoid femora. Folia Primatol. 26:139-150.
1983 The wrist of *Proconsul africanus* and the origin of hominoid postcranial adaptations, pp. 353-367 *in* New Interpretations of Ape and Human Ancestry, R. L. Ciochon and R. S. Corruccini, eds. New York: Plenum Press.

Mahalanobis, P. C.
1930 On tests and measures of group divergence. Jour. and Proc. Asiatic Soc. of Bengal 26:541-588.
1936 On the generalized distance in statistics. Proc. Natl. Inst. Sci., India 2:49-55.

Maier, W.
1970 New fossil Cercopithecoidea from the lower Pleistocene cave deposits of the Makapansgat limeworks, South Africa. Paleontol. Africana 13:69-107.
1971 Vergleichend-und funktionell-anatomische Untersuchungen an der Vorderextremitat von *Theropithecus gelada* (Ruppell 1835). Abhdl. Senckenb. Naturf. Ges., no. 527:1-284.
1972a Anpassungstyp und systematische Stellung von *Theropithecus gelada* Ruppell, 1885. Zeit. Morph. Anthrop. 63:370-384.
1972b The first complete skull of *Simopithecus darti* from Makapansgat, South Africa, and its systematic position. Jour. Human Evol. 1:395-405.
1973 Palaookologie und zeitliche Einordnung der sudafrikanischen Australopithecinen. Zeit. Morph. Anthrop. 65:70-105.

Manaster, B. J. M.
1975 Locomotor Adaptations within the *Cercopithecus, Cercocebus,* and *Presbytis* Genera: A Multivariate Approach. Ph.D. dissertation, University of Chicago. 190 pp.

Manly, B. F. J.
1986 Multivariate Statistical Methods: A Primer. London: Chapman and Hall.

Maples, W. R.
1972 Systematic reconsideration and a revision of the nomenclature of Kenya baboons. Am. Jour. Phys. Anthropol. 36:9-20.

Mein, P.
1979 Rapport d'activité du groupe de travail vertébrés mise a jour de la biostratigraphie du Neogene basée sur les mammifères. 7th Internatl. Congr. Medit. Neogene. Ann. Geol. Pays Hellen. 1979:1367-1372.

Miller, R. A.
1932 Evolution of the pectoral girdle and in the primates. Am. Jour. Phys. Anthropol. 17:1-56.

Morbeck, M. E.
1972 A Re-Examination of the Forelimb of the Miocene Hominoidea. Ph.D. dissertation, University of California, Berkeley. 205 pp.

Mori, A., and G. Belay
1990 The distribution of baboon species and a new population of gelada baboons along the Wabi-Shebelli River, Ethiopia. Primates 31:495-508.

Mosimann, J. E.
1970 Size allometry: Size and shape variables, with characterizations of lognormal and generalized gamma distributions. Jour. Am. Statist. Assoc. 65:930-945.

Napier, J. R., and P. R. Davis
1959 The forelimb skeleton and associated remains of *Proconsul africanus*. Fossil Mammals of Africa (Brit. Mus. Nat. Hist.) 16:1-69.

Napier, J. R., and P. H. Napier
1967 A Handbook of Living Primates. London: Academic Press. 456 pp.

Necrasov, O., Samson, P., and Radulesco, C.
1961 Sur un nouveau singe catarrhien fossile, decouvert dans un nid fossilfière d'Oltenie (R.P.R.). Ann. Stiintifice Univ. Al. I. Cuza, Iasi; Sec. II, stiinte naturale 7:401-416.

Oates, J. F.
1977 The guereza and man: How man has affected the distribution and abundance of *Colobus guereza* and other black colobus monkeys, pp. 420-467 *in* Primate Conservation, H.S.H. Prince Rainier III of Monaco and G. H. Bourne, eds. London: Academic Press.

O'Connor, B. L.
1975 The functional morphology of the cercopithecoid wrist and inferior radioulnar joints, and their bearing on some problems in the evolution of the Hominoidea. Am. Jour. Phys. Anthropol. 43:113-122.

Oxnard, C. E.
1967 The functional morphology of the primate shoulder, as revealed by comparative anatomical, osteometric, and discriminant functions techniques. Am. Jour. Phys. Anthropol. 26:219-240.
1968 Primate evolution—a method of investigation. Am. Jour. Phys. Anthropol. 26:289-302.
1969 Mathematics, shape, and function: A study of primate anatomy. Am. Scientist 57:75-96.
1973 Form and Pattern in Human Evolution: Some Mathematical, Physical, and Engineering Methods. Chicago: University of Chicago Press. 218 pp.
1978 Primate quadrupedalism: Some subtle structural correlates. Yrbk. Phys. Anthropol. 20:538-554.
1984 The Order of Man: A Biomathematical Anatomy of the Primates. New Haven: Yale University Press. 366 pp.

Pan Yuerong and N.G. Jablonski
1987 The age and geographical distribution of fossil cercopithecids in China. Human Evol. 2:59-69.

Partridge, T. C.
1973 Geomorphological dating of cave opening at Makapansgat, Sterkfontein, Swartkrans and Taung. Nature (London) 246:75-79.

Partridge, T. C.
1982 The chronological positions of the fossil hominids of southern Africa, pp. 617-675 *in* Prétirage, Ier Cong., Internat. Paléo. Humaine, Vol. II, H. de Lumley and M-A. de Lumley, eds. Nice: Cent. Nat. Rech. Sci.

Pickford, M.
1987a The chronology of the Cercopithecoidea of East Africa. Human Evol. 2:1-17.
1987b The diversity, zoogeography and geochronology of monkeys. Human Evol. 2:71-89.

Pienaar, U. de V.
1967 The small mammals of the Kruger National Park: A systematic list and zoogeography. Koedoe (7):1-25.

Pimentel, R. A.
1979 Morphometrics: The Multivariate Analysis of Biological Data. Dubuque, Iowa: Kendall/Hunt Publishing Co.

Preuschoft, H.
1971 Body posture and mode of locomotion in early Pleistocene hominids. Folio Primatol. 14:209-240.
1973 Body posture and locomotion in some East African Miocene Dryopithecinae, pp. 13-46 in Human Evolution, M. H. Day, ed. (Symp. Soc. Study Human Biol., vol. 11.)

Rao, C. R.
1948 The utilization of multiple measurement in problems of biological classification. Jour. Royal Statistical Soc. B-10:159-203.
1952 Advanced Statistical Methods in Biometric Research. New York: John Wiley & Sons. 292 pp.

Reyment, R. A., R. E. Blackith, and N. A. Campbell
1984 Multivariate Analysis, 2nd ed. London: Academic Press.

Rightmire, R.
1969 On the computation of Mahalanobis' generalized distance (D^2). Am. Jour. Phys. Anthropol. 30:157-160.

Ripley, S.
1979 Environmental grain, niche diversification, and positional behavior in Neogene primates: An evolutionary hypothesis, pp. 37-74 *in* Environment, Behavior, and Morphology: Dynamic Interactions in Primates. M.E. Morebeck, H. Preuschoft and N. Gomberg, eds. New York: Gustav Fisher.

Rohlf, F. J., and F. L. Bookstein
1987 A comment on shearing as a method for "size correction." Syst. Zool.36:356-367.

Rohlf, F. J., and R. R. Sokal
1965 Coefficients of correlation and distance in numerical taxonomy. University Kansas Sci. Bull. 45.

Rose, M. D.
1988 Another look at the anthropoid elbow. Jour. Human Evol. 17:193-224.

Rothenfluh, E.
1976 Uberprufung der Gewichsangaben adulter Primaten. Sem. Arbeit. Ph.D. dissertation, University of Zurich.

Sabater Pi, J.
1972 Contributions to the ecology of *Mandrillus sphinx* (Linnaeus 1758) of Rio Muni (Republic of Equatorial Guinea). Folia Primatol. 17:304-319.

Savage, D. E., and D. Russell
1983 Mammalian Paleofaunas of the World. Reading, Mass: Addison-Wesley. 432 pp.

Schultz, A. H.
1970 The comparative uniformity of the Cercopithecoidea, pp. 39-51 *in* Old World Monkeys: Evolution, Systematics, and Behavior, J. R. Napier and P. H. Napier, eds. New York: Academic Press.

Schultz, M.
1986 The forelimb of the Colobinae, pp. 559-669 *in* Comparative Primate Biology, Vol. 1: Systematics, Evolution and Anatomy, D. R. Swindler and J. Erwin, eds. New York: Alan R. Liss.

Senut, B.
1986a Nouvelle découvertes de restes post-crâniens de primates Miocènes (Hominoidea et Cércopithecoidea) sur le site Maboko au Kenya occidental. C. R. Acad. Sci. Paris 303, Ser. II:1359-1362.
1986b Long bones of the primate upper limb: Monomorphic or dimorphic? Human Evol. 1:7-22.

Shipman, P., W. Bosler, and K. L. Davis
1981 Butchering of giant geladas at an Acheulian site. Curr. Anthropol. 22:257-264.

Simons, E. L.
1965 New fossil apes from Egypt and the initial differentiation of the Hominoidea. Nature (London) 205:135-139.
1970 The deployment and history of Old World monkeys (Cercopithecidae, Primates), pp. 97-138 *in* Old World Monkeys: Evolution, Systematics, and Behavior, J. R. Napier and P. H. Napier, eds. New York: Academic Press.

Simpson, G. G., A. Roe, and R. C. Lewontin
1960 Quantitative Zoology. New York: Harcourt, Brace and World.

Sjovold, T.
1975 Some notes on the distribution and certain modifications of Mahalanobis' Generalized Distance (D^2). Jour. Human Evol. 4:549-558.

Smith, R. J.
1980 Rethinking allometry. Jour. Theor. Biol. 87:97-111.

Sneath, P. H. A., and R. R. Sokal
1973 Numerical Taxonomy: The Principles and Practice of Numerical Classification. San Francisco: W. H. Freeman. 573 pp.

Somers, K. M.
1986 Multivariate allometry and removal of size with principal components analysis. Syst. Zool. 35:359-368.

Somers, K. M.
1989 Allometry, isometry, and shape in principal components analysis. Syst. Zool. 38:169-173.

Steudel, K.
1982 Allometry and adaptation in the catarrhine postcranial skeleton. Am. Jour. Phys. Anthropol. 59:431-441.

1985 Allometric perspectives on fossil catarrhine morphology, pp. 449-475 *in* Size and Scaling in Primate Biology, W. L. Jungers, ed. New York: Plenum Press.

Steyskal, G. C.
1968 The size factor in numerical taxonomy. Syst. Zool. 17:210.

Strasser, E.
1988 Pedal evidence for the origin and diversification of cercopithecoid clades. Jour. Human Evol. 17:225-245.

Strasser, E. and E. Delson
1987 Cladistic analysis of cercopithecoid relationships. Jour. Human Evol. 16:81-99.

Struhsaker, T. T.
1975 The Red Colobus Monkey. Chicago: University of Chicago Press.

Struhsaker, T. T. and J. F. Oates
1975 Comparisons of the behavior and ecology of red colobus and black and white colobus monkeys in Uganda: A summary, pp. 103-123 *in* Socioecology and Psychology of Primates, R. H. Tuttle, ed. The Hague: Mouton.

Sundberg, P.
1989 Shape and size-constrained principal components analysis. Syst. Zool. 38:166-168.

Sweet, S. S.
1980 Allometric inference in morphology. Am. Nat. 20:643-652.

Szalay, F. S.
1981 Phylogeny and the problem of adaptive significance: The case of the earliest primates. Folia Primatol. 36:157-182.

Szalay, F. S., and M. Dagosto
1980 Locomotor adaptations as reflected on the humerus of Paleogene primates. Folia Primatol. 34:1-45.

Szalay, F. S., and E. Delson
1979 Evolutionary History of the Primates. New York: Academic Press. 580 pp.

Taub, D. M.
1977 Geographic distribution and habitat diversity of the Barbary macaque. Folia Primatol. 27:108-133.

Teaford, M.F.
1993 Dental microwear and diet in extant and extinct *Theropithecus*: Preliminary analysis, pp. 331-349 *in Theropithecus*: The Rise and Fall of a Primate Genus, N.G. Jablonski, ed. Cambridge: Cambridge University Press.

Temerin, L. A., and J. G. H. Cant
1983 The evolutionary divergence of Old World monkeys and apes. Am. Nat. 122:335-351.

Trofimov, B.A.
1977 Primate *Paradolichopithecus sushkini* sp. nov. from the Upper Pliocene of the Pamirs Piedmont. Jour. Palaeontol. Soc. India 20:26-32.

Turnquist, J. E.
1983 Influence of age, sex, and caging on joint mobility in the patas monkey (*Erythrocebus patas*). Am. Jour. Phys. Anthropol. 61:211-220.
1985 Passive joint mobility in patas monkeys (*Erythrocebus patas*): Rehabilitation of caged animals after release into a free-ranging environment. Am. Jour. Phys. Anthropol. 67:1-5.

Vogel, C.
1977 Ecology and sociology of *Presbytis entellus*, pp. 24-45 *in* Use of Non-human Primates in Biomedical Research, M. R. Prasad and T. C. Anand Kumar, eds. New Delhi: Indian National Academy of Sciences.

Von Koenigswald, G. H. R.
1969 Miocene Cercopithecoidea and Oreopithecoidea from the Miocene of East Africa, pp. 39-52 *in* Fossil Vertebrates of Africa, vol. 1, L. S. B. Leakey, ed. London: Academic Press.

Vrba, E. S.
1975 Some evidence of chronology and paleoecology of Sterkfontein, Swartkrans, and Kromdraai from the fossil Bovidae. Nature (London) 254:301-304.
1982 Biostratigraphy and chronology, based particularly on Bovidae, of southern hominid-associated assemblages: Makapansgat, Sterkfontein, Taung, Kromdraai, Swartkrans; also Elandsfontein (Saldanha), Broken Hill (now Kabwe) and Cave of Hearths, pp. 707-752 *in* Prétirage, Ier Cong., Internat. Paléo. Humaine, Vol. II, H. de Lumley and M-A. de Lumley, eds. Nice: Cent. Nat. Rech. Sci.
1985 African Bovidae: Evolutionary events since the Miocene. South Afr. Jour. Sci. 81:263-266.

Walter, R. C., and, J. L. Aronson
1982 Revision of K/Ar ages for the Hadar hominid site. Nature (London) 296:122-127.

Wiley, E. O.
1981 Phylogenetics: The Theory and Practice of Phylogenetic Systematics. New York: John Wiley and Sons. 439 pp.

Wolfheim, J. H.
1983 Primates of the World: Distribution, Abundance, and Conservation. Seattle: University of Washington Press. 832 pp.

Ziemer, L. K.
1978 Functional morphology of forelimb joints in the Woolly monkey *Lagothrix lagothricha*. Contr. to Primatol. 14:1-130.

Plates

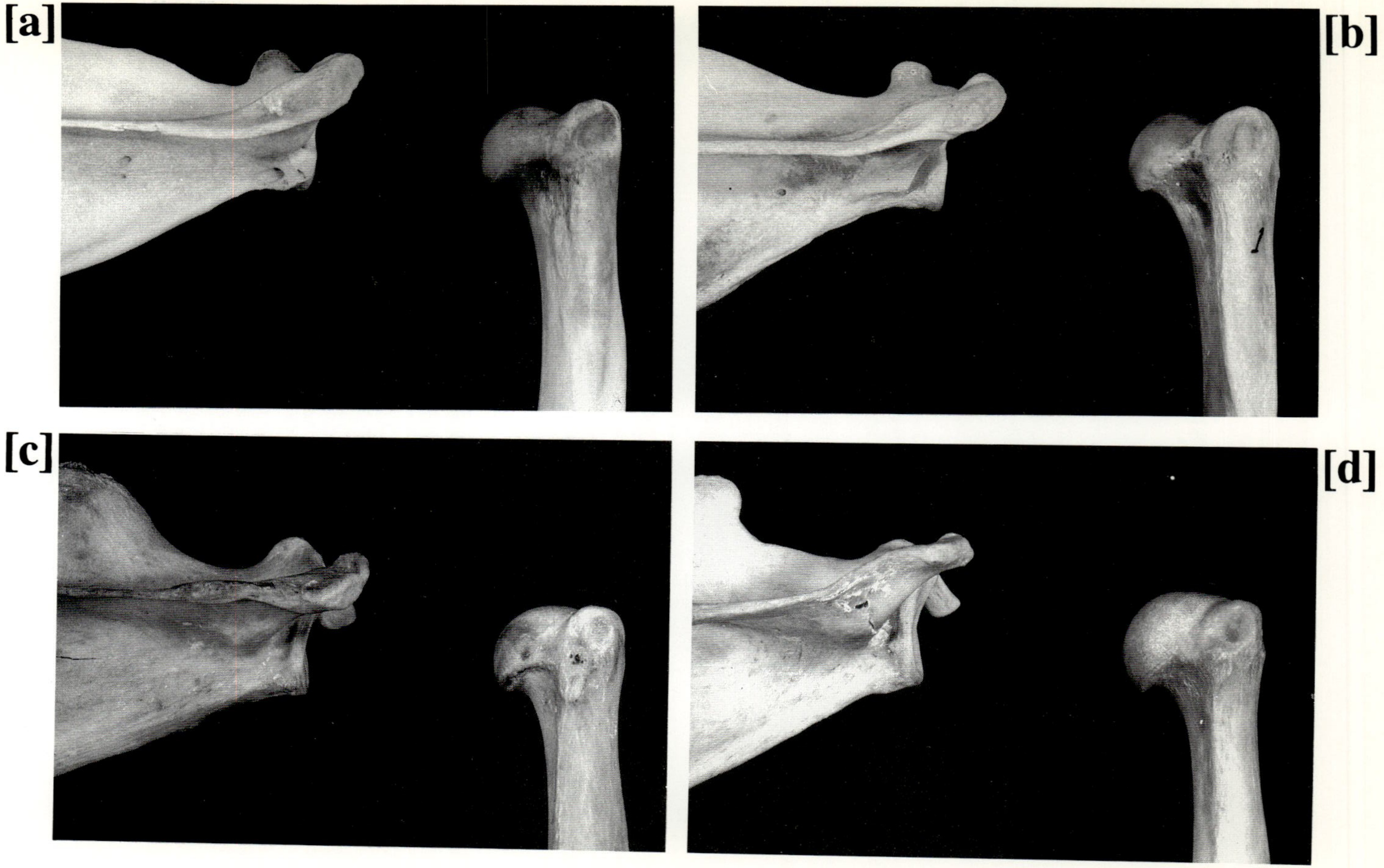

Plate 1. Comparative view of the shoulder joints of two cercopithecines, [a] *Theropithecus gelada*, [b] *Papio ursinus*, and two colobines [c] *Colobus badius*, [d] *Nasalis larvatus*, in the extant comparative sample. Specimens are not to same scale.

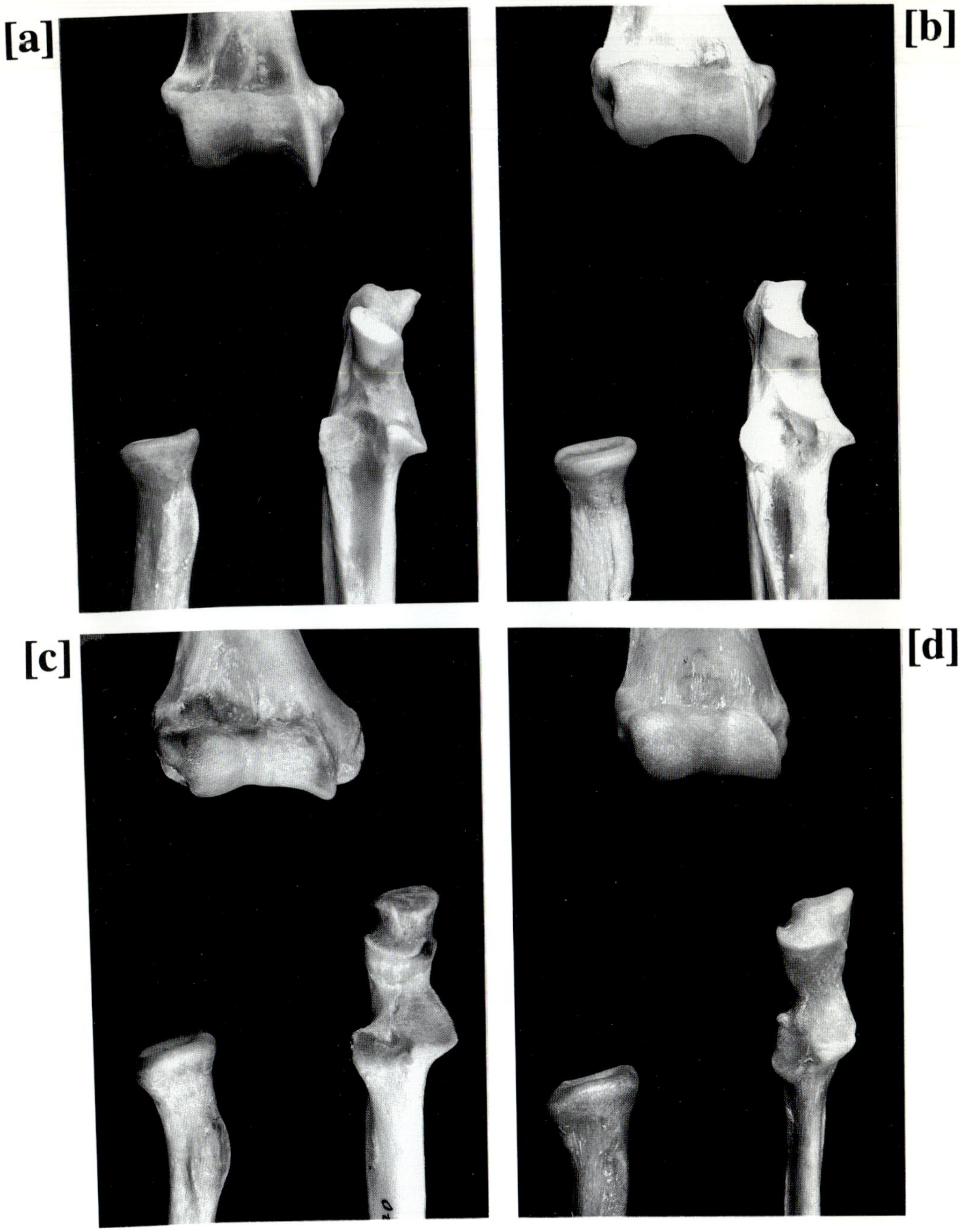

Plate 2. Comparative view of the elbow joints of two cercopithecines, [a] *Theropithecus gelada*, [b] *Papio ursinus*, and two colobines, [c] *Colobus badius*, [d] *Nasalis larvatus*, in the extant comparative sample. Specimens are not to same scale.

[a] [b]

[c] [d]

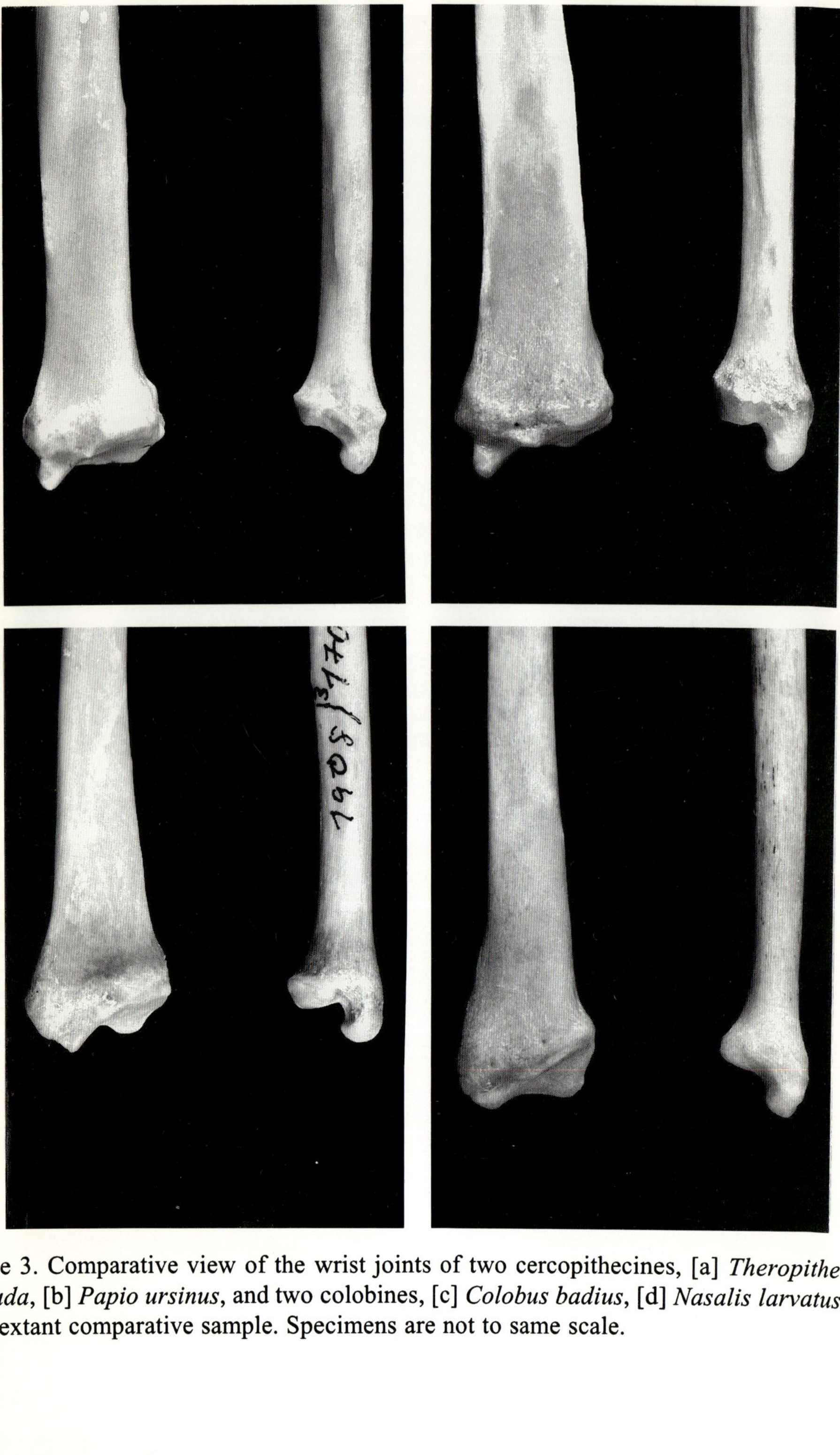

Plate 3. Comparative view of the wrist joints of two cercopithecines, [a] *Theropithecus gelada*, [b] *Papio ursinus*, and two colobines, [c] *Colobus badius*, [d] *Nasalis larvatus*, in the extant comparative sample. Specimens are not to same scale.

PROGRAMMING IN C FOR UNIX

ELLIS HORWOOD BOOKS IN COMPUTING SCIENCE

General Editors: Professor JOHN CAMPBELL, University College London, and BRIAN L. MEEK, King's College London (KQC), University of London

Series in Computers and Their Applications

Series Editor: BRIAN L. MEEK, Computer Centre, King's College London (KQC), University of London

Atherton, R.	**Structured Programming with COMAL**
Bailey, R.	**Functional Programming with HOPE**
Barrett, R., Ramsay, A. & Sloman, A.	**POP-11: A Practical Language for Artificial Intelligence**
Berry, R.E.	**Programming Language Translation**
Bishop, P.	**Fifth Generation Computers: Concepts, Implementations and Uses**
Brailsford, D.F. & Walker, A.N.	**Introductory ALGOL 68 Programming**
Bull, G.M.	**The Dartmouth Time Sharing System**
Burns, A.	**New Information Technology**
Burns, A.	**The Microchip: Appropriate or Inappropriate Technology**
Carberry, J.C.	**COBOL: A Practical Approach**
Chivers, I.D.	**A Practical Introduction to Standard PASCAL**
Chivers, I.D. & Clark, M.W.	**Interactive FORTRAN 77: A Hands-on Approach**
Clark, M.W.	**pc-Portable FORTRAN**
Cleal, D.M. & Heaton, N.O.	**Intelligent Systems: A User Perspective**
Cope, T.	**Computing using BASIC: An Interactive Approach**
Dahlstrand, I.	**Software Portability and Standards**
Davie, A.J.T. & Morrison, R.	**Recursive Descent Compiling**
de Saram, H.	**Programming in micro-PROLOG**
Deasington, R.J.	**A Practical Guide to Computer Communications and Networking, 2nd Edition**
Deasington, R.J.	**X.25 Explained: Protocols for Packet Switching Networks, 2nd Edition**
Eastlake, J.	**A Structured Approach to Computer Strategy**
Eisenbach, S.	**Functional Programming: Languages, Tools and Architectures**
Ellis, D.	**Medical Computing in Practice**
Ennals, J.R.	**Artificial Intelligence: Applications to Logical Reasoning and Historical Research**
Fernando, C.	**Cluster Analysis on Microcomputers**
Fossum, E. *et al.*	**Computerization of Working Life**
Gray, P.M.D.	**Logic, Algebra and Databases**
Harland, D.M.	**Concurrency and Programming Languages**
Harland, D.M.	**Polymorphic Programming Languages**
Hill, I.D. & Meek, B.L.	**Programming Language Standardisation**
Hogger, C.J. & Hogger, E.I.	**PROLOG for Scientists and Engineers**
Hutchins, W.J.	**Machine Translation: Past, Present, Future**
Hutchison, D.	**Fundamentals of Computer Logic**
Hutchison, D. & Silvester, P.	**Computer Logic: Principles and Technology**
Lester, C.	**Advanced Programming in PASCAL**
McKenzie, J., Elton, L. & Lewis, R.	**Interactive Computer Graphics in Science Teaching**
Matthews, J.	**FORTH**
Meek, B.L., Heath, P. & Rushby, N.	**Guide to Good Programming Practice, 2nd Edition**
Millington, D.	**Systems Analysis and Design for Computer Applications**
Moore, L.	**Foundations of Programming with PASCAL**
Moseley, L.G., Sharp, J.A. & Salenieks, P.	**PASCAL in Practice: Using the Language**
Moylan, P.	**Assembly Language for Engineers**
Narayanan, A. & Sharkey, N.E.	**An Introduction to LISP**
Paterson, A.	**Office Systems: Planning, Procurement and Implementation**
Pemberton, S. & Daniels, M.C.	**PASCAL Implementation**
Pesaran, M.H. & Slater, L.J.	**Dynamic Regression: Theory and Algorithms**
Peter, R.	**Recursive Functions in Computer Theory**
Phillips, C. & Cornelius, B.J.	**Computational Numerical Methods**
Ramsden, E.	**Microcomputers in Education 2**
Schirmer, C.	**Programming in C for UNIX**
Sharp, J.A.	**Data Flow Computing**
Smith, I.C.H.	**Microcomputers in Education**
Späth, H.	**Cluster Analysis Algorithms**
Späth, H.	**Cluster Dissection and Analysis**
Stratford-Collins, M.J.	**ADA: A Programmer's Conversion Course**
Teskey, F.N.	**Principles of Text Processing**
Tizzard, K.	**C for Professional Programmers**
Turner, S.J.	**An Introduction to Compiler Design**
Whiddett, R.J., *et al.*	**UNIX: A Practical Introduction for Users**
Whiddett, R.J.	**Concurrent Programming: Techniques and Implementations**
Yannakoudakis, E. & Hutton, P.	**Speech Synthesis and Recognition Systems**
Young, S.J.	**An Introduction to ADA, 2nd (Revised) Edition**
Young, S.J.	**Real Time Languages**

PROGRAMMING IN C FOR UNIX

C. SCHIRMER
Leader of Documentation
PM Targon, Nixdorf Computer AG, Paderborn, West Germany

Translation Editor:
KEITH TIZZARD
University of Exeter

ELLIS HORWOOD LIMITED
Publishers · Chichester

Halsted Press: a division of
JOHN WILEY & SONS
New York · Chichester · Brisbane · Toronto

English language edition first published in 1987 by
ELLIS HORWOOD LIMITED
Market Cross House, Cooper Street,
Chichester, West Sussex, PO19 1EB, England
The publisher's colophon is reproduced from James Gillison's drawing of the ancient Market Cross, Chichester.

Distributors:
Australia and New Zealand:
JACARANDA WILEY LIMITED
GPO Box 859, Brisbane, Queensland 4001, Australia
Canada:
JOHN WILEY & SONS CANADA LIMITED
22 Worcester Road, Rexdale, Ontario, Canada
Europe and Africa:
JOHN WILEY & SONS LIMITED
Baffins Lane, Chichester, West Sussex, England
North and South America and the rest of the world:
Halsted Press: a division of
JOHN WILEY & SONS
605 Third Avenue, New York, NY 10158, USA

This English edition is translated from the author's original German edition *Programmieren in C*, published in 1985 by Carl Hanser Verlag, Munich, West Germany,

British Library Cataloguing in Publication Data
Schirmer, C.
Programming in C for Unix. —
(Ellis Horwood books in computing science: series in computers and their applications)
1. C (Computer program language)
2. UNIX (Computer operating system)
I. Title II. Programmieren in C. *English*
005.4'2 QA76.73.C15

Library of Congress Card No. 86–21317

ISBN 0–7458–0098–X (Ellis Horwood Limited)
ISBN 0–470–20768–X (Halsted Press)

Printed in Great Britain by R.J. Acford, Chichester

Contents

The computer language C has been compared with a sleek sports car. The comparison is quite apt. Both are small, nimble, extremely powerful and reach their destination with a great deal of elegance and, nevertheless, at quite some speed. And just as the steering wheel of a sporty car is integrated into the mechanical functions by means of hard suspension and a powerful-sounding engine, so too can the method of operation of the computer be more or less experienced at first hand with the machine-oriented elements of the language, namely with the aid of a structured program text.

Sports car drivers have to ensure that they do not cut too many corners on their way. The same applies for C programmers. As with its sheet metal counterpart, the result could be a tangled-up mess, the best place for which being the scrap heap (that is to say, the waste paper basket).

Martin Stübs

Preface

As the UNIX® operating system becomes more widely used, the C programming language is also becoming more and more of a standard. And although C continues to have close ties with UNIX, there now exist C compilers and C standard libraries on many other operating systems as a result of this programming language's extreme portability, together with the increasing number of applications now available in C. C's efficiency and its programmer-friendliness both contribute to its popularity.

The book is based both on practical experience gained by using the programming language and on additional literature.

The book gives a comprehensive description of the C language and is suitable as a reference or study work. The most up-to-date version of the C standard libraries based on UNIX System V is included in full in the book. Throughout, the author has attempted to portray the high information content of the book as clearly as possible by using simple examples to show how each language element may be used in practice. This saves the reader from having to solve complicated program logic.

The book also examines the UNIX programming environment by considering the operation of this environment, tools, interpretation of error messages, testing of C programs, programming style, etc. A series of useful summaries completes the book.

August 1986 Claus Schirmer

® Registered trademark of AT&T in the USA and other countries.

1 Introduction

This section briefly describes the origin, significance and basic characteristics of the C language. There is also a subsection providing hints to the user of this manual.

1.1 The classification and significance of C

C was originally developed for an operating system that was specifically designed to be a powerful programming environment. The operating system, in whose development some of the same people were involved, has become the de facto standard for 16-bit processors and is considered the top candidate for the 32- and 64-bit microprocessors to come. Its name is UNIX (a registered trademark of Bell Laboratories).

Thus, C was originally an ancillary tool.

The increasing use of the UNIX operating system fostered the widespread use and popularity of the C programming language. 90 % of the UNIX operating system is written in C, the C compiler itself is written in C, and almost all UNIX utilities and all UNIX-based application software has been, and continues to be, written in C.

The C language soon developed from the UNIX system implementation language into a general system implementation language because it operates exactly on the objects with which the hardware usually works, namely characters, numbers and addresses, and because it provides a wealth of operators supporting any conceivable manipulation of the objects.

Higher-level data structures such as arrays, strings and records (termed "structures" in C) can be formed with C.

C, as opposed to Pascal, PL/1 or BASIC, does not provide language features for handling arrays or strings, but equivalent operations are provided by a standard library of functions or can easily be developed by the programmer.

Standard language constructs for defining the control flow such as tests, loops, branch tables and subroutines are, of course, available in C, and even the "goto", much scorned by software engineering fans, is available, though not required.

Furthermore, C offers (in contrast to BASIC, for example) powerful tools for structured programming: modular design through functions and separate compilation, which are the ideal prerequisites for easy-to-understand and compact programs.

Thanks to these advanced control and data structures, C is a high-level language while providing extreme efficiency by means of the above-mentioned reflection of the processor hardware. C is, of course, a compiled language, eliminating the time-consuming interpretation of the C source program at run time.

Input/output and file access do not form part of C, but there are comprehensive standard libraries of input/output, string handling, memory management and mathematical functions that have been implemented in all operating systems under which C can run. When considering the portability aspect, you can be sure that these functions are available under other operating systems. It is the large number of functions of the C standard libraries which make C a particularly powerful language.

The C language does not incorporate any elements for the concurrent execution of program segments. The control flow is always single-threaded. If multi-thread programming is required, you must access the operating system interfaces by a system call.

C is a type-oriented language, and each value of a data item must be of exactly one of the admissible data types. What Pascal and Ada programmers will miss in C is strong typing, which checks the consistency of the variable and constant types in each expression and each assignment. Omission of this feature facilitates type conversion, but requires more discipline on the programmers' part when handling the various types.
In addition the lint utility is provided which verifies correct C programming and detects type conflicts and other errors.

Another aspect giving flexibility to C is the preprocessor which can modify C source programs at compile time. The preprocessor's functions include macro handling and name substitution, conditional compilation and inclusion of header files.

Owing to the relatively small language set, C is easy to learn and requires very little coding effort by the programmer. The C compiler is compact and easily ported. The same applies to the standard library for input/output, string handling and memory allocation. All these features make for a very simple portability of the language.

Although the C language has not yet been standardized, there exists an effective de facto standard for C in the form of the "C Reference Manual" by Dennis M. Ritchie. Experience has shown that C programs are astonishingly portable and pose fewer portability problems than "standardized" languages such as Pascal and FORTRAN.

1.2 Program structure and basic characteristics

A C program consists of a collection of functions that can be held in one or more source files. Each source file is compiled separately. One and only one function is distinguished by the name "main", while all other functions can have any names and are peers in every other respect (from which it follows that functions must not be nested). A minimum program would consist only of the main function.

The functions of the C language can have the properties of procedures. However, functions and procedures are not distinguished, but are both classified as functions.

The control flow of a program begins in the function named "main" and will end there if the program ends normally. In the course of execution, the main program can transfer control by calling one or more of the other program functions which in their turn can call other functions. When a function is complete, it will return control to the calling function. Therefore, the program must always end in the main function.

The data objects available include characters, integers, floating point numbers and pointers as well as derived types in the form of arrays, structures, unions and functions.

Besides the basic objects you may use only structures as arguments for functions, but this does not represent a limitation as you will see in the "Pointers" section. When a function is called, the values of the function parameters are copied into the formal parameters of the function (call by value). This means that the function has no access to the objects themselves that form the function parameters (no call by reference). You will also learn in the "Pointers" section that this limitation can be overcome by using pointers as parameters.

Manipulation of the objects is supported by a comprehensive set of operators enabling you to write compact and simple expressions to compute values. By contrast to other high-level languages, C even provides bit manipulation operators as well as powerful assignment operators and logical operators (with logical values being represented by integer values).

1.3 Notes for the user of this manual

This introduction is followed by a chapter intended for the beginner who has never written a C program. More experienced programmers can skip the second chapter.

Following a summary of the syntax conventions for this manual, the C data types, storage classes and the representation of variables and constants are explained. The next chapter describes how the many operators are used to manipulate data and form expressions.

Another chapter describes the C statements. Functions play an important part in modular program design. A complete chapter is therefore dedicated to functions and program structure.

The handling of addresses (pointers) using C is somewhat complicated, but very efficient. All you need to know is explained in the chapter about "Pointers".

Another chapter deals with structures, followed by a chapter about input/output, which is supported by standard library functions.
The host of other standard library functions require so much attention (and space) that a complete chapter is devoted to them.

Then are described the interfaces with the UNIX operating system, followed by notes on the programming style that should be observed to obtain easy-to-read, easily maintainable and portable programs.

To enable the user to operate the C programming environment, the next chapter explains special Shell commands and the functionality of the preprocessor.

Following a summary of the error messages, the manual is rounded off by appendices containing summary lists and an index to the manual.

This manual is intended for programmers within the following target groups:

1. Experienced programmers wishing to learn and master the C language. These readers can skip the chapter "C programming for the beginner" and should try out on a computer the language elements they are not acquainted with while working through the other chapters.
2. Experienced C programmers can use this document as a reference manual. The index at the end of the manual is particularly useful for them.

3. Beginners without programming experience and not knowing the C language. These people should read the chapter "C programming for the beginner". Carefully read the chapter and think it over and try out the examples on a computer. Then study the remaining chapters, incorporating each language element at least once in a brief sample program.

2 C programming for the beginner

2.1 Your first C program

You have learned in the introduction that each C program must comprise at least a main function. What the main function looks like will be explained by a simple example. So that you can see that the program works we are going to place one statement in the program:
The function call printf (...) is used to display text on the screen.

The program looks as follows:

```
main()
{
    printf("Hello, John\n");
}
```

(You may, of course, use your own name instead of John.)

"main" is the function name. The two parentheses indicate that main is a function. The parentheses would normally include the formal parameters of the function. However, if there are no parameters as in this example, the parentheses must still be specified after the function name to distinguish it from other variable names.

Incidentally, you should know that you can pass parameters to the main function (viz. Chapter 8).

The body of the function includes the statements that are executed when the function is called. They are enclosed in braces whose meaning is similar to begin...end in Pascal.

Note:

Statements are usually ended with a semicolon ";" as in PL/1 or Ada.

Printf is a function call as can be seen from the subsequent parentheses. The argument (or parameter) of printf is "Hello, John\n", which the program will display on the screen if it is properly compiled and executed. The symbols "\n" represent a special character (a single character) generating a newline. Try out the effect of the character by leaving it out.

How can you start executing this program?

The first thing you do is to create a file using the editor. You can choose any file name ending with the qualifier ".c", for example "greeting.c". Type your program into this file and quit the editor. This will save the file (source file) on disk.

The compiler (and linker) are called with the command

```
cc greeting.c
```

If no error message is displayed when this command is executed, the program is successfully compiled and linked and can be executed. Unfortunately, the linker always assigns the name "a.out" to the executable file. If you want to prevent this file from being overwritten during the next compilation, rename it to

```
mv a.out greeting
```

Now you can execute the file. The call

```
greeting
```

loads the program into the memory and starts it. It will output the text

```
Hello, John
```

on the screen, and your first C program has successfully executed.

Just another remark about the linker. Larger C programs will usually consist of several parts that can be compiled separately. This feature facilitates the detection of errors and saves a lot of time when correcting errors because only the corrected program portion needs to be re-compiled. In this context it is the responsibility of the linker to combine the compiled program sections into a single machine code file and define consistent relative addresses for all program data.

2.2 Computations and function call

Let's write another program that uses the following typical C features:

- comments,
- variables,
- arithmetic, and
- functions.

Comments are enclosed in /* ... */. They may extend over several lines, but must not be nested. So be careful, when marking program sections for debugging purposes, of using comment symbols if the program sections already contain comments.

Variables must be defined before they are used for the first time. The following basic types are available:

Name	Meaning
char	Character
int	Integer
float	Floating-point number

Integers are complemented by the types unsigned int, short int and long int, and float is complemented by the type double, which refers to floating-point numbers with double precision. The value range of variables of a specific type (which depends on the number of bits available for representing a variable) is machine-dependent, which may cause portation problems.

In the C language, as in other high-level languages, values are computed using expressions. The assignment operator used in C is the "=", as opposed to ":=" in Pascal or Ada, just to save one character. To make a distinction, the relational operator for equal is coded as "==", using two characters because this operator is not so often used.

The following arithmetic operators are available:

Operator	Meaning
+	Addition
–	Subtraction
*	Multiplication
/	Division
%	Modulo (remainder of integer division)

Some other special operators will be introduced later. Note that C does not provide an exponentiation operator. However, this operator is fully substituted by the function "pow" in the standard library of mathematical functions.

The following example will give you an understanding of how user-written functions can be included:

```
main() /* Control and output computation of circular area */
{
    float dia, rad, ar, area();

    dia = 7.0;
    rad = dia / 2;
    ar = area(rad);
    printf("Radius: %f, Area: %f\n", rad, ar);
}

float area(r) /* Circular area computation */
float r;
{
    float f;

    f = r * r * 3.14159;
    return(f);
}
```

The program computes the area of a circle with a diameter of seven units.

You will notice immediately that the program consists of two parts, namely the functions main() and area(r). Execution of the (compiled) program begins with the first statement of the main function and continues sequentially. If program execution comes to a statement containing a function call – in this example the statement ar = area(rad); – the function is supplied with the parameter values (in this case, the radius rad), and the statements of the function are executed. If a return statement with a parameter is encountered during execution of the function, control is returned to the calling function (in this case, the main function), with the parameter value of the return statement being inserted for the function call.

This is what you should initially know about functions and their control flow.

What is interesting in the main function is the expansion of the printf function and, in general, the scope of variables depending on the place where they are declared.

The first parameter of the printf function is a control string that must contain places marked by % for the subsequent parameters. The % must be followed by a letter identifying the type of the associated parameter:

Identifying letter	Type
d	Decimal number
o	Octal number
c	Character
f	Floating-point number
etc.	

The % and identifying letter may be separated by length specifications and identifiers for left- or right-justified. For more details see the chapter "Input/output".

The control string is processed from left to right as follows: Normal characters except the % are transferred to the output on a one-to-one basis. When a % is encountered, the segment from the % to the identifying letter is replaced by the value of the next unprocessed parameter. Thus, the number and type of the subsequent parameters must correspond exactly to the number of % characters with matching identifying letters in the control string.

Example:

```
printf("Radius: %f, Area: %f\n",rad,ar);
```

generates the following output:

```
Radius: 3.500000, Area: 38.484478
```

2.3 General remarks on definitions

Called functions are declared in the calling function by the type of the value they return. The parentheses "()" after the name identify it as a function name; parameters are never specified in a declaration.

The definition of variables may be made in the function body after the opening brace. The scope of such a variable ends at the closing brace of that function. When the same function is called the next time, the variable is recreated in the memory and the old value has been lost. Variables of this type are described as automatic variables in C as in some other languages (e.g. PL/1).

Formal function parameters (in the example the "r" in the function "area") are defined immediately after the function name and before the opening brace, but are also treated as automatic variables for this function just like the variables defined after the brace.

The counterpart of these local variables are the global (or external) variables whose scope may embrace all functions belonging to the program. They are distinguished from local variables in that they are defined outside any program function.

Using the global variable rad, our sample program looks as follows:

```
float rad;     /* Radius globally defined */
main() /* Control and output computation of circular area */
{
    float dia, ar, area();

    dia = 7.0;
    rad = dia / 2;
    ar = area();
    printf("Radius: %f, Area: %f\n", rad, ar);
}

float area() /* Circular area computation */
{
    float f;

    f = rad * rad * 3.14159;
    return(f);
}
```

If you take advantage of the fact that expressions rather than variables can be used as function parameters, the program can be reduced to the following form:

```
float rad;     /* Radius globally defined */
main() /* Control and output computation of circular area */
{
    float dia = 7.0, area();

    rad = dia / 2;
    printf("Radius: %f, Area: %f\n", rad, area());
}

float area() /* Circular area computation */
{
    return(rad * rad * 3.14159);
}
```

2.4 Conditional execution and program loops

In the last example of this chapter we shall discuss two statements used to direct the control flow:

if statement and

while statement

These two statements have the same effect as in most other languages. You should pay particular attention to the notation.

Conditional execution of statements:

```
if (<expression>) <statement1>

[ else <statement2> ]
```

If <expression> is "true", <statement1> will be executed. If <expression> is "false", <statement2> will be executed, if present, and otherwise program execution continues with the next statement.

The absence of logical variables and values is made up for by numeric values: a logically true expression returns the numeric value 1, and a logically false expression, the value 0. Testing in if statements therefore works as follows: if the <expression> is other than 0, <statement1> is executed; if the <expression> is 0, <statement2> is executed (if present; otherwise control goes to the next statement). This extension can also be utilized to test numeric values, with "if" working properly for logical expressions.

A single <statement1/2> may be replaced by a compound statement which is a list of statements enclosed in braces.

Loop with test of continuation condition before execution:

```
while (<expression>) <statement>
```

In this case, the <expression> is first tested to see if it is not equal to 0 (or true). If so, the (compound) <statement> is executed. Then the <expression> is tested again, and the procedure continues in the same way until the value of <expression> is 0. Then control goes to the first statement after the while loop. If the <expression> was 0 at the beginning, control is immediately transferred without executing the <statement>.

Example:

```
main() /* Computation of circular area and circumference */
{
   float dia = 1.0;
   float rad, area(), circumf();
   char c;

   while (dia <= 10.0)
   {
      printf("Diameter: %f Circumference/area(c/a)? ", dia);
      c = getchar ();
      if (c == 'c')
         printf("Circumference: %f\n", circumf(dia));
      else
      {
         rad = dia / 2.0;
         printf("Area: %f\n", area(rad));
      }
      c = getchar();                         /* get CR key code */
      dia = dia + 1.0;
   }
}

float circumf(d) /* Computation of circumference */
float d;
{
   return(d * 3.14159);
}

float area(r) /* Computation of circular area */
float r;
{
   return(r * r * 3.14159);
}
```

The program will compute the circumference of the circle successively for each diameter of 1 to 10 units when "c" is entered. If another entry is made, it will compute the circular area. The function getchar() supplies a character from the input device. Like printf, it is contained in the standard library of input/output functions.

To pass the value to the program when the letter "c" or "a" is entered, depress the CR key. Note, however, that this causes the CR key code to be accepted into the input stream. To ensure that this code is not accepted as user input during the next "circumference/area" request, the second "c = getchar()" fetches it from the input stream into the program, where it is not evaluated.

You will learn later how this problem can be solved more elegantly or how the end of an input is recognized by the value of the CR key.

2.5 Summary

Following is a brief summary of the concepts, statements and other information presented so far.

Operating the programming environment:

Keying in the C program:

Use the editor to key the program into a file whose name ends in ".c" (e.g. greeting.c).

Compiling (and linking) the C program:

```
cc greeting.c
```

Renaming the compiled program:

```
mv a.out greeting
```

Executing the program:

```
greeting
```

C statements:

Outputting text and values:

printf("<control-string>", <output-values>);

Substitution characters in the control string:

%d	Decimal number
%o	Octal number
%c	Single character
%f	Floating-point number

Assignment:

<variable> = <expression>;

Definition:

<type> <variable list>;

Scope of variables:

Program-wide	when defined outside functions
Limited to function	when defined within a function

Function call (within expression or as a separate statement):

<function-name>(<parameter-list>)

Conditional statement:

if (<expression>) <statement1>
[**else** <statement2>]

while loop:

while (<expression>) <statement>

Comments:

/*...*/

Data types:

Name	Data type
char	Character
int	Integer
unsigned int	Unsigned integer
long int	Long integer (more digits)
short int	Short integer (fewer digits)
float	Floating-point number
double	Double-precision floating-point number

Numeric operators:

Operator	Meaning
+	Addition
–	Subtraction
*	Multiplication
/	Division
%	Modulo (remainder of integer division)

3 Syntax elements

This chapter introduces the elements of the C language required to form statements and write programs.

The rules governing the formation of the language elements are set forth using a simple metalanguage.

3.1 The metalanguage

In the following, the C language set is described using a metalanguage similar to the Backus-Naur-Form (BNF).

The metalanguage symbols and their meanings are as follows:

< > (left/right angle bracket)

Angle brackets enclose an abbreviated name of a set of language elements. Example: <statement> is an abbreviation for the set of all valid C statements.

If it is necessary to distinguish similar sets of language elements in the syntax description, this is done by subscripting the abbreviation within the angle brackets (e.g. <statement1>, <statement2>).

::= (... is defined as ...)

The language element to the left of the definition symbol is defined by the syntactical expression to the right of the symbol.

| (either ... or ...)

The language elements before and after this symbol are mutually exclusive. Example: <letter> | <digit> means that either a letter or a digit must be specified.

{ } (left/right brace)

Language elements specified within braces must be coded in this place (mandatory entry).

[ ] (left/right bracket)

Language elements specified within brackets are optional, i.e. they may either be specified or omitted (optional entry).

Brackets that belong to the C syntax and are therefore mandatory are printed in bold face (**[ ]**).

The following applies to both braces and brackets:

Entries specified one beneath the other are mutually exclusive.

Example: $\ldots \begin{Bmatrix} \text{<letter>} \\ \text{<digit>} \end{Bmatrix} \ldots$

means that either a letter or a digit must be entered at this point. (If the braces are replaced by brackets, either a letter or a digit may be entered.)

$\{\,\}_2^8[\,]^4$ (minimum/maximum number of multiple entries)

A superscript following a closing brace or bracket specifies the maximum number of times that a mandatory or optional entry can be coded at this point.

A subscript following a closing brace specifies the minimum number of times that a mandatory entry must be coded at this point.

If no subscript or superscript is specified, this is synonymous with "1".
The superscript "n" means any number of times.

Examples:

... $\{\text{<digit>}\}_2^4$ ...

means that no fewer than two and no more than four digits may be written in this place.

... $[\text{<letter>}]^4$ ...

means that the specification of letters at this position is optional, but if letters are specified, the maximum number is four.

All other characters and words in **bold** print outside the angle brackets are final symbols and must be written "as is".

Note:

If a syntactic element overflows a line, it is continued on the following line and indented.

3.2 List of syntax elements

- <letter> ::= A | B | C | ... | X | Y | Z | _ | a | b | c | ... | x | y | z

  <letter> is the set of letters of the alphabet.

- <digit> ::= 0 | 1 | ... | 8 | 9

  <digit> is the set of decimal digits.

- <odigit> ::= 0 | 1 | ... | 6 | 7

  <odigit> is the set of octal digits.

- <hdigit> ::= 0 | 1 | ... | 8 | 9 | A | B | C | D | E | F | a | b | c | d | e | f

  <hdigit> is the set of hexadecimal digits.

- <integer> ::= [±]{<digit>}^{n}

An integer is a sequence of decimal digits that may be preceded by a sign.

Examples: 9 −1234567890 815

- <octal-number> ::= [±]0[<odigit>]^n

An octal number is a sequence of octal digits with a leading 0 that may be preceded by a sign.

- <hexadecimal number> ::= [±]0$\left\{\begin{matrix} x \\ X \end{matrix}\right\}$[<hdigit>]^n

A hexadecimal number may be preceded by a sign, then continues with 0x (x may be lowercase or uppercase), followed by a sequence of hexadecimal digits.

- <decno> ::=

$$\left\{\begin{matrix} \text{<integer>} \\ \{\text{<digit>}\}^n . \\ . \{\text{<digit>}\}^n \\ \{\text{<digit>}\}^n . \{\text{<digit>}\}^n \end{matrix}\right\} \left[\left\{\begin{matrix} E \\ e \end{matrix}\right\} [\pm]\{\text{<digit>}\}_1^3 \right]$$

A decimal number is a number that may contain a decimal point.
In addition, the letter E or e and an integer can be used to multiply the value of the number by 10 to the power of the integer.
The exponent and the decimal point distinguish a decimal number from other numbers. This is why one of these elements must be specified.

Examples:

.99 1. 6.789E−5 0.05e3 345E6

- <char>

<char> is the overall set of ASCII characters.

- <ocode> ::= \{<odigit>}^{3}

  An octal code comprises the overall character set with each character represented octally in the ASCII code with a leading \.

- <noper> ::= + | – | * | / | %

  <noper> includes the operators for addition (+), subtraction (–), multiplication (*), division (/) and integer division remainder (%).

- <reloper> ::= < | > | == | >= | <= | !=

  The relational operators are from left to right: "less than", "greater than", "equal", "greater or equal", "less or equal" and "not equal".

- { <varname> | <fnctname> | <labelname> } ::= <letter> [<letter> | <digit>]^n

  A name (variable, function or label name) begins with a letter (including "_"), followed by letters or digits. Names are case sensitive. You should avoid the letter _ at the beginning of names because internal system names begin with this letter. Also note that names must never be identical with any of the following keywords:

auto	default	enum	if	short	typedef
break	do	extern	int	sizeof	union
case	double	float	long	static	unsigned
char	else	for	register	struct	void
continue	entry	goto	return	switch	while

  Examples:

  field_2 type a2 This_is_a_long_name

- <storage-class> ::= { [auto] | static | extern | register }

  [auto]
  static
  extern
  register

You can give a storage class specification when defining variables. Note the rules set forth in the following chapter.

- <type> ::= {

  char
  int
  float
  unsigned [int]
  short [int]
  long [int]
  unsigned short
  unsigned long
  double
  <structtype>
  <enumtype>
  <typedeftype>

  }

<type> is the data type specification for variable or function definitions. The data types have the following meanings from top to bottom: "character", "integer", "floating-point number", "unsigned integer", "short integer", "long integer", "unsigned short integer", "unsigned long integer", "double-precision floating-point number", "structure type", "enumeration type", "type name from typedef statement".

4 Types, storage classes, variables and constants

This chapter gives you an introduction to the data objects used by the C language. Note, however, that pointers (address variables), which also belong to the basic types of the C language, are discussed in a separate chapter because of their extreme versatility and power.

The basic data types are the elements of the higher-level data objects, of which the arrays are described in this chapter, while structures, unions and fields are dealt with in a separate chapter.

4.1 Types and definitions

4.1.1 Variable definition

Variables must always be defined before their first use. Definition statements can occur at the following points in a program:

- outside functions,
- following the function name before the beginning of the function (formal parameter definition),
- within a function before the first statement, or
- within a function and within braces before the first statement within the braces.

A definition has the following format:

[<storage-class>] <type> <variable$_1$>, ..., <variable$_n$>;

The <type> and <storage-class> are basically distinguished as follows: the type of a variable defines its storage requirements, while the storage class reflects its dynamic properties (life, scope).

Let us discuss types first. The following basic types are available:

char	Character
int	Integer
float	Floating-point number

Numbers can be of the following additional types:

short [int],
long [int],
unsigned [int],
unsigned short,
unsigned long,
double (= double-precision floating-point number).

- Data type 'char'
  - Any character of the ASCII character set
  - 8 bits long
  - Representation of contents in single quotes

- Data type 'int'
  - Integer numbers between -2^{15} and $2^{15}-1$ or -2^{31} and $2^{31}-1$
  - 16 or 32 bits long
  - Constants can be decimal, octal or hexadecimal

- Data type 'long [int]'
  - Integer numbers between −2147483648 and 2147483647
  - 32 bits long
  - Constants can be decimal, octal or hexadecimal (identified by an appended "L" or "l")

- Data type 'short [int]'
  - Integer numbers between −32768 and 32767
  - 16 bits long
  - Constants are decimal, octal or hexadecimal

- 'unsigned' data types

  - Sign bit belongs to the value
  - 'unsigned short': 16 bits long, integers between 0 and 65635
  - 'unsigned [int]': 16 or 32 bits long
  - 'unsigned long': 32 bits long, integers between 0 and 4294967295
  - Constants are decimal, octal or hexadecimal

- Data type 'float'

  - Rational numbers between approximately –0.17E38 and 0.17E38
  - Smallest value other than 0 that can be represented is 0.29E-38
  - Stored in floating-point representation with binary mantissa and exponent
  - Approx. seven significant decimal digits
  - Constants are fixed-point numbers or floating-point numbers with exponent

- Data type 'double'

  - Rational numbers in the same range as 'float'
  - Stored in floating-point representation with binary mantissa and exponent
  - Approx. 16 significant decimal digits
  - Constants are fixed-point numbers or floating-point numbers with exponent

The value range for a specific type of variable (which depends on the number of bits available for representing the variable) is machine-dependent, and this may cause portability problems. On 16-bit processors, the type 'int' is normally identical with 'short int', while on 32-bit processors it is identical with 'long int'.

4.1.2 Array definition

You can define not only single variables, but also arrays of elements of the same type. The most commonly used array is the one-dimensional character array, the string.

When defining an array, you write the size and number of dimensions, enclosed in square brackets, after the variable name. Each dimension requires a pair of brackets. The dimension size must be specified as a constant or a constant expression.

It is not possible, in C, to specify dimension sizes dynamically.

Examples:

Definition of a 10 character long 'char' array named "colour" and a 10 * 10 number array named "inv":

```
...
char colour[10];
float inv[10] [10];
...
```

Once you have defined the size of an array dimension by N, the individual elements are addressed by the subscripts 0 through N-1 enclosed in brackets when used in statements. You realize that the subscript can be understood as an offset in number of elements from the beginning of the array.

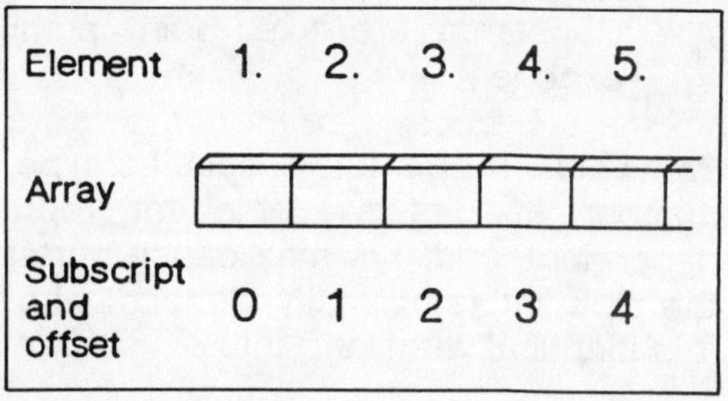

Array element addressing by specifying an element offset seems to be rather awkward (cf. PL/1), but is ideally suited for the use of pointers (see Chapter 8).

Note that, in contrast to most other languages, it is your responsibility to keep the subscript within the array limits. As explained above, C uses the subscript as an element offset from the beginning of the array. Negative subscripts or subscripts larger than the dimension will access any memory locations before or after the array.

The 'char' arrays, which may be more familiar to you under the name of "character strings", play a particularly important part because of their frequent use and the rich set of standard library functions that may be used with them. Strictly speaking, a string within C is a sequence of characters terminated by the end character '\0'. Most standard functions use the end character to detect the end and, hence, the current length of a string. When writing string processing routines yourself you will note that end characters are very helpful.

When using string variables in C, you are responsible for reserving 1 byte for the end character.

Example:

If you want to store a string of up to 10 characters in the 'char' array variable called name, the following definition is required:

```
char name[11];
```

Two other types – the pointers just mentioned and structures (comparable to the records of PL/1 and Pascal) – will be discussed in a separate chapter.

4.1.3 Enumeration type definition

Similarly to Pascal, the C language allows you to define types in which you can specify in the form of an enumeration which values the variables of this type can assume. The keyword for enumeration types is enum, which is optionally followed by a name for the enumeration type that is defined by the values following in braces.

Example:

```
...
enum lights {red, yellow, green};   /* Type definition */
...
```

Semantically, this definition only creates a new type named 'enum lights'. The data elements (variables) of this type are then defined and created as follows:

```
...
enum lights lights_bl, lights_ufw, lights_ofw;
...                                 /* Variable definition */
```

Note that the name "lights" is the enumeration **type** name and the names "lights_b1", "lights_ufw" and "lights_ofw" are variable names.

The type and data definitions can be combined by appending the variable names to the type definition.

Example:

```
...
enum lights {red, yellow, green}
   lights_bl, lights_ufw, lights_ofw;
...
```

The combined type and variable definition is often used with enumeration types that are not assigned a type name.

The values the variables can assume are internally stored as the constants 0, 1, 2 etc. (testing for this value is possible). If you wish to change this automatic identification, you can explicitly specify the value in the enumeration:

```
...
enum lights {red, yellow = 2, green = 4};
...                                /* Type definition */
```

4.1.4 Renaming types using typedef

Finally let us consider a construct used to assign meaningful names to the types described here (and some other types described later). The names can be used in subsequent variable definitions and with the type conversion operator.

Example:

```
...
typedef int LENGTH;
typedef enum lights LIGHTS;
...
LENGTH lome,max,min;
LIGHTS lights_bl, lights_ufw, lights_ofw;
...
```

(User-defined keywords are conventionally written in uppercase letters.)

Remember that typedef does not create a "new" type but just assigns an additional name to an already existing type.

4.1.5 Function definitions and declarations

Functions that are called by another function should always be declared in the calling function. The word "declaration" means that something (external) is made known, while "definition" means that an object is physically created with storage space being reserved and possibly initialized.

The function definition is the function itself. The declaration requirement for functions is eliminated when the function definition and its use are in the same source file and the definition appears before the use or when the function returns an 'int' value.

The format of a function declaration is similar to that of a variable definition:

[<storage-class>] <type> <function-name>();

and can therefore be combined with suitable variable definitions.

The type of the function is the type of the value that the function returns. Confusion with a variable definition is excluded by the "()".

Example:

```
...
int i, first(), sig, last();
...
```

This statement defines the two 'int' variables i and sig and declares the two functions first and last.

An additional type is provided for the use with functions. This type cannot be applied to other data elements. The type name is 'void'. If a function is defined and declared with this type, that function will not return a function value, i.e. behave like a Pascal procedure.

4.2 Variables, storage classes and initialization

Just one more remark about names (variable, function and label names). Names consist of letters, digits and the underscore "_", with the "_" being syntactically included in the set of letters. Internal names are any length, and all characters are significant; the length of external names depends on the machine and can be up to 6, 7 or 8 characters.

The first character of a name must be a letter (including "_"). However, you should avoid beginning names with "_" to preclude collision with internal names.

Names are case-sensitive. You must, of course, not use C language keywords.

List of the C reserved words:

auto	else	int	switch
break	entry	long	typedef
case	enum	register	union
char	extern	return	unsigned
continue	float	short	void
default	for	sizeof	while
do	goto	static	
double	if	struct	

Examples of valid names:

LENGTH   length   Length   This_is_a_long_name

4.2.1 Initialization

Only external and static variables are automatically initialized (to 0). If you want to initialize external variables to values other than 0, or initialize automatic variables, you must do this explicitly. The initialization must always accompany the definition and not a declaration. Its format is as follows:

```
<type> <variable> = <expr>;
```

For initialization of external and static variables <expr> must consist of constants. You should define each variable separately to facilitate the detection of errors. The following notation is also allowed:

```
int l=5, i, j=0, z;
```

Array initializations are separated by commas and enclosed in braces; where several dimensions exist, the initialization of each dimension is additionally enclosed in braces (aggregate initialization).

Note, however, that aggregate initializations are only permitted for globally defined arrays (or structures) and for local arrays or structures in conjunction with the storage class **static**.

Example:

```
int mat[2] [3] = { { 2, 3, 4 }, { 2, 3, 3 } };
```

If an array is initialized at the same time that it is defined, the length information can be omitted because it is derived from the initialization values. The preceding example is equivalent to

```
int mat[] [] = { { 2, 3, 4 }, { 2, 3, 3 } };
```

For the initialization of character strings refer to the subsection 4.3, "Constants".

Internal variables should always be initialized to prevent undefined values (garbage). Initialize them either at the time of definition like external variables or by a value assignment within the function.

4.2.2 Storage classes

Recall the syntax of a definition or declaration statement:

[<storage-class>] <type> <variable$_1$>, . . . , <variable$_n$>;

The storage class affects the scope, life and initialization of data elements. A distinction is made between local and global definitions: local variables are defined within a function or within a block in a function and are only known within this function or block (for "block" see "Compound statements and block structure"). Global variables are defined outside any function and can be referenced from the overall program.

If the program consists of several source files, external variables (i.e. variables defined in a different source file) must be declared (made known) in each source file in which they are to be used.

The following differences exist in the initialization of locally and globally defined variables, respectively: each externally defined variable is automatically initialized to 0, whereas internally defined variables must be explicitly initialized.

Each function is a global function definition.

The following storage class options may be specified in global definitions of variables or functions:

– The **static** option in the definition means that the variable or function is not available to the overall program, but only to the functions within the same source file. Static thus provides a means of information hiding by not allowing the overall program access to a variable that cannot be kept local (you just need to distribute the program to various source files).

The following storage class option can be specified for global declarations:

– The **extern** option means that the declaration refers to an externally defined variable or function that is to be available to the rest of the source file.

Four different storage classes are supported for internal definitions or declarations:

- If no option is specified, this is an automatic variable (storage class **auto**) which is re-created and re-initialized on each re-entry into the function or block.
- The storage class **static** means that the variable is only created on the first entry into the function or block, that it keeps its value when the function or block is exited, and that it can be re-accessed with its old value upon re-entry. Explicit initialization in the definition of a local static variable is only performed on the first execution of the function or block containing the variable.
- The **extern** option means that this is the declaration of an externally defined variable.
- Variables identified by **register** are a subset of the automatic variables (local variables not defined with static or extern, or formal parameters of a function). The register option is a recommendation to the runtime system to use a register for storing the variable. When extremely frequently referenced variables are defined in this way, this will usually accelerate program execution.

  If the type of a register variable is 'int', the type may be omitted in the definition. The following definitions are synonymous:

  ```
  register int i;
  register i;
  ```

  In general, register definitions are only accepted with the data types 'char', 'int' or (any) pointer type, but no error message is issued in other cases.

Note:

The address operator (&) must not be used for variables defined with the register storage class (see Chapter 8).

Overview of C storage classes

Storage class	Scope	Life	Auto. initial.	Aggregate init. allowed
register	Block	Block	No	No
auto	Block	Block	No	No
static intern	Block	Program	To 0	Yes
static extern	Source file	Program	To 0	Yes
extern	Program	Program	To 0	Yes

4.3 Constants

The two constant classes of C are

- Literal constants and
- Symbolic constants

Literal constants are represented by their value (e.g. 10, 'a', "text"). Symbolic constants have a name that represents their value. The name and value are defined by a preprocessor statement (cf. the subsection "The C compiler preprocessor"). Wherever the syntax requires constants, you may choose to specify a literal or a symbolic constant (e.g. dimension size for arrays).

The notation of literal constants, which also applies to the definition of symbolic constants, is explained in the following.

'int', 'long', 'short' and 'unsigned' constants

Integer constants are integer numbers without a decimal point. Integer numbers may also be represented in octal or hexadecimal notation. The notation is

0[<odigit>]^{n} and

0x[<hdigit>]^{n} or
0X[<hdigit>]^{n}

Example:

10, 012, 0xa, 0Xa, 0xA and 0XA represent the same number.

Short int constants have the same notation, while long int constants must be followed by an L or l if they are less than 32768 and greater than –32769.

Example:

0xaL (10 as a long constant in hexadecimal notation)

'float' and 'double' constants

Fixed-point constants must contain a decimal point. Also allowed is the scientific exponential notation (floating-point notation). The floating-point notation does not require a decimal point.

Examples of valid fixed- and floating-point constants:

123.45 7.0 123.456e7 0.12E3 123e45 3. .45

Fixed-point and floating-point constants are always represented in decimal.

The distinction between 'float' and 'double' applies only to the storage mode of variables. The storage of constants and the processing of all floating-point numbers is always in the 'double' type.

'char' constants

A character constant is enclosed in single quotes.

Examples:

'u', 'a'

Note the representation of the following characters and special characters:

'\n'	New line
'\t'	Tab character
'\b'	Backspace character
'\r'	CR character
'\f'	Top of form
'\0'	Null character
'\\'	Backslash (\)
'\''	Quote
'\{<odigit>}^{3}'	Character represented by octal code

String constants

A string constant is enclosed in double quotes:

"I am a string"

A " character within the string constant must be represented by \". The compiler will reserve a field of the proper size and store the string in it. In addition, it will automatically append the end character '\0' to any string constant so that the program can easily recognize the end of string.

String constants may be used to initialize character arrays, and in this case you may even omit the braces. The following initializations are synonymous:

```
char colour[] = { 'r', 'e', 'd', '\0' };
char colour[] = { "red" };
char colour[] = "red";
```

5 Operators and expressions

Operators enable you to modify variables. Expressions consisting of operators, variables, constants and function calls produce new values and results.

We are not going to discuss the syntax for forming expressions because there are no major differences from other languages. Primary expressions include variables, constants, array and structure elements as well as function calls that are connected by operators to form higher-level expressions.

The operators are governed by fixed rules of precedence that can be overridden by parenthesizing. Expressions enclosed in parentheses are always evaluated first.

The evaluation of parenthesized expressions is governed by the same rules of precedence. Therefore, when several pairs of parentheses are nested, the parenthesized expressions are evaluated from the innermost to the outermost. Redundant pairs of parentheses are ignored.

5.1 Type conversion in expressions

Except for the unary and the conditional expression operators, all other operators are binary and are used in the following notation:

<expression1> <operator> <expression2>

If the data types of the operands of a binary operator are different, the value of the "weaker" data type is converted to the data type of the other operand while retaining its value. The type of the result is the same as the operand types after conversion.

These conversions are governed by some simple rules:

There are only four executable data types, namely 'int', 'long', 'unsigned' and 'double'. The values of other types are automatically converted to a value of an executable type.

Values of the types 'char' and 'short' are internally converted to 'int' values. 'Float' values in expressions are always converted to 'double' values.

After these conversions, when different data types are connected, they are converted (in the specified order):

- If one of the operands is a 'double' type, the other operand is converted to 'double'. The result is of type 'double'.
- If one of the operands is a 'long' type, the other operand is converted to 'long'. The result is of type 'long'.
- If one of the operands is an 'unsigned' type, the other operand is converted to 'unsigned'. The result is of type 'unsigned'.

Example:

```
main()    /* Type conversion in expressions */
{
    double f, g;

    f = 10 / 3;
    g = 10 / 3.0;
    printf("f = %.3f, g = %.3f\n", f, g);
}
```

produces the following output:

```
f = 3.000, g = 3.333
```

In the first division, both operands are 'int' constants that do not require conversion, and the fractional positions are truncated in an integer division. In the second division, the value 3.0 is a 'double' constant. This causes the other operand 10 to be converted to the 'double' type, producing a real division.

5.2 Unary operators

Operator	Meaning
*	Contents of the storage field to which a pointer is pointing (see Chapter 8)
&	Address of a variable (see Chapter 8)
–	Negative value (minus sign)
!	Negation (0, if value ≠ 0; 1, if value = 0)
~	Bitwise complement
++	Increment
––	Decrement
(<type>)	Forced type conversion
sizeof	Size of the subsequent data element or type in bytes

The absence of logical values in the C language and their representation by numeric values is the reason for the peculiar use of the not operator (!): numeric values not equal to 0 mean "true", and the value 0 means "false".

Bitwise complement means that the individual 0 bits of the operand are set to 1, and the 1 bits are set to 0 (cleared). This operator is only allowed on values of the 'char' and 'int' (including all variations of signed, unsigned, short and long) types.

The increment and decrement operators are, of course, particularly useful for integer type operands.

5.3 Arithmetic operators

Operator	Meaning
+	Addition
–	Subtraction
*	Multiplication
/	Division
%	Modulo (remainder of integer division)

The division operator truncates any fractional positions from the result if both operators are of the integer types ('char', 'short', 'int', 'unsigned' or 'long').

The following rule of thumb applies to the modulo operator: the sign of the result is the same as the sign of the first operand. The expression a % b is evaluated as follows:

```
a – a / b * b
```

Examples:

```
 5 % 3 equals 2, because 5 / 3 = 1, remainder of 2
 6 % 3 equals 0
–5 % 3 equals –2
 5 % –3 equals 2
–5 % –3 equals –2
```

5.4 Logical operators

Operator	Meaning
<	Less than
<=	Less than or equal
>	Greater than
>=	Greater than or equal
==	Equal
!=	Not equal
&&	Logical AND
||	Inclusive OR

The logical operators return the 'int' value 1 or 0 depending on whether the logical operation is "true" or "false", respectively.

Truth table of the AND and OR operators:

a	b	a AND b	a OR b
0	0	0	0
0	1	0	1
1	0	0	1
1	1	1	1

The AND and OR operations do not always evaluate both operands. If, in an AND operation, the first operand is "false", the overall expression is "false", and the second expression is not evaluated. The same applies to OR when the first expression is "true". This can lead to problems when you use, for example, the following expression:

```
++m || ++n
```

If the lefthand operand is non-zero, the righthand operand will not be evaluated so that n is not incremented!

5.5 Bitwise logical operations

Operator	Meaning
&	Bitwise AND
^	Bitwise exclusive OR (XOR)
|	Bitwise inclusive OR
>>	Bitwise right shift in the value of <expression1> by the value of <expression2> places (the value of <expression2> must not be negative)
<<	Bitwise left shift (analogous to right shift)

The first three operators compare the operands bit by bit, producing the appropriate result bits. The operands of all bitwise operators must be of the 'char' or 'int' types.

Truth table of the XOR operator compared to the OR operator:

a	b	a XOR b	a OR b
0	0	0	0
0	1	1	1
1	0	1	1
1	1	0	1

Note:

Another bitwise operator is the unary operator "binary complement (~)". See subsection 5.2, "Unary operators".

Example:

```
main()    /* Bitwise operators */
{
    int a = 02436, b = 065;    /* Octal initialization */

    printf("%06o & %06o = %06o\n", a, b, a&b);      /* AND */
    printf("%06o | %06o = %06o\n", a, b, a|b);      /* OR  */
    printf("%06o ^ %06o = %06o\n", a, b, a^b);      /* XOR */
    printf("~%06o = %06o\n", a, ~a);                /* NOT */
}
```

This program produces the following output:

```
002436 & 000065 = 000024
002436 | 000065 = 002477
002436 ^ 000065 = 002453
~002436 = 175341
```

The bitwise shift is not a cyclic shift, so note that bits shifted out at the end of the operand are lost with zeros being inserted at the other end.

When integer values are shifted right, ones will be inserted only if it is a negative value (so that the sign is retained).

Examples:

5 << 2 produces 20
5 >> 2 produces 1
–5 << 2 produces –20
–5 >> 2 produces –2

5.6 Conditional expression operator

<expression1> ? <expression2> : <expression3>

Result: <expression2>, if <expression1> is true;
<expression3>, if <expression1> is false.

Example:

Assigning the maximum of two values:

```
max_a_b = (a >= b) ? a : b;
```

Note that, as with the other operators, the result is the stronger of the data types when the data types of <expression2> and <expression3> are different.

5.7 Assignment operators

= += −= *= /= %= >>= <<= &= ~= |=

The meanings of these operators are explained in subsection 6.1, "The assignment".

5.8 Precedence of operators and execution sequence of operations

<table>
<tr><th>Operator</th><th>Execution from/to</th></tr>
<tr><td>() [] -> .</td><td>left / right</td></tr>
<tr><td>! ~ ++ -- - (<type>) * & sizeof</td><td>right / left</td></tr>
<tr><td>* / %</td><td>left / right</td></tr>
<tr><td>+ -</td><td>left / right</td></tr>
<tr><td><< >></td><td>left / right</td></tr>
<tr><td>< <= > >=</td><td>left / right</td></tr>
<tr><td>== !=</td><td>left / right</td></tr>
<tr><td>&</td><td>left / right</td></tr>
<tr><td>^</td><td>left / right</td></tr>
<tr><td>|</td><td>left / right</td></tr>
<tr><td>&&</td><td>left / right</td></tr>
<tr><td>||</td><td>left / right</td></tr>
<tr><td>? :</td><td>right / left</td></tr>
<tr><td>= += -= etc.</td><td>right / left</td></tr>
<tr><td>,</td><td>left / right</td></tr>
</table>

Example:

Evaluation of an expression that might represent the conditions in an if statement:

```
Expression:   -(n++ * 3) != -a + b % 2
Values:       -(2++ * 3) != -4 + 5 % 2
1st step:     -(2   * 3) != -4 +   1     ; n = 3
2nd step:     -     6    !=   -3         ; n = 3
3rd step:                 1              ; n = 3
```

6 The C statements

The C language has only very few statements. The definition statement was introduced in Chapter 5, "Types, storage classes, variables and constants". All other types of statement will be discussed in this chapter.

6.1 The assignment

Assignments were introduced in Chapter 2, "C programming for the beginner".

The normal assignment operator is "=" as is shown in the following assignment:

```
a = b + 3;
```

As in other languages, the righthand expression is evaluated first, i.e. the value of b + 3 is computed. Then this value is written to the memory location of the lefthand variable.

The lefthand argument can be a 'char', 'short', 'int', 'long', 'unsigned', 'float' or 'double' variable or an element of an array, structure, union or bit field. Furthermore, the lefthand side may consist of pointer variables of any type or complete structure variables. Not allowed are complete arrays. The class of syntax elements that may appear to the left of the "=" in an assignment are termed **lvalue**.

6.1.1 Value of an assignment

The special feature of an assignment statement in C is that the assignment has a value after it has been executed. The value is the one that is assigned to the lefthand variable. If, in our example, b was equal to 5, the assignment returns the value 8. This value can be used in other expressions. The resulting expression is sometimes hard to read, for example

```
c = ( a = b + 3 ) - 6;
```

but it is valid. The first thing this statement does is to assign the value of b + 3 to the variable a. Then 6 is subtracted from the assignment value and the result assigned to c. You can see that this single statement changes the values of both a <u>and</u> c.

This feature is often utilized for tests in the if or while statements, in particular to test for the end of input.

Example:

```
while ((c = getchar()) != EOF)   ...
```

EOF must have previously been assigned the (machine-dependent) end of file character (this can be done in the header file of the standard input/output library <stdio.h>; viz. the chapter "Input/output"). The operator "!=" means "not equal".

In this statement, the function getchar is executed first. It returns a character that is written to the variable c. The assignment

c = getchar()

is thus set to the value of this character, which is then tested for end of file. If it is not the end of file character, the subsequent loop will be executed; if it is end of file, processing continues with the statement following the loop.

It is necessary to parenthesize c = getchar() because the relational operator "!=" has a higher precedence than the assignment operator "=".

6.1.2 Type conversion in assignments

In the same way, during the evaluation of expressions, the assignment of a value to a variable to the left of the operator "=" may require a type conversion if the type of the variable and the type of the expression value are different. Since conversion in assignments follows different rules from those for expressions, all the possible conversion situations with a description of each conversion are summarized in the following table (where "signed int" corresponds to the types 'char', 'short', 'int' and 'long', and "integer" corresponds to all integer types including 'char').

<Variable> (lvalue)	<Expression> (rvalue)	Other conditions	Conversion
signed int	signed int	variable type is inferior to expression type	The high-order bits are truncated, leaving the value possibly changed
signed int	signed int	Variable type is superior to expression type	The sign bit is multiplied, transferring the value correctly
unsigned	signed int	Variable type is inferior to expression type	The high-value bits are truncated, leaving the value possibly changed
unsigned	signed int	Variable type is superior to expression type	The high-order bits are padded with zeros, leaving the value possibly changed
signed int	unsigned	Variable type is inferior to expression type	The high-order bits are truncated, leaving the value possibly changed
signed int	unsigned	Variable type is superior to expression type	The high-order bits are padded with zeros, transferring the value correctly

<Variable> (lvalue)	<Expression> (rvalue)	Conversion
float, double	integer	Zeros are assumed as fractional positions, transferring the value correctly
integer	float, double	The fractional positions (in fixed-point notation) are truncated, leaving the value possibly changed
float	double	The value is rounded so that seven significant digits are kept correctly
double	float	The low-order bits of the mantissa are padded with zeros, transfering the value correctly

This table clearly shows how dangerous it is to use different types in an assignment because only in a very few cases is the value transferred correctly.

Example:

```
main()    /* Type conversion in assignments */
{
    unsigned short u;
    long l1 = 1000000,   l2 = -1;
    char ch = -1;

    u = l1;
    printf("l1 = %ld,   u = %u\n", l1, u);
    u = l2;
    printf("l2 = %ld,   u = %u\n", l2, u);
    u = ch;
    printf("ch = %d,   u = %u\n", ch, u);
}
```

This program produces the following output:

```
l1 = 1000000,   u = 16960
l2 = -1,   u = 65535
ch = -1,   u = 255
```

6.1.3 Compound assignment operators

You probably know from other programming languages the "paradoxical" statement

```
x = x + 3;
```

which means that the value of x is incremented by 3 and then written back to the location x.

The C language provides the special assignment operator "+=" for this situation, allowing you to use the following notation for the statement:

```
x += 3;
```

If this notation is used, one memory access may be saved depending on the implementation.

Similar compound assignment operators exist for the operators

```
- * / % >> << & ^ |
```

Example:

```
...
int m = 8, n = 6;

m *= 2 + 3;
printf("m = %d\n", m);
m += n += 4;
printf("m = %d,  n = %d\n", m, n);
...
```

produces the following output:

```
m = 40
m = 50,  n = 10
```

Note that the righthand operand of the compound assignment operator is as though it were parenthesized.

```
m *= 2 + 3;
```

Is the same as

```
m = m * (2 + 3);
```

and not the same as

```
m = m * 2 + 3;
```

The second part of the example illustrates that multiple compound assignments are processed from right to left just the same as simple assignments.

In earlier C versions, the notation of compound assignment operators was reversed:

=+, =−, =* etc.

This notation is still allowed by some compilers, for reasons of compatibility. It should, however, be avoided because in some cases (e.g. x=−3;) it is not clear whether this is a compound assignment operator or a simple assignment operator and a unary operator.

6.1.4 Increment/decrement assignments

Another type of assignment is also a C specialty and, besides, does not even look like an assignment because there is no assignment operator in it:

```
x++:
++x;
x--;
--x;
```

These assignments represent the increment and decrement operations for integers (++ for increment, — for decrement) you know from Assembler languages. Whether you write it as a prefix or postfix is only relevant if the value of the assignment is also used in an expression:

```
y = ++x;
y = x++;
```

If x was 5 in the beginning, it will be 6 upon execution of each of the above statements. The value of the assignment ++x is 6, i.e. the value of x after incrementation; the value of x++, on the other hand, is 5 because x is not incremented until the value has been transferred.

Note that an incrementation in an expression containing the AND or OR operator is not always executed because the result is often known upon evaluation of the first operand, eliminating the need for C to process the second operand. Example:

```
++m || ++n
```

6.2 Compound statements and block structure

Syntax:

```
{
    [ <definitions> ]
    <statement-1>
    <statement-2>
    .
    .
    .
    <statement-n>
}
```

The compound statement can be substituted in other statements at any point where any statement is to be inserted (example: if statement).

In addition, C enables you to use the compound statement to form blocks of statements with re-defined variables that are only known to that block. If there exist variables with the same name outside the block, they will remain unaffected, but cannot be referenced within the block.

Example:

```
main()
{     /* outer block */
    int i=2, j=4;

    printf("Outer block: i = %d, j = %d\n", i, j);
    {     /* inner block */
        int i=5;

        printf("Inner block: i = %d, j = %d\n", i, j);
    }
    printf("Outer block: i = %d, j = %d\n", i, j);
}
```

The program produces the following output:

```
Outer block: i = 2, j = 4
Inner block: i = 5, j = 4
Outer block: i = 2, j = 4
```

Within the inner block, the variable i of the outer block is overridden by a variable of the same name, and this is why i = 5 in the printf statement of the inner block. The variable j is not overridden in the inner block, and therefore it is known to the inner block as defined in the outer block and retains that value.

6.3 The if statement

Conditional execution of statements

Syntax:

if (<expression>) <statement1>

[**else** <statement2>]

If the expression is different from 0, statement1 will be executed; if it is equal to 0, statement2 will be executed (if present; otherwise processing continues with the next statement). This feature can be used for testing numeric values, and will also work for logical expressions.

<Statement1/2> can be a compound statement, i.e. a sequence of individual statements enclosed in braces.

Note that the keyword "then" used in some other languages before <statement1> is not allowed in C.

Be cautious when using nested if statements. The "else" always refers to the last "if" unless a different relationship is established by braces. You should make a habit of enclosing statements that belong together in braces to provide easy-to-read code.

Example:

```
...
if((c = getchar()) != 'm')
   if(c == 'a')
   {
      avg = (a + b) / 2.0;
      printf("Average = %f\n",avg);
   }
   else
      if(c == 'p')
      {
         pro = a*b;
         printf("Product = %d\n",pro);
      }
      else printf("Illegal entry!\n");
else
{
   max = a>b ? a : b;
   printf("Maximum = %d\n",max);
}
...
```

In this program segment braces and indentations are used to clarify which else refers to which if.

If more than two cases need to be distinguished, the if statement can be used in the following format:

if (<expression-1>)
    <statement-1>
else if (<expression-2>)
    <statement-2>
    .
    .
else if (<expression-n>)
    <statement-n>
[**else** <statement-n+1>]

The statement associated with the first non-zero expression is executed. If no expression is unequal to 0 (or "true"), the statement under else is executed or, if this is not present, nothing is executed.

6.4 The switch and break statements

Multi-way branch

Syntax:

```
switch ( <expression> )
{
      case <constexpr-1>: ...
      case <constexpr-2>: ...
      .
      .
      .
      case <constexpr-n>: ...
      [ default           : ... ]
}
```

The ellipsis symbol ... can be replaced with nothing or one or more statements. Only the last ... in the switch statement must be replaced with at least one statement.

The constant expressions listed in case may be a result of the type 'char', 'short' or 'int' (= executable type 'int'). Note that in a compound expression each operand must be a constant or a symbolic constant (see subsection 14.4.1, "Substitution of names (constant definition)"). <Expression> must also yield an 'int' value.

When the value of the expression matches one of the listed constant expressions, this point of the switch statement is the entry point, and all subsequent statements within the switch statement will be executed, including those listed under other constant expressions. If none of the values of the constant expressions matches the value of <expression>, the entry point is at "default", if present, and otherwise nothing is executed. Default should, but does not have to, be the last case.

Note that each switch statement must contain at least two cases.

Example:

```
main()    /* Switch demonstration */
{
    char c = 'b';

    switch(c) {
        case 'a': printf("Case a)\n");
        case 'b': printf("Case b)\n");
        case 'c': printf("Case c)\n");
    }
}
```

Since the value of C in case 'b' causes control to be transferred to the switch block, the program produces the following output:

```
Case b)
Case c)
```

The entry point list of the switch statement can be made a real control switch by using the break statement.

Syntax:

break;

The break statement causes the current switch, for, while or do while block, including the statements enclosed in braces, to be exited. If several such blocks are nested, the innermost loop or switch statement is left.

If, in a switch statement, you want exactly those statements to be executed whose expression value matches the corresponding constant, the last statement of the statement sequence controlled by a case must be the break statement.

Semantically, a break statement is not required on the last case, but it is a good precaution against errors when the switch statement is extended later.

Example:

```
...
switch (c) {
    case 'c': cir = 2 * rad * 3.14159;
              printf("Circumference: %f\n",cir);
              break;
    case 'a': ar = rad * rad * 3.14159;
              printf("Area: %f\n",ar);
              break;
    default:  printf("Incorrect input!\n");
              break;
  }
...
```

If c contains a 'c', the circumference is computed and output, if it contains 'a', the area is output, and in the event of all other characters, the message "Incorrect input!" is output.

6.5 The while statement

Loop with test for continuation before execution

Syntax:

while (<expression>)
<statement>

The <expression> is tested to see if it is non-zero or "true". If so, the statement is executed. Then the expression is tested again, and the procedure continues until the value of the expression is 0 or "false". Then control is transferred to the next statement. If the expression was equal to 0 at the beginning, control is transferred immediately without the statement being executed.

Example:

```
...
while(i++ < lim)
{
    sum += i;
    num++;
}
...
```

6.6 The for statement

Loop with test for continuation before execution

Syntax:

for (<expression1>; <expression2>; <expression3>)
<statement>

The for loop is an enhancement of the while loop in that you can specify initialization values before the first loop execution as well as an expression that is executed after each loop execution. For is normally used when the loop is controlled by a control variable.

The classic for loop is:

```
for (i = 1; i < n; i++) ...
```

The for loop can nevertheless be used with the same semantics as the while loop. Syntactically, the for statement corresponds to the following statements including while:

```
<expression1>;
while ( <expression2> ) {
        <statement>
        <expression3>;
}
```

This paraphrase clearly shows the execution sequence of the for statement:

From the point of view of the syntax, the first statement to be executed is <expression1>;. This is called an <expression> because the ";" is explicitly written while the semicolon is automatically included in an expression <statement>.

Then the continuation condition <expression2> is tested to see if it is true (>0). If so, the (compound) <statement> is executed, and then <expression3>; is executed which is normally used to change an operand of the continuation condition <expression2>. This statement is the end of the loop, and the continuation condition is tested again before another execution. If the continuation condition is not met (=0), the loop is exited or, at the first time, not executed at all.

Example:

```
...
for (i = init; i <= lim; i++)
{
    sum += i;
    num++;
}
...
```

Generalizing the for statement

The three expressions within the for statement are optional (but note that all semicolons are syntactically mandatory). If <expression1> or <expression3> is omitted, initialization or the corresponding statement in the loop is eliminated. If <expression2> is omitted, the continuation condition is always "true" with the loop executing infinitely. In this situation you must exit the loop by a (conditional) break or return statement to avoid a processor deadlock.

Also note that <statement> may be an empty statement consisting only of the ";".

Conversely, <expression1> and <expression3> may consist of several expressions separated by commas. This results in several initializations or several expression statements at the end of loop. <Expression2> should be a simple expression, but you can connect several conditions by logical operators.

Examples:

Infinite loop:

```
...
for (;;)
   ...;
...
```

Loop without loop body:

```
...
for (i=0,j=0; i< n; s[i] != ' ' ? (t[j++]=s[i]) : 0, i++)
   ;
...
```

(Suppressing spaces while transferring strings)

6.7 The do while statement

Loop with test for continuation after execution

Syntax:

do
    <statement>
while (<expression>);

As opposed to the while and for statements, this statement causes the loop to be executed at least once, which is sometimes (though seldom) meaningful.

This statement is the semantic equivalent of the repeat until loop from Pascal, but note that in C execution continues when the <expression> is "true", while in Pascal the loop terminates in this case.

Example (itoa):

```
itoa(n, s)      /* Place integer in string */
char s[];
int n;
{
    int i = 0;

    do
        s[i++] = n % 10 + '0';
    while ((n /= 10) > 0);
    s[i] = '\0';
    reverse (s);
}
```

This function uses another function named "reverse" which reverses the contents of the string s.

6.8 The continue statement

Initiate another loop execution

Syntax:

continue;

The continue statement must be controlled by a for, while or do while loop but, in contrast to the break statement, cannot be applied to a switch block. It causes the current loop execution to be terminated immediately and, in the event of while or do while, causes the continuation condition to be tested immediately, which may be followed by another loop execution. With a for statement, the <expression3> is executed before that (see subsection 6.6, "The for statement").

When multiple loops are nested, the continue statement only affects the innermost loop.

Examples:

```
...
for(j = 0; j < lim; j++)
{
    if(j == 4)
        continue;
    printf("%d ", j);
}
putchar('\n');
...
```

```
...
while(j++ < lim)
{
    if(j == 4)
        continue;
    printf("%d ", j);
}
putchar('\n');
...
```

Note that in contrast to this, the break statement discussed before always causes the loop to be exited.

Example:

```
...
for(j = 0; j < lim; j++)
{
    if(j == 4)
        break;
    printf("%d ", j);
}
putchar('\n');
...
```

6.9 The return statement

Exit a function

Syntax:

return [<expression>];

The return statement returns control and a function value to the calling function. If the value of <expression> is of a type different from that with which the function was defined, type conversion is performed. Specification of a value can be omitted if the function value is not evaluated by the caller. If it is nevertheless tested, its contents are treated as garbage (without an error message).

If a function does not include a return statement, control is returned upon execution of the last statement. The function value is then also undefined.

Example:

```
main()    /* Output the maximum of two integers */
{
    int a, b;

    printf("Enter 2 numbers:  ");
    scanf("%d %d", &a, &b);
    if(a != b)
        printf("Maximum: %d\n", max(a, b));
}

max(x, y)
int x, y;
{
    if(x > y)
        return(x);
    return(y);
}
```

The termination statements you have available can be summarized as follows:

continue	Termination of a loop execution and continue with the next loop execution
break	Exiting a loop or a switch statement
return	Exiting a function and returning the function value

6.10 The function call statement

Syntax:

<fnctname>([<actual-parameter-list>]);

Functions are called in expressions with the returned function value being used again. If the function value is irrelevant or if the function is of the 'void' type, the function can be called by a separate statement. This should only be done if the function modifies external variables or has other side effects.

6.11 Labels and the goto statement

Branch to a symbolically referenced point

Syntax:

goto <label>;

<label> ::= <name>:

Program control is transferred to the statement following the <label>. Each <label> must be followed by a statement, but note that one statement may be preceded by several labels.

The goto statement should and can be avoided in every situation. The only nearly justified case for using goto is to exit deeply nested loops and compound statements in a single action in the event of errors or program abort.

6.12 The dummy statement

Syntax:

;

A useful example of the dummy statement was given under the extended for statement. The dummy statement can, of course, be used wherever the syntax requires a <statement>. Another example would be where the end of a compound statement is to be marked by a label. You would code a dummy statement after the label because each label must be followed by a statement.

7 Functions and the program structure

Functions in C are the means used to modularize and simplify programs. When complex processes and computations are performed, you do not usually want to know how the result is obtained; the only thing you want to know is the result itself. And this is exactly what a function call does: the computation of the result is accomplished by the function external to the calling module, and the result is made available to the calling module. The result does not necessarily have to be a single function value as you will see in Chapter 8.

It is up to the calling function to decide whether the function value is evaluated or whether only the other effects of a function are utilized. Thus, functions in C are not only used to determine values, but also provide the same capabilities as procedures in Pascal.

The limitation to the structure category of "functions" simplifies programming in C, like many other features of the language, but at the same time it requires more attention with respect to the transfer and use of values in various program modules.

7.1 Notation and position within the program

A function has the following general structure:

```
[<storage-class>] [<type>] <name>( [<formal-parameter-list>] )
[<definition-of-the-formal-parameters>]
{
    [<definition-of-the-local-variables>]

    <statements>
}
```

The only admissible storage class is "static". Functions of this class cannot be accessed from other source files.

<Type> is the type of the function value (returned value) and may be omitted if it is 'int'. The admissible types for function values include only the basic types as well as pointer and structure types, and in addition the function-specific type 'void' which means that the function does not return a value. <Name> is the name of the function that must be specified to call it from another function.

Example of a typical function:

```
double cylvol(rad, height)          /* Cylinder volume */
double rad, height;
{
    double ar;

    ar = rad * rad * 3.14159;
    return(ar * height);
}
```

A function can be defined anywhere in the program outside other functions. The definition is therefore treated as external, just like a variable definition outside functions.

Functions must, like external variables, be declared in other functions that use them. The "extern" option, however, is only required when the function was defined in a different source file. Parameters are omitted in the declaration, and the function name is only followed by a pair of parentheses.

Examples:

```
...
double sqrt(), average();
char *find(), lim();
...
```

There is only one situation in which a function needs no declaration because it is implicitly declared. The following conditions must be met:

- The definition and use of the function must be contained in the same source file.
- The function definition must appear before its use.

Conversely, each non-declared function is taken to be a function that returns an 'int' value.

A function is called in the following way:

<function-name>([<actual-parameter-list>])

The call can be issued either in the form of an expression if the function value is used again, or as a standalone statement. In the latter case, no function value can be evaluated.

Examples:

```
ar = area(rad);

if(area(rad) * height < 33.33) ...

volume = vol(height, area(rad));

reverse();
```

In the first three examples, the function "area" is called within an expression. Note that it is even possible for a function call to be the parameter of another function call (third example). In the fourth example, any function value that may be returned is not used.

Incidentally, the main program named "main" is also a function that is called by the operating system.

7.2 Control flow

When you have called your program by a command, the operating system will start it as a separate process. The control flow begins at the "main" function of your program. The main function retains control until it encounters a function call in a statement. The function call transfers control to the called function until that function either calls another function or returns control to the caller.

There are three varieties of returning control:

- Within the function, control encounters a return <expression>;. Control is returned to the caller, and the value of <expression> is returned as the function value.
- Within the function, control encounters a return;. Control is returned to the caller. The function value is undefined (garbage).
- Within the function, control encounters the end of the function. Control is returned to the caller. The function value is undefined.

Since any function will eventually return control to the caller, control will finally go back to the main function. If the program ends there or if the system function "exit" is called, the process is ended (note, however, that exit can be called in any function).

7.3 Transferring values to functions

When you need to supply values, you can either use external variables whose contents are available to any function within a program, or you can transfer values via function parameters.

The transfer via parameters is straightforward and easy to control and is therefore preferable to the other method using external variables for transferring values. External variables should only be used where several functions must change the same data objects.

Parameters are separated by commas and enclosed in parentheses and then appended to the function name. In connection with function calls, we speak of actual parameters because each parameter is formed by an expression whose (actual) value counts. Therefore, a constant may definitely be an actual parameter.

In connection with function definitions, on the other hand, we speak of formal parameters. They are always represented by variables, but have neither a value nor a memory location before the function is executed. The type of the n-th actual parameter must be identical with the type of the n-th formal parameter, or you will be transferring undefined values. This implies that the number of each type of parameter must be identical. With respect to the parameter types you should, however, take into account that the non-executable types 'char', 'short' and 'float' are automatically converted to the (executable) types 'int' and 'double', respectively.

If the types of corresponding parameters do not match even after automatic conversion, it is your responsibility to convert the type of the actual parameter using the unary operator (<type>).

Example:

```
main()
{
    double sqan, sqrt();
    int number;

    ...
    sqan = sqrt((double)number);
    ...
}
```

The mathematical function "sqrt" requires a parameter of the 'double' type. Since the square root of an 'int' variable is to be computed, its value must be converted beforehand.

When the function is called, the values of the actual parameters are copied into the corresponding formal parameters of the function. This type of parameter supply is termed

"call by value".

Note that the function has no access to the objects themselves that form the actual parameters (no "call by reference"). This limitation can, however, be overcome by transferring addresses of variables. This is described in Chapter 8.

The limitation that the contents of arrays cannot be transferred as parameters can also be overcome by supplying their addresses (see Chapter 8).

7.4 Recursive functions

Functions may be used recursively, i.e. a function may call itself directly or indirectly. This would only be reasonable if on each call a state is changed so that at some time a uniquely defined final state is reached.

Recursive programming of functions will generally save program code and may improve readability, but it will never accelerate program execution. So you should think twice before using recursive functions.

Example:

```
revers()    /* Read string and output in reverse order */
{
    int c;

    if ((c = getchar()) != '\n')
    {
        revers();
        putchar(c);
    }
}
```

The character c is re-created and re-read at each call level. When the function encounters the end of input ('\n'), the last read character (except the '\n' itself) is printed, control goes back one level, and the last but one character is printed etc.

8 Pointers

A pointer is a data item that can accommodate the address of another memory item. Addresses can be formed by simple or compound data items, or by functions. Even a pointer itself can store the address of a pointer.

C provides, in contrast to comparable languages, totally free and easy manipulation of pointers, which are therefore of paramount importance in the C language. Pointers play an important part in C not only for indirect addressing that you may know from assembly languages, but above all in the addressing of arrays and in the transfer of function arguments.

8.1 Notation and the use of pointers

When using pointers, you will very often see and use the two unary operators * (contents of) and & (address of). Let ptr be a pointer variable for a specific type, and var, a normal variable of the same type. Then the following assignments would be meaningful:

```
var = *ptr;

ptr = &var;
```

The first line assigns the contents of the memory location pointed to by the pointer ptr to the variable var. The second line enters the address of the variable var into the pointer variable ptr.

The operator & can legally be used only for variables and array elements, while it is illegal for compound expressions or constants. Note also that the address must not be formed from register variables. The * operator may, of course, only be applied to pointer variables or expressions.

Pointer variables may point only to variables of a specific type, or you will produce undefined results. Therefore, pointer variables are defined with reference to the type to whose elements they can point.

Example:

```
int *iptr1, *iptr2;
```

To make a distinction from normal variables, the * operator precedes the variable, which is also correct from the semantic point of view: the contents of the object to which the pointer variable iptr1 points is of the integer type.

Pointer variables that point to a pointer of a specific type are defined with a dual * operator.

Example:

```
main()    /* Addressing using pointers */
{
    int i = 18, *ip, **ipp;

    ip  = &i;
    ipp = &ip;
    printf("ip   = %lo, &ip = %lo, ipp = %lo\n", ip,&ip,ipp);
    printf("*ipp = %lo, *ip = %d\n", *ipp, *ip);
    printf("&i   = %lo, i   = %d\n", &i, i);
}
```

The program produces the following output:

```
ip   = 73777424, &ip = 73777420, ipp = 73777420
*ipp = 73777424, *ip = 18
&i   = 73777424, i   = 18
```

The contents of the addresses that contain pointer variables can be used in calculations like normal variables, and constructs like *iptr can even be coded to the left of the "=".

Examples:

```
*iptr = length * 3.8;
*iptr *= *iptr2;
```

Pointer variables themselves can be used to a limited extent in arithmetic operations:

- Pointer variables may be assigned to other pointer variables (of the same type). Caution: if the pointer types do not match, you will not get an error message!
- Pointer variables that do not contain a defined value should be assigned a value of NULL.
- Pointer variables can be incremented or decremented.
- Pointer variables may be compared (==, !=, <, >, <=, >=). This includes the test for no contents (== NULL).
- Pointers may be increased or decreased by integer values (ptr +/– ivar).
- The difference between two pointers may be determined (ptr1–ptr2).

The special feature in incrementing or decrementing pointers is that the increment or decrement (1 integer) is not 1 byte, but, depending on the type to which the pointer refers, the overall length of a variable of that type ("scaled pointer arithmetic"). If, for example, an 'int' pointer is incremented, it will point to the next 'int' variable as compared to the previous state. You will learn in the next subsection how this feature is used in dealing with arrays.

8.2 Pointers and arrays

Access to array elements via pointers is usually faster and more efficiently programmed than access via array subscripting.

Let arr[10] be an 'int' array and ptr a pointer to variables of type 'int'. If you place in ptr the address of the first element of arr using

```
ptr = &arr[0];
```

then its contents can be accessed by either

```
arr[0]   or   *ptr
```

Likewise, arr[1] can be accessed by

```
*(ptr + 1)
```

because the pointer is incremented exactly by the length of an 'int' variable and the array elements are of the 'int' type. Any array element arr[i] then is *(ptr + i) because ptr + i is the address of arr[i].

Example:

```
char    carr[4] = "ABC";
double darr[3] = {1.2, 3.4, 5.6};
main()    /* Scaled pointer arithmetic */
{
   char   *cptr = &carr[0];
   double *dptr = &darr[0];

   for ( ; *cptr; cptr++, dptr++)
   {
      printf("*cptr: %c, cptr: %lo, ", *cptr, cptr);
      printf("*dptr: %g, dptr: %lo\n", *dptr, dptr);
   }
}
```

The program produces the following output:

```
*cptr: A, cptr: 40000004, *dptr:       1.2, dptr: 40000010
*cptr: B, cptr: 40000005, *dptr:       3.4, dptr: 40000020
*cptr: C, cptr: 40000006, *dptr:       5.6, dptr: 40000030
```

The relationship between pointers and arrays is actually even closer than suggested by the above notation. The array name is converted by the compiler to an address constant pointing to the first array element so that

```
ptr = &arr[0];
```

can also be written as

```
ptr = arr;
```

However, this assignment is not always necessary because, as you know, arr itself can be used as a pointer.

Thus, the following notations are equivalent:

```
arr[i]    and  *(arr + i),

&arr[i]   and  arr + i.
```

Rarely used, but allowed, is the subscripting of pointers. If ptr = arr; was assigned, then ptr[i] is equivalent to arr[i]. The only difference between array names and pointer variables is that an array name is not a variable but an address constant. Constructs like

```
arr++       or

arr = ptr;
```

are illegal.

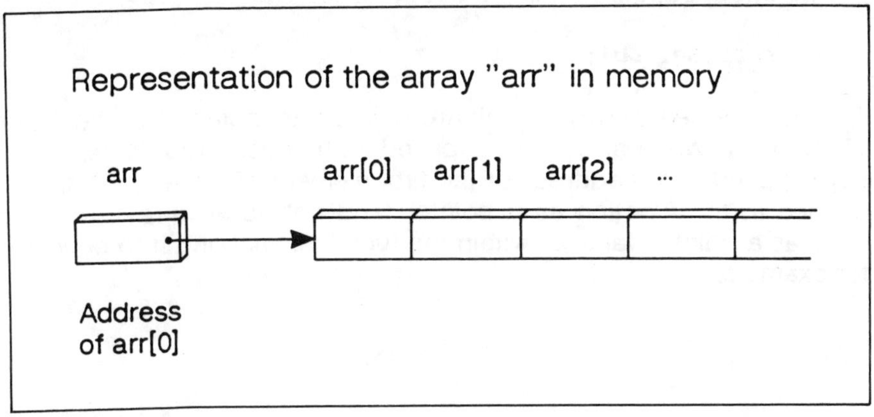

8.3 Pointers as function arguments

Since function parameters are in C transferred by copying their values ("call by value") it is not possible to change the contents of variables once transferred.

You can use pointers to bypass this limitation: you can transfer the addresses of the variables rather than the variables themselves. Then you can use these addresses to access the contents of the variables and change them.

The principal application for this feature is when a function produces more than one value (the function value) and supplies the values to the caller.

Example:

```
swap(p, q)    /* Swap two integers */
int *p, *q;
{
    int aux;

    aux = *p;
    *p = *q;
    *q = aux;
}
```

When this function is called, you must not specify integer variables as function arguments, but their addresses: if the values of a and b are to be swapped, the function must be called by

```
swap(&a, &b);
```

As you have seen, array names are pointers (constants) to the first element of the array. When arrays are supplied as function arguments, what is actually supplied is the address of the first element of the array. If the pointer is defined in the function as a pointer (and not as an array), it can even be used as a pointer variable within the function, enabling it to be incremented, for example.

Example:

```
strlen(s)    /* Function value, length of string s */
char *s;
{
   int n = 0;

   while(*s++)
      n++;
   return(n);
}
```

You can also define s by

```
char s[];
```

This is treated just like char *s;. (The length of a one-dimensional array can be omitted in a formal function parameter because it is accepted from the actual parameter at the time of call.)

8.4 Pointer arrays

Pointer arrays are mainly used where data objects of variable lengths are to be handled. The first address of each data object is stored in the array. You can walk through the entire data object by incrementing a pointer array element; you can access the next data object by incrementing the array subscript.

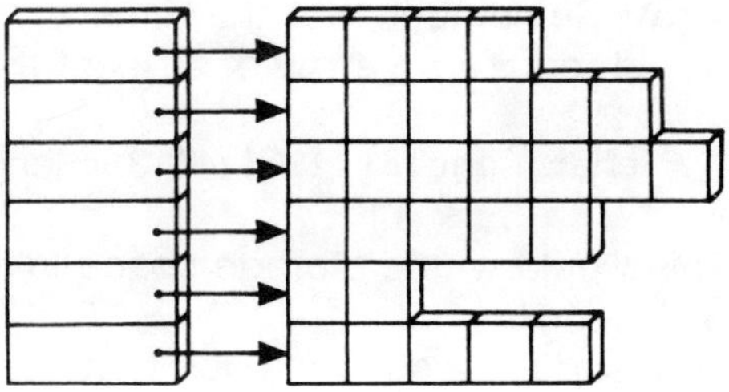

If the contents of the objects pointed to by the pointer array are not defined by assignments, they may be explicitly initialized. As in the case of normal arrays, you may omit the array length because it is derived from the initialization values.

Example:

```
char *day_name(n) /* Function value: pointer to the
                                      n-th week-day name */
int n;
{
    static char *name[] = {
        "illegal day",
        "Monday",
        "Tuesday",
        "Wednesday",
        "Thursday",
        "Friday",
        "Saturday",
        "Sunday"
    };
    return((n < 1 || n > 7) ? name[0] : name[n]);
}
```

The "static" option is required to enable the pointer array to be initialized in the function while preventing it from being re-initialized on each entry into the function.

Instead of this pointer array you might have used a two-dimensional 'char' array. Remember that the dimension lengths are constant in this case; this would require an array of at least the following size:

`char name[8] [13];` (13 = length of "illegal day")

This would waste storage space in most of the rows.

You should also consider that hash index access can easily be implemented with pointer arrays.

8.5 Pointers to functions

When discussing definitions, we noticed some similarities between functions and variables. In the same way, pointers can contain variable addresses or function addresses. It is, of course, not meaningful to do computations with function addresses or change them, but it is sometimes appropriate to transfer function addresses as function parameters.

The definition of pointer variables pointing to functions is a logical step in the C language, but the result is rather hard to read:

```
<type> (*<name>) ();
```

The return value of the function (shown by "()") which is the contents (*) of the address held in the pointer name <name> is of type <type>.

The parentheses around *<name> are required because if you omitted them (<type> *<name>()) you would define a function with the name <name> whose return value would be a pointer to variables of type <type>.

As in arrays, the function name itself is an address constant which points to the beginning of the function upon compilation.

Example:

A universal sort program that can arrange number arrays in numeric sequence and strings in alphabetic sequence.

The only thing that most sort algorithms do with array elements is compare and exchange them. Because numbers are in this example also stored in strings, the exchange function can be used in both cases. Note, however, that the numeric and alphabetic compare will in some cases return different results (example: consider the two numbers 9 and 10; 9 is less in a numeric compare, while 10 is less in an alphabetic compare).

You can solve this problem by writing one alphabetic and one numeric compare function. Depending on whether numbers or strings are to be sorted, the main program supplies to the sort function the pointer of either the compare function for numbers or that for strings.

Definition in main:

```
...
int strcomp(), numcomp();
...
```

Calling the sort function:

```
...
if (numeric)
    sort(lineptr, nlines, numcomp);
else
    sort(lineptr, nlines, strcomp);
...
```

Beginning of sort function:

```
sort(v, n, comp)    /* Sort function */
char *v[];
int n;
int (*comp)();
{
    ...
```

Within the sort function, you can now always call the compare function using

```
(*comp)(..., ...);
```

and the proper compare function will be used each time.

8.6 Function arguments of the main function

The main function of each program may contain two formal parameters: argc and argv. With these two parameters you can (theoretically) supply any number of parameters to the main program because argv is an array of pointers pointing to the actual function parameters, and argc is their number.

The way the function parameters are transferred depends on your specific operating system. In the Shell, when a program is called, the parameters are appended to the program name separated by spaces. In the program call

```
greeting john walter
```

greeting is the program name, and john and walter are two function parameters.

In the sense of the formal parameters of the main function, however, greeting is the first parameter and john and walter are more parameters. The number of parameters (argc) is 3, where

```
*argv[0] = "greeting"
*argv[1] = "john"
*argv[2] = "walter"
```

Example:

```
main(argc, argv)    /* Repeat arguments */
int argc;
char *argv[];
{
   while (--argc > 0)
      printf(("%s%c", *++argv, (argc > 1) ? ' ' : '\n');
}
```

(The program name itself is not printed.)

Another parameter of the main function is permitted under the operating system UNIX and the Shell. It is envp, which is an array of pointers to the environment of the user executing the program.

The array of pointers is terminated by a NULL pointer and contains entries of the form

<name>=<value>

Envp must be defined in the same way as argv in either of the following ways:

```
char *envp[];
```

```
char **envp;
```

Note that the names argc, argv and envp are commonly used, but are not required to be these particular identifiers.

Example:

Displaying the user environment

```
main(argc, argv, envp)    /* User environment */
int argc;
char **argv, **envp;
{
    while(*envp)
    {
        printf("%s\n", *envp);
        envp++;
    }
}
```

9 Structures

A structure in C is a combination of several variables of different types related by processing requirements or just having a logical relationship. The structures of the C language are named by the keyword "struct" and behave like the "records" of the languages PL/1, Pascal and Ada.

9.1 The use of structures

9.1.1 Defining structure types and structures

The keyword for structures is struct, optionally followed by a name for the structure type that is subsequently specified in braces.

Example:

```
...
struct date {
   int  day;
   char mon_name[3];
   int  year;
};    /* Type definition */
...
```

Semantically, this declaration just creates a new type named 'struct date'. Data items (structures) of this type are then defined and created by

```
...
struct date birthday, exam, wedding;
...                          /* Variable definition */
```

Note that the name "date" is the structure **type** name, and the names "birthday", "exam" and "wedding" are structure names (i.e. variable names).

You may combine the type and data definitions by appending the variable names to the structure type definition.

Example:

```
...
struct date {
    int  day;
    char mon_name[3];
    int  year;
} birthday, exam, wedding;
...
```

A combined structure type and variable definition is often done with structure types that are not assigned a structure type name.

When you define a structure type, you can specify structure elements of the type "structure". This means that structures can be nested. The structure types must, of course, not be identical.

9.1.2 Initializing structures

Structure initializations can be appended to data definition:

```
...
struct date birthday = { 18, "jan", 1956 };
...
```

Note that a structure initialization is an aggregate initialization so that you can initialize immediately in the definition only those structures that are defined externally or with the storage class "static".

9.1.3 Referencing structure elements

Operands in an expression cannot be entire structures, but only individual structure elements. Only in assignments is it allowed to specify entire structure variables to the left and right of the = sign.

The structure elements are referenced individually by

<structure-name>.<element-name>

Thus, the year of birth in the above example would be

```
birthday.year
```

or the first letter of the month of birth would be

```
birthday.month[0]
```

As opposed to array variables, entire structure variables can be transferred to and from functions both as function parameters and as function values.

9.1.4 Pointers to structures

Pointers to structures are an ideal means of creating incarnations of structures of a structure type. You can also use pointers to access structure elements.

Example:

The definition of a pointer that can point to a structure of the type 'struct date' looks as follows:

```
...
struct date *pbirthday; /* Structure pointer variable */
...
```

Subsequently you must assign to the pointer variable the address of the desired structure:

```
...
pbirthday = &birthday;
...
```

Address formation is one of the few operations permitted on entire structure variables.

Since after this sample assignment *pbirthday is identical with birthday, elements are referenced in expressions by

```
(*pbirthday).year
```

A special notation is designed for this purpose: (*pbirthday).year is equivalent to

```
pbirthday->year
```

General notation:

<structure-pointer>-><element-name>

Also note that the incrementation of a pointer to a structure increases the address by the size of the entire structure and not by the size of a single structure element. Therefore, this operation is only useful for arrays of structures.

9.1.5 Arrays of structures

Structure arrays are defined as follows, as you would expect:

```
...
struct date anniversary[10];    /* Array of structures */
...
```

An (aggregate) initialization of a structure array looks as follows:

```
...
struct date anniversary[] = {
    {  8, "Mar", 1953 },
    { 20, "Aug", 1949 },
    { 25, "Jul", 1952 },
    {  2, "Aug", 1955 },
    { 19, "Sep", 1950 } };
...
```

As with initialized arrays, you can omit the dimension size because it can be derived from the initialization values.

Note that, in common with all aggregate initializations, this initialization is only supported for an external definition or a definition with the storage class "static".

9.2 The sizeof operator

If you want to know, for some reason, the size of data items of a specific type (this may, for example, be the incrementation size of pointers of that type), you can use the sizeof operator.

Syntax:

> **sizeof(<type>)** or
> **sizeof** <expression>

This operator enables you to test the size of the data items of a specific type, or of variables or even of expressions. The value of the above expression is the size in bytes of its type.

This operator is most often applied to structures because their size is often difficult to determine due to the different data types and, besides, does not equal the sum of the sizes of the structure elements because these may on some systems have to begin at word boundaries. You can just as well apply sizeof to entire arrays, array elements or any expression because these also belong to a specific type.

For arrays (and this includes strings), the sizeof operator returns the defined size of the array even if it is only partially filled.

Example:

```
...
struct {
    int length, height;
    char descr[15], *descript;
} article;
...
fread(&article, sizeof(article), 1, adptr);
...
```

The input/output function fread (see Chapter 10, "Input/output") requires, as the second parameter, the size of the data element to be read, whose address is specified as the first parameter.

9.3 Self-referencing structure types

It is allowed in C to define structure types that contain pointers to structures of their own structure type. This provides you with an excellent means of managing lists or trees.

Example (list of strings):

```
...
struct list_elem {          /* List element structure */
   char *contents;          /* Element contents       */
   ...                      /* Other properties       */
   struct list_elem *successor;    /* Successor       */
};
...
```

It is not permitted for a structure to contain an element of the same structure type, but the line

```
struct list_elem *successor;
```

just defines a pointer to a structure of the same type, and this is, indeed, permitted.

The standard library function "malloc" that requests from the operating system the storage space for a list element, and pointer variables that point to the beginning and the end of a list provide you with a comprehensive list management mechanism.

The procedure for trees is similar. The difference is that a tree node in general contains several pointers to other tree nodes (two pointers in binary trees). Unused pointers contain a null value.

9.4 Fields

Fields are a special variation of structures that allow several bit variables to be stored in a location of an integer variable (e.g. flag tables).

The definition of fields is similar to that of structures and additionally specifies for each field element the length in bits.

Example:

```
...
struct {
   unsigned debug_flag: 1;
   unsigned err_flag  : 1;
   unsigned err_code  : 5;
} status;
...
```

This defines the variable "status", which contains three fields. The individual fields are referenced in the same way as in structures, for example:

```
status.err_code
```

and computations with fields are performed in the same way as with short unsigned integer variables.

Example:

Setting flags:

```
...
status.debug_flag = status.err_flag = 1;
...
```

Deleting flags:

```
...
status.debug_flag = status.err_flag = 0;
...
```

Testing flags and fields:

```
...
if(status.debug_flag) ...
...
if(status.err_code == 037) ...
...
```

Remember the following restrictions when using fields:

A field must fit into an integer variable, the individual fields have no sign, and the & operator (address of) must not be applied to fields.

9.5 Unions

Unions are defined syntactically in much the same way as structures, but they occupy only the storage space required by the longest element because they can at any one time contain only one of their defined elements.

Thus, unions are used to define a storage location to accommodate variables of different types. At the same time you should, however, define a variable or field element in which you can record which type is currently stored in the union.

Example:

```
...
int vtvp;                    /* Type of stored value */
union ifs {                  /* Union type declaration */
  int    ivalue;
  double dvalue;
  char   svalue[5];
} vvalue;                    /* Variable definition */
...
```

You can picture this union as follows:

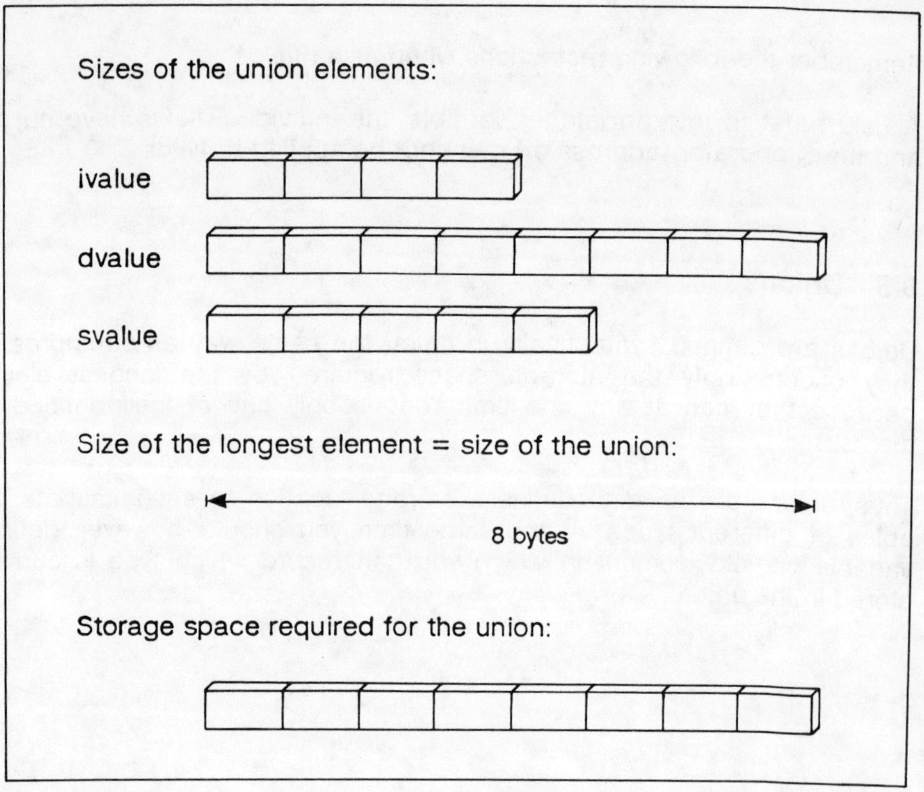

You can now assign to the union vvalue a value by assigning the value to the appropriate union element depending on its type. The most recent previous assignment is cancelled by the new assignment even if it referred to another union element.

Retrieving a value from a union is only reasonable if you access that union element that was last assigned a value. In order not to leave this to chance, you should remember the element in vtype.

Union elements are referenced the same way as structure elements:

<union-name>.<element-name> or

<union-pointer>-><element-name>

Examples of unions:

```
...
if (vtyp == INT)
    printf("%d\n", vvalue.ivalue);
else if (vtyp == FLOAT)
    printf("%f\n", vvalue.fvalue);
else if (vtyp == STRING)
    printf("%s\n", vvalue.svalue);
...
```

(For the definition of the constants INT, FLOAT and STRING refer to Chapter 14, "Operating the programming environment".)

10 Input/output

Input/output and file access are, strictly speaking, not supported by the C language, but there exists a standard library of input/output functions that is available in all operating systems under which C can run. In any portability considerations you can proceed on the assumption that these functions are available under other operating systems.

We wish to point out that the UNIX operating system offers an additional input/output interface that has been included in the C language set for the C programming systems running under UNIX. Utilization of these functions will, however, jeopardize portability.

The UNIX input/output interface is described in Chapter 12, "Interfaces with the UNIX operating system".

10.1 Standard input/output library and <stdio.h>

Each program wishing to use the input/output functions of the standard library must include the standard input/output header file at the beginning of each source file with

```
#include <stdio.h>
```

This header file has some constants and macros defined which are required during input/output. In addition, it declares all library input/output functions, eliminating the need for your program to declare these functions. A few of the most commonly used input/output functions (e.g. printf) do not need to be declared even if the header file is not present.

On some systems a "-ls" must be specified as a parameter in the cc command to include explicitly the standard input/output library.

Example:

```
cc erf.c ctex.c form.c -ls
```

Try out on your system whether this parameter is required or not on the cc command.

10.2 Input/output from/to standard input/output

10.2.1 Character input/output using getchar and putchar

getchar()

returns the next character from the "standard input".

putchar(<char>)

writes a character supplied as a parameter to the "standard output". Standard input, unless redefined, is the keyboard of your terminal; standard output is the screen of your terminal. How the standard input or output can be redirected is discussed in the "Shell" manual.

10.2.2 Formatted output using printf

The function printf is used for the formatted output of data (to standard output). Printf is called as follows:

printf(<control-string>, <expression-1>, ..., <expression-n>);

The first parameter of the printf function is a control string which must contain place holders, marked by %, for the subsequent parameters. The % must be followed by a letter identifying the type of the associated parameter:

Identifying letter	Type
d	Decimal number
o	Octal number
x, X	Hexadecimal number
u	Unsigned decimal number
c	Character
s	String
f	Fixed-point number without exponent
e, E	Floating-point number in exponential notation
g, G	e or f, whichever is shorter

x and X are different in that they output the characters a-f and A-F, respectively, for the hexadecimal digits; e and g write an "e" before the exponent in a floating-point number, while E and G write an "E".

Between the % and the identifying letter can occur size specifications, identifiers for left- or right-justified notation and other modifiers in the following sequence:

- A minus sign for left-justified notation (default: right-justified).
- A plus sign for sign output even if the value is positive.
- A space character for output of a space if the value is positive (default: nothing).
- A number sign (#) for octal output with leading 0, hexadecimal output with leading 0x or 0X, output with decimal point with the identifying letters e, E, f, g and G even if no fractional positions follow, or forcing after-decimal zeros with g and G (which are omitted otherwise).
- A number for the minimum output field length for the value of the parameter; if the representation of the value is shorter than this specification, the value is preceded by an appropriate number of spaces (if the number begins with 0, it will be preceded with zeros). If the representation of the value is longer than this specification, the output field length is automatically increased.
- (With s, f, g or G only) a separating point.
- (With s, f, g or G only). With s, the number of string characters to be displayed, and with f and g, the number of fractional positions.
- An "l" for 'long int' (in connection with d, o, x or X).

The identifying letters e, E, f, g and G cause six fractional positions to be output by default.

The control string is processed from left to right as follows: normal characters except % are transferred to the output "as is" in the control string. When a % symbol is encountered, the part from the % symbol to the identifying letter is replaced by the value of the next parameter not yet processed.

From this it follows that the number and type of the subsequent parameters must match exactly the number of % signs with the corresponding identifying letters in the control string. Type conversion is never performed with the printf function. If you want to change the type of a value, you must force its conversion to the other type.

Example:

The statement

```
printf("Area: %f, Material: %d\n", ar, (int)(ar+1));
```

produces the following output:

```
Area: 38.484478, Material: 39
```

The following example demonstrates the various output modes using printf:

```
main()    /* printf demonstration */
{
    float x1 = 12.34E-5, percent, part = .18;
    int n = 6, m = 3;
    long l = 1234567;
    char a = 'a', b = 'b';

    percent = 100. * part;
    printf("Sum of n + m: %d,   l = %ld\n", n+m, l);
    printf("x1 = %3.7f,  x1 = %.8E\n", x1, x1);
    printf("%.2f is %.0f%%\n", part, percent);
    printf("a = \'%c\', b = \'%c\'\n", a, b);
    printf("\"Not now!\" said Maria\n");
}
```

The program produces the following output:

```
Sum of n + m: 9, l = 1234567
x1 = 0.0001234, x1 = 1.23399994E-04
0.18 is 18%
a = 'a', b = 'b'
"Not now!" said Maria
```

Note that the % sign is output where the control sequence "%%" is coded. Where single or double quotes are to be output, precede them with a \.

The output of negative integer values is only possible with the identifying letter d; with o, x and X, the bit pattern of the memory location is represented in octal or hexadecimal notation, respectively.

With each identifying letter you must specify the associated parameters in the form of direct variables, constants or expressions whose values are to be output. The only exception is the identifying letter s which must be given a ('char') pointer as a parameter. This can be built as an array name, string constant, pointer variable or expression with a value of the type 'char *'. Starting at the location to which the pointer refers, characters are output until the end character '\0' is encountered or the maximum number of characters to be displayed (if specified) is reached.

The special properties of string output are illustrated by the following example.

Example:

The string "Hammerfest" is to be output.

Format	Output produced
`%6s`	`Hammerfest`
`%-6s`	`Hammerfest`
`%12s`	`  Hammerfest`
`%-12s`	`Hammerfest`
`%12.6s`	`      Hammer`
`%-12.6s`	`Hammer`

10.2.3 Formatted input using scanf

The function scanf is the counterpart of printf: it causes formatted input via the standard input. Scanf is called as follows:

scanf(<control-string>, <pointer-1>, ..., <pointer-n>);

The first parameter of the scanf function is a control string which must contain place holders, marked by %, for the subsequent parameters. The % must be followed by a letter identifying the type of the associated parameter.

The control string is processed as follows from left to right: spaces, tab and newline characters are ignored both in the control string and in the input. Other characters except % are expected in the input "as is" in the control string.

When a % sign is encountered, an input field is expected with the format specified after the %, and its value is transferred to the next unprocessed parameter.

From this it follows that the number and type of the subsequent parameters must match exactly the number of % signs with the corresponding identifying letters in the control string. Scanf does not perform any type conversion.

Each of the subsequent parameters must be a pointer variable because only pointers enable the contents of variables in a function to be changed (see Chapter 8). One of the most common mistakes in C (which the author himself has committed several times) is the specification of normal variables as parameters of the scanf function.

Identifying letter	Type
d	Decimal number
o	Octal number
x	Hexadecimal number
c	Character
s	String
e, f	Fixed- or floating-point number
[...]	String consisting of specified characters

The identifying letter "u" (for 'unsigned') is also allowed, but it allows no sign. Both can be used for inputs of type 'int' and 'unsigned'. Hexadecimal numbers must be input without leading 0x or 0X.

The identifying letters "e" and "f" are both suited for the input of fixed- and floating-point numbers (in contrast to their meaning in printf).

The difference between "s" and "[...]" is as follows: on normal string input with "s", characters are accepted until a space, tab or newline character is encountered. On input with "[...]", characters are accepted as long as they are contained in the set of characters enumerated in the brackets. The first input character that is different from any of the enumerated characters terminates the input. Space, tab and newline characters are not ignored in this case. If the first character within the brackets is a "^", the effect is reversed, causing the input to be terminated on the first input character that matches any of the characters within the brackets.

Between the % and the identifying letter can be coded length specifications and identifiers for value suppression:

- A number for the maximum field length.
- A "*" to ignore the value (which means that none of the subsequent parameters corresponds to this place holder).
- An "l" for 'long int' (in connection with d, o or x) or for 'double' (in connection with e or f).
- An "h" for 'short int' (in connection with d, o or x).

An input field is limited by the length specified in the control string or by a space, tab or newline character.

Examples:

```
scanf("%d %f %s %d", &i, &x, name, &j);
```

with the following input line

```
123     12.34e1       Andrew       123.456
```

assigns the value 123 to i, the value 123.4 to x, the value "Andrew" to name and the value 123 to j ("." does not belong to an 'int' value. ".456" remains in the standard input and is handled by the next input command.)

```
scanf("%2d %f %*d %2s", &i, &x, name);
```

with the input line

```
123        1234        Andrew
```

assigns the value 12 to i (field length 2), the value 3.0 to x (end of field defined by the space after 3), skips 1234 and assigns the value "An" to name (field length 2).

Note:

Scanf terminates the transfer when the input does not match the control string or when execution of the control string is complete.

10.2.4 Formatting in memory

We want to mention here that there are two functions that are similar to printf/scanf and that format strings in the memory rather than formatting values from/to standard input/output. These functions are

sprintf(<output-string>, <control-string>, <expression-list>**);**

sscanf(<input-string>, <control-string>, <pointer-list>**);**

They have the same effect as printf/scanf except that they have a different target and source, respectively.

An important application of these two functions is the conversion of strings containing images of numeric values into real numeric values and vice versa.

Sprintf also provides a simple means of initializing, assigning and concatenating strings.

Example:

```
char *sptr1 = "FirstPart";
char *sptr2 = "SecondPart";
main()        /* sprintf demonstration */
{
   char str[60];

   sprintf(str, "%s", "Initial");      /* Initialization */
   printf("%s\n", str);
   sprintf(str, "%s", sptr1);          /*     Assignment */
   printf("%s\n", str);
   sprintf(str, "%s%s", sptr1, sptr2); /* Concatenation */
   printf("%s\n", str);
}
```

produces the following output:

```
Initial
FirstPart
FirstPartSecondPart
```

10.3 File access

Besides the redirection of standard input and output to files performed by the operating system, it is possible to access files directly.

10.3.1 UNIX file philosophy

Under the UNIX operating system, input and output are always performed by reading or writing files. Even the standard input and output discussed above and all peripheral devices as well as the pipes used for inter-process communication are seen as files by the operating system and can therefore be consistently programmed in a device-independent manner.

In contrast to most other popular operating systems, UNIX has only a single type of file – the sequential file. Each file consists of a non-structured sequence of characters (bytes). If you attempt to read beyond the end of file, you will get the end of file character (EOF – predefined in <stdio.h>). Files are dynamically expanded. Space allocation or preformatting is therefore not required.

Reading and writing are also performed sequentially and are controlled by an internally managed file pointer which, after each read or write operation, points to the next character to be processed (writing is usually performed at the end of file). There is a special command, for disk files only, which allows the file pointer to be positioned to any point, thus enabling random access.

In contrast to the unstructured files, UNIX provides a powerful, hierarchically structured file system. Sophisticated access is supported by the (logical) search for files in the file system, and fast access is supported by the physical search within a large file.

Order of file access

Before you can access a file, it must be opened. However, there is no rule without exception: the standard input, the standard output and the standard error output are automatically opened by the Shell, saving you three commands.

Then the read and write accesses to the file are performed together with any file pointer positioning.

A file should always be closed after the last access because otherwise it will remain open until the end of program, it will occupy storage space and will not be protected from inadvertent overwriting.

10.3.2 Opening files

A predefined function is available for opening files. The opening function returns a pointer to a file (more precisely: to a file label in memory) which is declared or defined as the type FILE in <stdio.h>. The function itself makes available the standard library of input/output functions.

You must define a pointer variable to FILE and open the file.

You need the following definition:

```
FILE *<filepointer>;
```

You may choose any name for the <filepointer>.

The (permanent) name of the opening function is fopen. The code to open a file looks as follows:

```
<filepointer> = fopen(<filename>, <mode>);
```

<Filename> is the name of the file to be opened. <Mode> is a string with the following contents:

"r"	read
"w"	write
"a"	append
"r+"	read and write
"w+"	read and write after deletion
"a+"	read and write at end of file

If a file that does not exist is opened for writing or appending, it is created at that time. If a file is opened with "w" or "w+", its old contents will be deleted.

Opening a file with "a" or "a+" prevents any overwriting of the existing file contents. Irrespective of where the file pointer currently is, writing is always at the end of file (data is appended) and the file pointer repositioned.

When a file is opened with "r+" or "w+", you can alternately read or write. If, however, you want to write after a read access, you must first explicitly position the file pointer (except when it is at the end of file). The file pointer must also be explicitly positioned when a write operation is followed by a read access.

If you attempt to open a file for reading that does not exist or for which there is no access permission, you will get an error status and, in <filepointer>, the NULL pointer. NULL is also defined in <stdio.h>.
You should check the return value of fopen in every case.

Example:

```
...
#include <stdio.h>
FILE *datptr;
...
if((datptr = fopen("names", "r")) == NULL)
{
    printf("File names cannot be opened\n");
    exit(1);
}
...
```

At the time a program is started, three files are automatically opened (so you do not need to open them in your program) and can be referenced by the following file pointers (which you do not need to declare either):

stdin	Standard input (usually your keyboard)
stdout	Standard output (usually your screen)
stderr	Standard error output (usually your screen)

10.3.3 Character input/output using getc, putc and ungetc

Getc and putc have the same effect for files as getchar and putchar have for the standard input and output. The statement

<charvar> = **getc**(<filepointer>);

transfers into the 'char' variable <charvar> the next character from the file to which the file pointer refers.

The function call

putc(<char>, <filepointer>)

on the other hand, outputs 'char' to the file to which the file pointer refers. (<char> may be any expression which produces a character.)

Note:

Getc and putc are implemented in the standard library as macros and not as genuine functions. The expansion of the following expressions would therefore not be deterministic:

```
getc(*f++), putc(c,*f++)
```

For such cases the (somewhat slower) genuine functions named **fgetc** and **fputc** are available to you whose syntax and semantics are identical with those of getc and putc.

<charvar> = **fgetc**(<filepointer>);

fputc(<char>, <filepointer>)

ungetc

The function ungetc is a special feature of the C language. You can use it to put an already read character back into the input file. On the next sequential read from the file, this character will again be the first character read. One character can be put back into each file at any one time.

The function value of ungetc is EOF if the character itself was EOF or if it is not possible for other reasons to put the character back.

Syntax:

ungetc(<char>, <filepointer>**)**

Note:

Ungetc can be used in conjunction with any input functions (except getc).

Access to memory words

Syntax:

<intvar> = **getw(**<filepointer>**)**;

putw(<intvalue>, <filepointer>**)**

These two functions are used to transfer memory words (whose length is machine-dependent) to/from a file. Since a memory word can accommodate exactly one integer number, the function value of getw and the parameter of putw are 'int' values.

When using either of these functions be sure not to refer to word addresses.

10.3.4 Line input/output using gets, puts, fgets and fputs

Syntax:

gets(<string>**)**

fgets(<string>, <max-number>, <filepointer>**)**

puts(<string>**)**

fputs(<string>, <filepointer>**)**

gets and fgets

Gets reads the next entire input line from the standard input file (stdin) up to the newline character into the <string>. An EOF character will also end the process.

Fgets reads the next input line from a file including the newline character into the <string>. The number of characters is, however, limited to <max-number>−1, which is usually the length of the <string>. Both functions append an end character ('\0') to the transferred characters. The function value of gets and fgets is usually the pointer to the <string>. At end of file, the function value is NULL.

puts and fputs

Puts writes the <string> to the standard output file stdout, while fputs writes the <string> to the file that is referred to by the <filepointer>. The <string> does not need to contain a newline character. The function puts automatically appends a newline character to the transferred string, while fputs does not.

Both functions return EOF when they fail.

10.3.5 Field input/output using fread and fwrite

Syntax:

fread(<address>, <field-length>, <no-of-fields>, <filepointer>)

fwrite(<address>, <field-length>, <no-of-fields>, <filepointer>)

You use fread to read <no-of-fields> fields with a length of <field-length> bytes into <address>.

Fwrite has the reverse effect. It writes the contents of the fields that you have previously stored at <address> to the file referred to by the <filepointer>.

Both functions terminate when any error occurs (fread also terminates at EOF), and both functions return the number of the actually transferred fields.

10.3.6 Formatted input/output using fprintf and fscanf

Fprintf and fscanf have the same effect for files as printf and scanf have for the standard output and standard input, respectively. The file pointer that points to the file to which the command refers must be specified in fprintf and fscanf as the first parameter before the control string:

fprintf(<filepointer>, <control-string>, <expression-list>);

fscanf(<filepointer>, <control-string>, <pointer-list>);

10.3.7 Random access: positioning and testing the data pointer

The functions presented in the preceding subsections operate sequentially, but you can manipulate the file pointer that is managed in the process. This possibility enables you to access files randomly.

Syntax:

fseek(<filepointer>, <offset>, <reference-point>**)**

rewind(<filepointer>**)**

fseek

The first parameter is the file pointer you obtained when opening the file. The <offset> is specified relative to the <reference-point> in number of bytes and must be a 'long int' constant, variable or expression. The <reference-point> can be specified in one of the following ways:

<reference-point>	Meaning
0	Beginning of file
1	Current file pointer
2	End of file

The <reference-point> must be of type 'int'.

Example:

```
...
fseek(fp, 0L, 2);          /* End of file */
...
```

This statement positions the file pointer at the end of file.

The return value of fseek is non-zero if positioning failed, and zero otherwise.

rewind

The rewind function positions the file pointer to the beginning of file and is thus identical to

```
fseek(<filepointer>, 0L, 0);
```

ftell

The function

```
ftell(<filepointer>)
```

is used to determine the current position of the file pointer. The function returns a 'long int' value giving the offset in bytes of the file pointer from the beginning of file.

10.3.8 User-defined input/output buffers and unbuffered input/output

Every file input/output including input/output from/to standard input and standard output are buffered for reasons of efficiency. Only the standard error output (stderr) is not buffered because errors are to be displayed immediately when they occur.

If you want to provide the input or output buffer yourself, for example to examine it from time to time, you must define a 'char' array (string) of the length BUFSIZ (BUFSIZ is a constant defined in <stdio.h>) and call the function "setbuf":

```
setbuf(<filepointer>, <pointer>)
```

<pointer> refers to the 'char' array (name of the string).

The function must be called after a file is opened and before the first read/write access.

If input/output is to be accomplished totally unbuffered, you must call the setbuf function with a <pointer> value of NULL.

10.3.9 Flushing the output buffer

Syntax:

fflush(<filepointer>)

Fflush immediately writes all buffered data for the file referred to by the <filepointer> to the file. The file remains open.

Note:

This function is implicitly called when files are closed.

10.3.10 Closing files

Syntax:

fclose(<filepointer>)

Fclose cuts the connection from the file pointer to the file. At the same time, if output is buffered, the buffer is flushed. The file pointer variable can be reused to open another file.

Note:

All open files are automatically closed at the end of a program.

The closing of one file and the subsequent opening of another file can be accomplished in a single command:

freopen(<filename>, <mode>, <filepointer>)

The file to which the <filepointer> refers is closed, and then the file named <filename> is opened. The <filepointer> points to the new file and is supplied as the return value of the function.

The old file is closed in any case even if opening of the new file failed. In the event of a failure, the function returns a NULL value.

10.4 Buffering in temporary files

Data that is produced by one program to be reused later or further processed by another program can be buffered in "temporary files" which are removed when the data has been extracted from them.

The system supports two different tools for temporary files:

- The system-supported creation and opening of a temporary file that is automatically removed at the end of the process;
- The creation of a file name that does not yet exist in the system. You can use this name to create and open a file in your program. You are responsible for removing this file.

10.4.1 System temporary files

The function

tmpfile()

creates a temporary file. The return value of this function is the file pointer to a temporary file. You can use the file pointer in connection with the usual input/output functions. Once created, the file is open for reading and writing.

When the current process ends, the file is automatically removed.

10.4.2 User-managed temporary files

The most difficult problem you are faced with when managing temporary files yourself is not to use the name of an existing file because this would destroy the contents of the existing file.

The C standard library therefore provides three functions for finding names:

<pointer> = **mktemp(**<string>**)**;

<pointer1> = **tmpnam(**<pointer2>**)**;

<pointer> = **tempnam(**<directory>, <prefix>**)**;

mktemp

When using the "mktemp" function, you must specify in <string> a mask for the file name for which the function returns a pointer. The contents of <string> must begin with the desired file name followed by six X's. The mktemp function replaces the six X's with a letter and the current process ID. The letter is selected so that duplicated file names are avoided.

tmpnam

When using the function "tmpnam", you can specify in <pointer> the address of a string variable in which the temporary file name is to be returned. The same pointer (i.e. the string itself) is in this case the return value of the function. The string variable must at least have the length L_tmpnam (defined in <stdio.h>).

If, however, <pointer2> has a value of NULL, an internal static memory area for the temporary file name is created and its address returned in <pointer1>. The next call of tmpnam will, however, destroy the contents of this memory area.

Tmpnam returns a complete path name. The temporary file is listed in the directory defined by P_tmpdir in <stdio.h> (usually /tmp or /usr/tmp).

tempnam

The function "tempnam" allows the directory to be specified as a parameter (string variable). If, however, <directory> is NULL or not a valid path name to the directory, the directory defined by P_tmpdir is also used in this case. If that directory is also not defined or not accessible, then "/tmp" is substituted for <directory>.

This order of search is overridden if the user calling (the program) has in his environment defined the Shell variable TMPDIR whose value is the path name of his temporary file directory and which is supplied instead of <directory> to the function tempnam.

The second parameter <prefix> is either NULL or a pointer to a 'char' array containing up to five characters with which the temporary file name is to begin.

The return value of the function is a pointer to the file name or, if it fails, NULL.

Further handling of the temporary files

Once you have obtained a file name for your temporary file using one of the above three functions, you can create and open the file with the function fopen. The opening mode must be neither "r" nor "r+". Now you can write or read the file with the usual input/output functions.

It is your responsibility to remove the temporary file after use. This can be done via the Shell or from C using the system call "unlink" (see Chapter 12, "Interfaces with the UNIX operating system"). There is no standard library function for removing files.

10.5 Data transfer between processes via pipes

Besides the data transfer between programs via temporary files, the UNIX operating system provides a much more elegant method of transferring data from one program to another: pipes.

A pipe causes the standard output of one program to be used as the standard input for another program.

Besides simplifying programming, this gives you the advantage of faster execution because both programs run concurrently and synchronize each other via the pipe (the data transfer channel).

In this connection, the programs are also called "processes" because a separate process is started for each executing C program and because this mechanism can just as well be used for commands, command procedures or mixtures of these three types, which all start separate processes.

Note:

> The pipe programming discussed here is required if either or both processes involved are C programs and the pipe is to be started not by a Shell command but by the C program.

10.5.1 Opening pipes and starting the second process

Syntax:

popen(<command-string>, <mode>**)**

This command creates and opens a pipe and starts the Shell command, command procedure or C program specified in <command-string> (the contents of <command string> must syntactically correspond to a Shell command line). The direction of transfer from or to the command is defined by a one-character string in <mode>:

<mode>	Meaning
"r"	Your program reads the standard output from command
"w"	Your program writes the standard input to command

The return value of the function is a file pointer that you can use immediately for reading from or writing to the pipe by means of the usual input/output functions.

Caution:

If the call in <command string> is a C program, the input or output corresponding with the pipe must, of course, not be programmed as a pipe because this program references the standard output or input.

10.5.2 Closing pipes

A pipe opened with popen should always be closed with

pclose(<file pointer>**)**

because only this function waits until the corresponding process has ended and returns its end of process status.

10.6 Testing for status and errors

Although with most input/output functions the return value indicates the success or failure, you can explicitly test to see if any error has occurred during input/output. In addition, you can test input files to see if the EOF (end of file) character was read:

ferror(<filepointer>)

feof(<filepointer>)

The return value of each function is 0 if no error and no EOF character has occurred, otherwise it is non-zero.

Caution:

The two status codes are automatically set but not cleared.

If, for example, an error has occurred, the error status is set to a value other than 0 (irrespective of the current state). If your program tests the error status only sporadically, you will not know during which of the recent input/output statements the error has occurred. It is therefore recommended to test after each input/output command.

After each test that detected a status other than 0, you must clear the status code with the function

clearerr(<filepointer>)

This function clears to 0 both the error and the EOF status.

11 Standard library functions

Besides the input/output functions described in the preceding chapter there are a number of other standard library functions that make programming in C easier and much more comfortable.

These functions can be classified as follows:

- Functions for the classification of characters and for the conversion of characters and numbers
- Functions supporting string programming
- Mathematical functions
- Functions for managing and controlling data and data structures
- Memory management functions
- System interfaces with the standard library
- Functions providing you with information about your system environment

11.1 Conversion and classification functions

11.1.1 Classification of characters

Before any character classification function can be used in a program file, the following preprocessor statement must be included:

```
#include <ctype.h>
```

All classification functions are implemented as macros, eliminating the need for function declaration.

Syntax:

```
<classification-function>(<char>)
```

The function value table for each function is as you might expect:

Function value	Meaning
=0	Character not contained in the class
≠0	Character is contained in the class

Each function has a parameter of the type 'int' or 'char'. The functions are applicable to all characters and to 'int' values that would be valid ASCII codes of characters (0 through 127) as well as EOF. The function isascii is defined for all 'int' values.

To avoid undefined results, you should first call the function isascii if you are not sure; if this function executes successfully, you can call any other function.

Function	Character class
isalpha	Letters
isupper	Uppercase letters
islower	Lowercase letters
isdigit	Decimal digits
isxdigit	Hexadecimal digits
isalnum	Letters and digits
isspace	Space, tab, CR, newline, vertical tab and form feed characters
iscntrl	Control characters (codes 00 through 037, 0177)
ispunct	Punctuation characters (neither letters/digits nor control characters)
isprint	Printable characters (codes 040 through 0176)
isgraph	Graphic characters not including space
isascii	ASCII characters (codes 00 through 0177)

11.1.2 Conversion functions

11.1.2.1 Character conversion functions

You need the same preprocessor statement that was mentioned above in connection with the classification functions:

#include <ctype.h>

Syntax:

toupper(<char>)

_toupper(<char>)

tolower(<char>)

_tolower(<char>)

toascii(<intvalue>)

Toupper and _toupper return the lowercase equivalent of an uppercase letter; if a character other than a lowercase letter is supplied, toupper will leave it unchanged, while _toupper will produce an undefined result.

_toupper is implemented as a macro and therefore executes faster, but on the other hand _toupper only produces usable results if a lowercase letter is supplied as the parameter. The function toupper, on the other hand, will accept 'char' and 'int' values from −1 through 255 as parameter (in values greater than 128 the high-order bit will be ignored).

Tolower and _tolower return the uppercase equivalent of lowercase letters. The differences between the two functions are the same as those discussed for toupper and _toupper.

The function toascii sets all bits in the parameter value except for the seven low-order bits to 0 so that a valid ASCII character is returned.

11.1.2.2 Converting strings to numbers

Converting strings to integers

atoi(<string>)

atol(<string>)

strtol(<string>, <pointer>, <base>)

The function "atoi" converts a (decimal) integer contained in <string> into an 'int' value. The 'int' value is returned as the function value. Any leading space, tab or newline characters in the <string> are ignored, and leading zeros in the integer do not affect the result. The digits may be preceded by a sign.

The end of the number is detected by the first character other than a digit.

The function "atol" has the same effect as atoi except that it returns a 'long int' value (this is important for the declaration!). It should therefore be used for long integers.

The function "strtol" provides you with a conversion tool that is more powerful in two ways:

- You can convert an integer with any base between 2 and 36 to a 'long int' value.
- The function returns the address of a pointer referring to the first character after the integer in the <string> (so you can continue processing the string).

Since the <pointer> itself is changed by the function, it must be defined with a dual * operator (or with a single * operator, and then its address must be specified as a parameter).

The contents of the <pointer> must be unequal to (char **)NULL, or it will not be changed.

If the <base> was specified as 16, the function ignores not only leading zeros but also a leading "0x" or "0X".

You may also specify a <base> = 0; in this case the base will be derived from the initial digits of the integer:

- A leading 0 initiates an octal number (<base>: 8).
- A leading 0x or 0X sets the <base> to 16.
- In every other case the base is set to 10.

If it is not possible to form an integer value from any of the functions, the value 0 will be returned, and in the case of strtol the **<pointer> refers to the beginning of the <string>.

Converting strings to floating-point numbers

strtod(<string>, <pointer>)

atof(<string>)

The function "strtod" converts a fixed- or floating-point number contained in <string> to a 'double' value. The 'double' value is returned as the function value. Leading space, tab or newline characters within the <string> are ignored, and leading zeros do not affect the result. The digits may be preceded by a sign. The number may contain a decimal point and an exponential part initiated by "e" or "E".

The end of the number is detected by the first character other than a digit or e/E. If the value of <pointer> is other than (char **)NULL, an indirect pointer to the first character following the number stored in the string is returned in it.

If the <string> does not conform to this convention, the function value is 0, and the <pointer> points (indirectly) to the beginning of the <string>. If a value produces an overflow, the predefined constant HUGE will be returned.

The function "atof" is equivalent to

strtod(<string>, (char **)NULL).

Note:

You should also note the conversion of strings to numbers using sscanf.

11.1.2.3 Converting numbers to strings

ecvt(<double-value>, <no-of-digits>, <dec-point>, <sign>**)**

gcvt(<double-value>, <no-of-digits>,<string>**)**

The function "ecvt" converts the value of the 'double' expression supplied in <double-value> to a string of <no-of-digits> significant digits, with the low-order digit possibly rounded.

The decimal point is not stored in the string, but its position relative to the first digit in the string is furnished in the 'int' pointer <dec-point>. A negative value means that the decimal point is that many positions before the first significant digit.

Likewise, the sign is not stored in the string but passed in the 'int' pointer <sign> as follows:

*<Sign>	Meaning
=0	positive number
≠0	negative number

The return value of the function is a 'char' pointer to a static memory area in which the function stores the string of significant digits.

In the function "gcvt" you specify a <string> in which the floating-point number is to be stored (the function value being the pointer to this <string>).

In contrast to ecvt, gcvt supplies an immediately printable floating-point number, possibly signed, with a decimal point and, if necessary, with an exponent. You use the parameter <no-of-digits> to specify how many significant digits the number is to have. Note that the <string> size must be longer because it is not only the digits that are transferred.

Note:

You should also see how numbers are converted to strings using sprintf.

11.2 String handling functions

Although there is no "string" data type in the C language, strings play a special role in C as they do in other languages, which is quite natural because of the frequent use of this data structure. Remember that strings in C are 'char' arrays terminated by the null character '\0'.

A comprehensive set of functions makes string handling so easy that you will certainly not miss the data type "string".

Programs wishing to use these functions must include the header file <string.h>:

#include <string.h>

The functions are listed and briefly described in the following.

strcat(<string1>, <string2>**)**

strncat(<string1>, <string2>, <number>**)**

These two functions concatenate <string2> to the end of <string1>, with strncat concatenating no more than <number> characters. Since the two functions do not check <string1> for overflow, it must be defined large enough. The function value is a pointer to <string1>.

strcmp(<string1>, <string2>**)**

strncmp(<string1>, <string2>, <number>**)**

These two functions compare <string1> and <string2> lexicographically, with strncmp comparing only up to the <number>th character. The following table gives the function value:

Condition	Function value
<string1> < <string2>	<0
<string1> = <string2>	=0
<string1> > <string2>	>0

strcpy(<string1>, <string2>**)**

strncpy(<string1>, <string2>, <number>**)**

Strcpy copies <string2> to <string1>. The copy ends when the null character has been transferred from <string2>. The function strncpy copies exactly the specified <number> of characters. If <string2> is shorter, zeros will be padded in on the right. Neither strcpy nor strncpy perform an overflow check, so be sure to dimension <string1> large enough.

strlen(<string>**)**

The function "strlen" returns the current length of <string> (excluding the zero character at the end) as an 'int' value.
You can use the sizeof operator to test the defined length of the array which holds the string.

strchr(<string>, <character>**)**

strrchr(<string>, <character>**)**

These functions return a pointer to the first (strchr) or last (strrchr) occurrence of <character> in <string>. Since the zero character is counted in the <string> in this case, it can also be interrogated. If the <character> does not occur within the <string>, the function returns the NULL pointer.

strpbrk(<string1>, <string2>**)**

This function returns a pointer to the first character in <string1> to match any character from <string2>. If there are no matching characters, you will get a NULL pointer.

strspn(<string1>, <string2>**)**

strcspn(<string1>, <string2>**)**

These functions return 'int' values specifying the length of the initial part of <string1> that consists only of characters from <string2> (strspn) or that consists only of characters different from <string2> (strcspn).

strtok(<string1>, <string2>)

The function "strtok" considers <string1> as a sequence of fields that are separated from each other by one or more characters from <string2>. You can retrieve the fields by several successive calls of the function strtok which each time returns a pointer to the field in question and automatically appends a null character to the field.

Note that it is only allowed in the first manipulation of <string1> to specify this string as a parameter. In each subsequent manipulation of <string1> you must specify a NULL pointer because the position is remembered internally.

When no more field is found in <string1>, the function value is NULL.

11.3 Mathematical functions

The mathematical functions are collected in a standard library. Each program wishing to use one of these functions must contain the statement

#include <math.h>

and the option **-lm** in the cc command to ensure that the standard library is available during compilation. In addition, a header file is used to declare all functions as well as the constant HUGE which is returned by some functions when the function value is too large to be represented or infinite.

Mathematical functions always give results with the greatest possible precision by returning function values of the 'double' type.
A few functions return 'int' or other values because of their special properties. These special cases are indicated in the following descriptions.

11.3.1 Arithmetic functions

Absolute value

abs(<int-value>**)**

fabs(<double-value>**)**

Abs and fabs leave a positive value unchanged and convert a negative number into a positive number of the same amount. Abs is applicable to integers, whereas fabs is applicable to floating-point numbers; the function values are of the 'int' or 'double' type, respectively.

Modulo function

fmod(<double-value1>, <double-value2>**)**

This function returns the same result as the modulo operator ("%"), but does so for 'double' values as parameters and gives a 'double' result. The result always has the same sign as <double-value1>.

If <double-value2> = 0, the function value is <double-value1>.

Gamma function

gamma(<double-value>)

Besides the usual function declaration, you must declare the external 'int' variable signgam:

```
extern int signgam;
```

The function value is ln | Γ (x) | (x = <double-value>) so that, if you want to obtain the mathematical gamma function, you must apply the exp function to the result. The sign is supplied in signgam.

Example:

```
...
y = gamma(x);                /* Standard function */
if(y > MAXG)
    error();
y = exp(y) * signgam;        /* Gamma function value */
...
```

On successful execution of this program segment, y will contain the value Γ(x).

Bessel functions

j0(<double-value>)
j1(<double-value>)
jn(<int-value>, <double-value>)
y0(<double-value>)
y1(<double-value>)
yn(<int-value>, <double-value>)

These functions are the Bessel (cylindrical) functions of the first and second type.

J0, j1 and jn are the Bessel functions of the first type of zeroth, first or n-th order, respectively. Y0, y1 and yn are the Bessel functions of the second type of zeroth, first and n-th order, respectively. The functions of the second type require a positive argument.

11.3.2 Transcendental functions

exp(<double-value>)

log(<double-value>)

log10(<double-value>)

pow(<double-value1>, <double-value2>)

sqrt(<double-value>)

Description of the functions:

- Exp is the exponential function to the base e (e = 2.71828128...) and returns the value

  e^x (x = <double-value>).

- The function log returns the natural logarithm (to base e) of <double-value> and is the inversion of exp:

  $_e\log(x) = \ln(x)$ (x = <double-value>).

  <double-value> must be positive.

- The function log10 returns the logarithm to the base 10 of <double-value>:

  $_{10}\log(x) = \lg(x)$ (x = <double-value>).

  <double-value> must be positive.

- The function pow is the substitute for the exponential operator used in some languages and returns the value:

  x^y (x = <double-value1>, y = <double-value2>).

  If x is negative, y must be an integer number.

- Sqrt returns the square root of <double-value>:

  $\sqrt{x}$ (x = <double-value>).

  <double-value> must not be negative.

11.3.3 Rounding functions

floor(<double-value>)

ceil(<double-value>)

Floor returns the next smaller and ceil the next greater integer to <double-value> as a 'double' value.

11.3.4 Trigonometric functions

sin(<double-value>)

cos(<double-value>)

tan(<double-value>)

asin(<double-value>)

acos(<double-value>)

atan(<double-value>)

atan2(<double-value1>, <double-value2>)

hypot(<double-value1>, <double-value2>)

sinh(<double-value>)

cosh(<double-value>)

tanh(<double-value>)

Sin, cos and tan are the normal sine, cosine and tangent functions. The parameter <double-value> must be specified as a radian measure.

Asin, acos and atan are the inverse functions arc sine, arc cosine and arc tangent. Asin and acos are only defined over the range $[-1,1]$; the function values of the three functions are in the ranges $[-\frac{1}{2}\pi, \frac{1}{2}\pi]$, $[0,\pi]$ and $[-\frac{1}{2}\pi, \frac{1}{2}\pi]$, respectively.

The function atan2 returns the value

atan(y / x)   (y = <double-value1>, x = <double-value2>),

whereas hypot returns the Euclidian distance

$$\sqrt{x^2 + y^2}$$

These two functions facilitate, for example, the translation of orthogonal coordinates into polar coordinates.

Sinh, cosh and tanh return the hyperbolic sine, hyperbolic cosine and hyperbolic tangent, respectively, of their argument.

11.3.5 Normal distribution functions

erf(<double-value>)

erfc(<double-value>)

Erf is the distribution function of the normal distribution with E = 0 and σ = 1 (error function):

$$\frac{2}{\sqrt{\pi}} \int_0^x e^{-t^2} dt \qquad \text{(with x = <double-value>).}$$

Erfc is the function 1 − erf(x). This construct allows the function value for large values of x to be output with a much greater precision.

11.3.6 Pseudo random number generator

rand()

srand(<unsigned-value>)

Rand generates ('int') random numbers in the range from 0 through $2^{15} - 1$. After 2^{32} calls of the function the same sequence of random numbers is repeated.

The function "srand" allows you to reset the random number generator to a specific starting point at any time. Without srand, the rand function generates the first random number from the value 1. Srand has no function value (type 'void').

11.3.7 Floating-point number manipulation

frexp(<double-value>, <int-pointer>**)**

ldexp(<double-value>, <int-value>**)**

modf(<double-value>, <double-pointer>**)**

These three functions return results of the 'double' type.

Frexp returns as its function value the mantissa, and in *<int-pointer> the exponent, of the number <double-value> if it is represented as

<double-value> = mantissa * pow(2, exponent)

The exponent is selected so that the absolute value of the mantissa is between 0.5 and 1.

Conversely, ldexp returns the function value

<double-value> * pow(2, <int-value>).

The function modf splits the <double-value> into an integer portion returned in *<double-pointer> and a fractional portion returned as the (signed) function value.

11.3.8 Error handling for mathematical functions

The functions of the mathematical standard library call the function "matherr" when an error is detected. You can utilize this feature by writing your own function with the same name and having the following function header:

```
int matherr(exptr)
struct exception *exptr;
```

The system supplies to this function a pointer to an exception structure containing the characteristics of the error. The structure type is declared in <math.h> and looks as follows:

```
struct exception
{
   int type;
   char *name;
   double arg1, arg2, retval;
};
```

The variable type describes the type of error and is an element of the following constant set:

Value	Name	Meaning
01	DOMAIN	Domain of definition error
02	SING	Argument uniqueness
03	OVERFLOW	Function value too long
04	UNDERFLOW	Function value too short
05	TLOSS	Total loss of significance
06	PLOSS	Partial loss of significance

The pointer "name" points to the name of the function that has caused the error. Arg1 and arg2 are the arguments with which the function was called. Retval is the return value of the function before it was changed by matherr.

If the user-defined matherr function returns the function value 0, the system in addition issues an error message and sets the external variable "errno" (see subsection 11.6.4, "System error messages").

If you have not defined your own matherr function, a standard function is called whose behaviour is described by the following table. In any case, EDOM or ERANGE (domain of definition or value range violation) is stored in errno and program execution continued.

Function errno	DOMAIN EDOM	SING EDOM	OVERFLOW ERANGE	UNDERFLOW ERANGE	TLOSS ERANGE	PLOSS ERANGE
Bessel function	–	–	–	–	M, 0	+
y0, y1, yn (x≤0)	M, –H	–	–	–	–	–
exp	–	–	H	0	–	–
log, log 10 (x≤0)	M, –H	M, –H	–	–	–	–
pow	–	–	±H	0	–	–
pow (x neg., y frctnl.)	M, 0	–	–	–	–	–
sqrt	M, 0	–	–	–	–	–
gamma	–	M, H	H	–	–	–
hypot	–	–	H	–	–	–
sinh, cosh	–	–	±H	–	–	–
sin, cos, tan	–	–	–	–	M, 0	+
asin, acos, atan2	M, 0	–	–	–	–	–

M means that a message is issued, and 0/H means that 0 or HUGE is returned as the function value. In the event of the partial loss of significance, as much as possible of the value is returned.

11.4 Managing data and data structures

11.4.1 Searching, updating and sorting lists

lsearch((char*) <pointer>, **(char *)** <base>, <no-of-pointers>,
<length>, <comp-fnct>**)**

lfind((char*) <pointer>, **(char *)** <base>, <no-of-pointers>,
<length>, <comp-fnct>**)**

qsort((char *) <base>, <number>, <length>, <comp-fnct>**)**

bsearch((char*) <pointer>, **(char *)** <base>, <number>,
<length>, <comp-fnct>**)**

The lsearch function linearly searches a list for an element to which the <pointer> refers. If an identical element exists, the function returns a pointer to that element. Otherwise, the element is added as the last element to the list and its address returned as the function value.

You can accommodate any type of element in your list because the syntax requires that pointers to list elements be explicitly converted to 'char' pointers. This is why you must explicitly convert the pointer returned to you as the function value into a pointer to your list element type.

<Base> points to the first element of the list, and *<no-of-pointers> stores the number of list elements (as an 'unsigned' value). A pointer is required here because the value may be incremented by 1.

To specify a compare function for your list elements, you must enter in <comp-fnct> the name of a function to which you pass two pointers to list elements and which returns the following values:

Condition	Function value
List-elem1 ≠ list-elem2	≠0
List-elem1 = list-elem2	=0

If you need to manage a list of strings, the function "strcmp", for example, will meet these requirements.

The compare function does, of course, not have to compare every byte of the list elements. There are special compare functions with which you can suppress portions of the list elements during the compare operation or compare only specific characteristics.

The versatility of the compare function thus makes list management with lsearch universally applicable.

The function "lfind" has the same effect as lsearch except that it does not enter an element that is not found but returns the NULL pointer.

The function "qsort" sorts your list in the quicksort mode. There is no function value (type 'void').

Note that you do not supply the <number> of list elements as a pointer but specify it directly, and that the compare function whose name you pass does not only detect mismatches but must also be able to decide which element is greater or smaller:

Condition	Function value
List-elem1 < list-elem2	<0
List-elem1 = list-elem2	=0
List-elem1 > list-elem2	>0

The function "bsearch" is used to perform a very fast binary search through your sorted list to find a specific element that you pass as the contents of <pointer>.

The function value returned will be a pointer to the list element, if it exists, and NULL otherwise. The function value is returned as a 'char' pointer, so be sure to convert it explicitly to a pointer to your list element type.

Note:

> If you have sorted your list with qsort, you must, of course, specify the same compare function <comp-fnct> for bsearch.

11.4.2 Managing hash tables

Hash tables support the fast access to large lists or memory areas in which a sequential search would be too time-consuming and sorting inappropriate because of constant updating.

To use hash table support, you must include a special header file:

```
#include <search.h>
```

This makes the following functions available to you:

```
hcreate(<number>)

hsearch(<entry>, <action>)

hdestroy()
```

With hcreate you create a hash table, specifying the maximum possible number of entries in <number>. Each entry consists of a pointer to a key characterizing the data item (e.g. a disk address in the virtual memory) and a pointer to the data item itself. This data type ENTRY is predefined in <search.h>:

```
struct ENTRY
{
    char *key;
    char *data;
};
```

You can, of course, use data types other than 'char'; if so, the pointer type must be explicitly converted when entering it.

Hcreate returns an 'int' value as the function value, which is 0 if the creation of the hash table has failed.

Entering and searching in the hash table is accomplished by the function "hsearch" to which is passed the <action> of either ENTER or FIND. ENTER and FIND are constants of the enumeration type ACTION predefined in <search.h> (they must not be enclosed in quotes).

For <entry> you supply the entry to be found or entered. The function value is a pointer to the found or added entry, and if no entry is found or the table is full, NULL is returned.

The function "hdestroy" clears the hash table. Hdestroy does not return a function value (type 'void').

Caution:

> Because only one hash table may be active at any one time, hdestroy must precede hcreate to create a new hash table.

11.4.3 Managing binary trees

If you want to manage data with the help of binary trees, you must also include the header file <search.h>:

#include <search.h>

This makes the following functions available to you:

tsearch((char *)<data-ptr>, **(char **)**<root-ptr>, <comp-fnct>**)**

tfind((char *)<data-ptr>, **(char **)**<root-ptr>, <comp-fnct>**)**

tdelete((char *)<data-ptr>, **(char **)**<root-ptr>, <comp-fnct>**)**

twalk((char *)<root-ptr>, <action>**)**

Tsearch is used to search for a tree node. If the node does not exist, it will be mounted in the proper position in the tree. Tsearch is also used to create the root for a new binary tree.

<Data-ptr> can be any type of pointer to a structure you have defined, and (char *) is used to cast this pointer explicitly to a 'char' pointer. In <root-ptr> you must supply an indirect pointer (address of a pointer) to the root node. If this pointer is NULL, the node you supplied in <data-ptr> is created as the tree root.

The compare function <comp-fnct> for your tree nodes must be specified as the name of a function that you supply with two pointers to tree nodes and which returns the following values:

Condition	Function value
Tree-node-1 < tree-node-2	<0
Tree-node-1 = tree-node-2	=0
Tree-node-1 < tree-node-2	>0

This function creates an ordering structure on your data which forms the basis for (fast) tree access.

The return value of the function tsearch is a ('char' type) pointer to the data element found or inserted. You may have to cast this pointer to your pointer type.

The function "tfind" has the same effect as tsearch except that, if the element is not found, it is not entered, but the NULL pointer returned instead.

Tdelete deletes a node from the tree structure. This may relocate the root node (if that node is deleted).

The parameters are the same as those for tsearch. The return value is a ('char') pointer to the predecessor node of the deleted node. If the node cannot be found in the tree, a NULL value is returned.

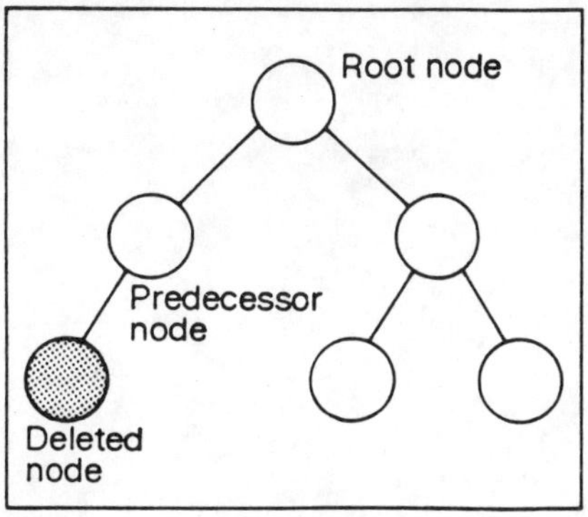

The twalk function allows you to walk through the binary tree from left to right and execute a function on each node.

In <root-ptr> you supply a (direct) pointer to the root node of the (partial) tree you want to walk through. Any node will do.

<Action> is a pointer to a function that is called each time a node is encountered. This function must be called with three arguments. The first argument is the address of the just visited tree node. The second parameter is a value of the enumeration type VISIT defined in <search.h> with the following constants:

Constant	Meaning
preorder	First encounter of the node
postorder	Second encounter of the node
endorder	Third encounter of the node
leaf	The node has no successors

The third parameter is the depth of the tree node (beginning with 0 for the root node).

The function to which <action> refers and the function twalk itself are of type 'void'.

11.5 Memory management functions

The functions described in the preceding subsection used to manage data structures in the form of lists, binary trees etc. internally use memory allocation functions which the user can access directly, if desired.

11.5.1 Allocating memory space

malloc(<length>**)**

calloc(<number>, <el-length>**)**

Malloc furnishes you with a ('char') pointer to a contiguous memory area of size <length> (in bytes).

The function calloc does even more: it allocates a memory area in which <number> data elements of size <el-length> will fit. In addition, it initializes the memory area to zeros.

All parameters to these functions must be supplied as 'unsigned' values.

If no contiguous memory area of sufficient size is available, each function will return a NULL pointer.

11.5.2 Expanding/contracting memory space

realloc(<pointer>, <length>**)**

This function is used to expand or contract to the new <length> a memory area whose <pointer> was returned to you by malloc, calloc or realloc. The function value is a 'char' pointer to the new memory area.

If the old memory area can not be expanded but a sufficiently large area is available in another location, the contents of the old area are automatically copied to the new area.

This function will also return a NULL pointer when sufficient space is not available.

11.5.3 Freeing memory space

Syntax:

free(<pointer>)

You can free a memory area acquired by malloc, calloc or realloc by executing the function free with the <pointer> returned by the allocate function used.

The memory area concerned is then available for re-allocation, although its contents are not automatically cleared.

Free does not return a function value (type 'void').

11.6 System interfaces with the standard library

11.6.1 Generating the IOT signal

The function

abort();

sends a SIGIOT signal to the process of the current program. SIGIOT is a software interrupt signal. The process will be aborted and a core file created with the current memory contents unless the signal is trapped or ignored (system call "signal").

The function "abort()" must be declared with the type 'int' and, in the event of non-abort, returns the same value as the system call "kill" (see "System calls" manual).

11.6.2 Preparing an execution profile

Syntax:

monitor(<address1>, <address2>, <'short'-array>, <size>, <number>)

"Monitor" is an interface with the Shell command "prof" and the system call "profil".

"Monitor" prepares for the recording of a histogram of periodically measured values of the program counter and the number of calls of specific functions indicated in <'short'-array>. <Size> is the size of this array. The lowest address of measured values is at <address1>, and the highest address just below <address2>. The two addresses must point to functions that return 'int' values (cast explicitly, if necessary).

Caution: <address1> must not be 0!

A maximum of <number> call counters can be recorded. Only the calls to functions that were compiled with the -p option of cc are recorded. <Size> and <number> must be specified with type 'int'. The standard library functions also maintain call counters when "cc -p" is used.

The <size> of <'short'-array> should only be smaller by a factor of 3 or 4 than the distance between <address1> and <address2>, and particularly so if compact and very frequently used routines are concerned.

For <address2> you may specify the external variable "etext" that points to the first memory address above the program code. This causes all function calls to be recorded upward of <address1>.

Example:

The following call provides you with a profile of the entire program:

```
...
extern etext;
...
monitor((int (*)())2, etext, sharray, length, no_of_fnct);
...
```

Note:

Besides "etext" you can declare and use two other external variables containing the first address after the initialized or non-initialized data area. They are edata and end (declarations: extern edata; extern end;).

Monitoring is ended and the results are written to the file mon.out by specifying 0 for all parameters in the function call:

```
monitor((int (*) ())NULL, 0, 0, 0, 0);
```

To evaluate the file mon.out, use the Shell command "prof".

Note:

An executable program generated by "cc -p" automatically contains calls of monitor with default parameters; you therefore do not need to explicitly to call monitor except when you want to monitor the profile of a program in a more detailed way.

11.6.3 Encryption

crypt(<string>, <modifier>)

The "crypt" function provides you with a password encryption facility.

The parameter <string> is the password to be encrypted (or any other string to be encrypted).
<Modifier> is a two-character string formed of the following letters and other characters: { 0-9, a-z, A-Z, . and / }. This string enables you to execute the encryption algorithm in any one of 4096 modes.

The function returns a pointer to the encrypted string and so must be declared as follows:

```
char *crypt();
```

Caution:

Each time crypt is called, the same memory location is used again, with the preceding value being overwritten.

11.6.4 System error messages

You use the function "perror" to output to the standard error output an explanation and the system error message concerning the latest error that has occurred on a system or library function call.

The following declaration is required:

```
void perror();
```

Syntax of the function call:

```
perror(<string>)
```

This function call first outputs the <string> (to the standard error output), followed by a colon, a space, the system message and a newline character.

You would normally place the name of the program or subroutine that caused the error in the <string>.

The operating system returns the error number of the last occurred error in the external variable "errno". When a system function has successfully executed, this variable is not changed and retains the last error number.

Declaration of "errno":

```
extern int errno;
```

You can obtain the system error messages via the following external table:

```
extern char *sys_errlist[];
extern int sys_nerr;
```

With "errno" specified as a subscript, you can retrieve the system error message (without newline character) from sys_errlist. You should make sure in advance that the subscript is not greater than "sys_nerr" because "sys_nerr" is the highest message number in the table.

11.6.5 Checking assertions

Syntax:

assert(<expression>**)**

To use this function, a special header file must be included:

#include <assert.h>

If, when this function is called, the <expression> (of type 'int') is false (0), the standard error output displays:

Assertion failed: <expression>, file <name>, line <number>

and the program aborts. <Name> is the name of the source file, and <number> the line number of the assert call.

The preprocessor statement

#define NDEBUG

before "#include <assert.h>" or the compiler option **-DNDEBUG** prevents assert calls from being compiled.

11.6.6 Causing a program to sleep

Syntax:

sleep(<seconds>**)**

The argument must be specified as an 'unsigned' value.

This function causes execution of your program to stop for the number of <seconds> at the point where the function is called. When the time has elapsed, execution is resumed with the subsequent statement.

Owing to the time-sharing characteristics of the operating system, the actual sleep time may be somewhat shorter or longer than the specified time. If the actual sleep time is shorter than the desired time, you will get the difference as an ('unsigned') return value of the "sleep" function.

11.6.7 Calling the Shell

Syntax:

system(<string>)

With this function you call the Shell from your program. <String> must contain a Shell command line as you would input it on the terminal in the command mode.

Program execution is interrupted until the Shell command has been executed. The function "system" returns the process end status of the Shell command ('int' value).

If no new process can be created or executed, the function system will return −1.

11.6.8 Swapping bytes

Syntax:

swab(<char-ptr1>, <char-ptr2>, <number>);

The function copies <number> bytes from the position pointed to by <char-ptr1> to the position pointed to by <char-ptr2>, swapping the bytes with even and odd byte addresses.

The <number> (type 'int') should be positive and even.

Swab does not return a function value (type 'void').

11.6.9 Directory scan

Syntax:

ftw(<string>, <fnct-ptr>, <depth>)

To use this function, a special header file must be included:

#include <ftw.h>

The function "ftw" scans the entire hierarchy of the directory tree whose root directory name is specified by <string>. For each encountered entry, ftw calls the function pointed to by <fnct-ptr>. In the process, it is supplied with a pointer to the entry name, a pointer to a stat structure and an 'int' value. The stat structure contains information about the entered object and is described in detail in the "System interfaces" manual. The 'int' value can assume the following constant values (defined in <ftw.h>):

Constant	Value	Meaning
FTW_F	0	File
FTW_D	1	Directory
FTW_DNR	2	Read-protected directory
FTW_NS	3	Undefined entry

In the case of FTW_DNR, the directory is not examined any longer, and in the case of FTW_NS, the stat structure contains garbage.

Ftw returns the value 0 when the entire directory tree has been processed successfully. If an execution of <fnct-ptr> returns a non-zero value, ftw will abort and return this ('int') value as its function value. If ftw fails, it will return −1.

<Depth> can be used to limit the number of hierarchy levels starting at the specified root directory. <Depth> must not be greater than number of files that you can still open at that time because ftw creates a file description for each level.

11.7 Miscellaneous information

11.7.1 CPU time

Syntax:

clock();

The function "clock" returns (as a 'long' value) the CPU time in microseconds used since this function was first called. The CPU time returned is the sum of the user and system CPU times for the current process and all completed child processes that were started by the function "system" or by a system call and whose completion was checked with the system call "wait".

Note that the time units used for measuring are normally longer than microseconds so that the last digit may be slightly inexact.

11.7.2 Login and terminal names

Syntax:

cuserid(<'char'-pointer>)

getlogin()

logname()

ctermid(<'char'-pointer>)

The function "cuserid" returns a 'char' pointer referring to the login name, represented as a string, of the user who has initiated the current process.

If the <'char'-pointer> you supply is NULL, the login name is stored in a static memory area whose address is returned. Otherwise, the <'char'-pointer> you supplied will point to the login name when the function is complete. It is assumed that it points to a 'char' array with a length of at least **L_cuserid** (defined in <stdio.h>).

If the login name cannot be found, the function will in the former case return a NULL pointer, whereas in the other case it places the end character '\0' in the first byte to which <'char'-pointer> refers.

The function "getlogin" also returns the login name, but searches for it in the file **/etc/utmp**. In conjunction with the function getpwnam you will in this way find the proper entry in the file **/etc/passwd** if a user has several login names. If getlogin is called from a process that is not assigned to a terminal it will return a NULL pointer.

The function "logname" returns a pointer to the login name from the user environment, i.e. the value of your Shell variable LOGNAME.

The function "ctermid" returns a 'char' pointer referring to the path name, represented as a string, of the terminal on which the current process is executing. You can use this string, for example, in the fopen function.

If the <'char'-pointer> you supply is NULL, the terminal name is stored in a static memory area whose address is returned. Otherwise, the <'char'-pointer> you supplied will point to the terminal name when the function is complete. It is assumed that it points to a 'char' array with a length of at least **L_ctermid** (defined in <stdio.h>).

The function

ttyslot()

returns (as an 'int' value) the subscript of the current program user's entry in the file **/etc/utmp**. The function also checks to see if the standard input/output and error output are linked with terminals. A return value of 0 indicates that no terminal name is found.

11.7.3 Date and time of day

To use the date and time functions, you must include a special header file:

#include <time.h>

This makes the following functions available to you:

ctime(<'long'-pointer>**)**

localtime(<'long'-pointer>**)**

gmtime(<'long'-pointer>**)**

asctime(<'struct tm'-pointer>**)**

tzset()

The function "ctime" converts the number of seconds that have elapsed since January 1, 1970, 0 hours Greenwich Mean Time and pointed to by <'long'-pointer>. The function value is a 'char' pointer referring to a 26-character string containing the current date and time in the following form:

Tue Aug 20 11:53:07 1985\n\0

The functions "localtime" and "gmtime" return pointers to structures of the type 'struct tm' which is defined in the above-mentioned header file. The current date and time are stored in this structure. The type 'struct tm' has the following format:

```
struct tm {
    int tm_sec;      /* Seconds (0 - 59) */
    int tm_min;      /* Minutes (0 - 59) */
    int tm_hour;     /* Hours   (0 - 23) */
    int tm_mday;     /* Day     (1 - 31) */
    int tm_mon;      /* Month   (0 - 11) */
    int tm_year;     /* Year - 1900 (0 - 99)      */
    int tm_wday;     /* Weekday     (Sunday = 0) */
    int tm_yday;     /* Day of year (0 - 365)    */
    int tm_isdst;    /* Summertime               */
}
```

Note that the months are counted from 0 through 11. Tm_isdst is non-zero in the summer time period.

Gmtime returns the Greenwich Mean Time, while localtime converts this time to your time zone.

The time zone difference (in seconds) is held in the external variable timezone which you can access directly provided you have made the following declaration:

extern long timezone;

The function "asctime" converts a tm structure to the 26-character output format mentioned in connection with the function ctime (function value: 'char' pointer).

If you have defined the variable TZ in your user environment, asctime uses its contents to overwrite the standard time zone. TZ contains a three-letter abbreviation for your time zone followed by a number representing the difference from the Greenwich Mean Time in hours.

The function "tzset" (type 'void') that is called by asctime but is also available on its own, uses TZ to set the above-mentioned external variable timezone as well as the external variable "tzname". Tzname must be declared as follows:

extern char *tzname[2];

Caution:

The function values of all these functions point to locations in the static memory area. Any values stored there are overwritten each time a function is called.

11.7.4 Current directory

The function "getcwd" gives you a 'char' pointer to the path name of the current directory.

Syntax:

getcwd(<'char'-pointer>, <size>)

If the <'char'-pointer> you supply is NULL, the function "malloc" is used to allocate a memory area of the size specified in <size>, the path name is placed in it and its address returned. You can use this address later in the "free" function. Otherwise, the <'char'-pointer> you supply will, upon execution of the function, point to the path name of the current directory. It is assumed that it points to a 'char' array whose length is at least <size>.

11.7.5 Parameters from the user environment

getenv(<name>)

putenv(<string>)

If there exists in the user environment a string

<name>=<value>

the function "getenv" returns to you a 'char' pointer to the assigned value. Otherwise you will get a NULL pointer.

For the function "putenv" the contents of the string must be of the form "<name>=<value>". At the time the function is called, the old value of <name> is set to <value>, or the variable <name> with this value is accepted into the user environment.

Note that after the call the memory area of <string> becomes an integral part of the user environment, so be sure not to use an automatic variable.

11.7.6 Entries in the group file

The file **/etc/group** contains the user groups of the system.

To be able to use the group functions, you must include a special header file:

#include <grp.h>

This makes the following functions available to you:

getgrent()

getgrgid(<gr-identifier>)

getgrnam(<name>)

setgrent();

endgrent();

The functions "getgrent", "getgrgid" and "getgrnam" return pointers to structures of the type 'struct group' (defined in <grp.h>). Each line in the file **/etc/group** corresponds to one such structure:

```
struct group {
    char *gr_name;       /* Group name       */
    char *gr_passwd;     /* Group password   */
    int  gr_gid;         /* Group identifier */
    char **gr_mem;       /* Group members    */
};
```

Getgrent returns on its first call a pointer to the first structure in /etc/group, and on each subsequent call the pointer is advanced by one structure so that the entire file is walked through. When the end of file is reached, the function returns a NULL pointer.

The function getgrgid searches the file /etc/group for an entry whose group identifier matches <gr-identifier>. If such a structure is found, its address is returned, and a NULL pointer otherwise.

Similarly, the function getgrnam searches the file /etc/group for an entry in which the group name matches <name>. If such a structure is found, its address is returned, and a NULL pointer otherwise.

Setgrent sets the internal file pointer back to the beginning of the file /etc/group so that another search can be performed.

Endgrent should be called to close the file when processing of the file is complete.

11.7.7 Entries in the passwd file

The file **/etc/passwd** lists all users known to the system and their login information.

In order to be able to use the passwd functions, you must include a special header file:

#include <pwd.h>

This makes the following functions available to you:

getpwent()

getpwuid(<user-identifier>**)**

getpwnam(<name>**)**

setpwent();

endpwent();

getpw(<user-identifier>, <string>**)**

The functions "getpwent", "getpwuid" and "getpwnam" return pointers to structures of the type 'struct passwd' (defined in <pwd.h>). Each line in the file **/etc/passwd** corresponds to one such structure:

```
struct passwd {
   char *pw_name;     /* Login name             */
   char *pw_passwd;   /* Password (encrypted)   */
   int  pw_uid;       /* User identifier        */
   int  pw_gid;       /* Group identifier       */
   char *pw_comment;  /* Comment                */
   char *pw_gecos;    /* Historical field       */
   char *pw_dir;      /* Login directory        */
   char *pw_shell;    /* Start program          */
};
```

Getpwent returns on its first call a pointer to the first structure in /etc/passwd, and on each subsequent call the pointer is advanced by one structure so that the entire file is walked through. When the end of file is reached, the function returns a NULL pointer.

The function getpwuid searches the file /etc/passwd for an entry whose user identifier matches <user-identifier>. If such a structure is found, its address is returned, and a NULL pointer otherwise.

Similarly, the function getpwnam searches the file /etc/passwd for an entry in which the login name matches <name>. If such a structure is found, its address is returned, and a NULL pointer otherwise.

Setpwent sets the internal file pointer back to the beginning of the file /etc/passwd so that another search can be performed.

Endpwent should be called to close the file when processing of the file is complete.

The function "getpw" has the same effect as the function getwuid except that the line found in the file /etc/passwd is not returned as a structure but placed character by character in <string>. Therefore, you do not need to include <pwd.h> to use getpw, and when you have used getpw you do not need to call the function endpwent.

Getpw returns (the 'int' value) 0 if an entry matching the supplied user identifier is found, and non-zero otherwise.

11.7.8 Option letters of the Shell command line

Syntax:

getopt(argc, argv, <opt-string>)

This function provides you with the next option letter from the Shell command line which started the current program. As you know from the subsection "Function arguments of the main function", the arguments supplied by the Shell are stored in an argument array, and their number is stored in a variable. You must define these two data items as follows:

```
char **argv;
int argc;
```

<Opt-string> is a string containing the admissible option letters. An option that is followed by an option argument when called must be followed by a colon in this string. In this case, you must in addition declare the external variable "optarg":

```
extern char *optarg;
```

which, after getopt has been called, points to the option argument of the option returned by getopt.

When all options have been examined (up to the first real argument), the function getopt returns the value EOF.

When the function getopt detects an option that is not contained in <opt-string>, it returns the character '?' and produces an error message that is output on stderr.

If, while getopt is being called, you want to check which argument of the command line is examined next, you must declare the external variable "optind":

```
extern int optind;
```

Before the first call of getopt, optind is at 0.

Example:

```
main(argc, argv) /* Program with 4 possible options */
int argc;
char **argv;
{
    int c;
    extern int optind;          /* Argument index */
    extern char *optarg;        /* Option argument */
    .
    .
    while ((c = getopt(argc, argv, "lbf:F:")) != EOF)
        switch(c) {
        case 'l': if (bflg) errflg++;
                  else lflg++;
                  break;
        case 'b': if (lflg) errflg++;
                  else bproc();
                  break;
        case 'f': lfile = optarg;
                  break;
        case 'F': ffile = optarg;
                  bufsiz = 512;
                  break;
        case '?': errflg++;
        }
    if(errflg) {
        fprintf(stderr, "usage: . . . ");
        exit(2);
    }
    .
    .
}
```

Within the while loop, all options of the program call are evaluated. The options l and b are mutually exclusive, and the options f and F must each be supplied with a file as an option parameter. Other options besides l, b, f, and F are not allowed.

11.7.9 Names list of executable files

To use the nlist function, you must include a special header file:

#include <a.out.h>

Syntax:

nlist(<filename>, <'struct nlist'-array-pointer>**)**

The function "nlist" checks the names list of the executable file whose name is supplied in the string <filename>, selects a number of values and stores them in an array of type 'struct nlist' to which the pointer <'struct nlist'-array-pointer> points. <'struct-nlist'-array-pointer> must be defined as follows:

struct nlist *<'struct nlist'-array-pointer>**[**<length>**]**;

The nlist structures consist of the elements "variable name", "variable type" and "variable value". You must specify the names of those variables whose type and value you want to obtain. The array is terminated by a structure in which the variable name begins with '\0'.

If a variable name is not found in the names list of the file, the type and value are returned as 0.

If the file cannot be read or contains no name list, all type entries are set to 0.

In the event of errors, the function value is ('int') −1, and 0 otherwise.

11.8 Protecting programs by passwords

The function "getpass" enables you to enter a password which is not displayed while being entered. The password is designed to protect your program from unauthorized use.

Syntax:

getpass(<prompt>)

Getpass displays the <prompt> on the screen and expects an input that is not displayed on the screen. The input is terminated by the newline or the EOF character.

The function value is a 'char' pointer to the entered password. A maximum of 8 characters are accepted and are terminated by an end character. If your terminal cannot be accessed, the function returns a NULL pointer.

Example:

```
char *text = "Password: ";
main()        /* Password-protected program */
{
   char *passptr, *getpass();

   if(strcmp(passptr = getpass(text), "Cream"))
   {
      printf("Not authorized !! \n");
      exit(2);
   }
   .
   .
}
```

At the beginning, the program requests input of a (blanked) password. If the entered password is other than "Cream", a message is issued and the program aborted.

12 Interfaces with the UNIX operating system

The standard library of input/output functions provides a uniform input/output interface for C under different operating systems. Each operating system must provide a special input/output interface simulating the functions of the standard library. The interface with the UNIX operating system is briefly described in the following.

A complete description of all system calls supported by C is given in the "System Calls" manual.

12.1 File access

Under UNIX, all inputs and outputs are considered as file accesses. Files in this sense include both conventional disk files and peripheral devices such as, in particular, the display terminal. You probably know from the Shell documentation that even inter-process communication, accomplished via pipes, is handled like file access.

Access to a file does not require the full file name on each transaction. The file name is only specified once when the file is opened or created. A file description is created in the memory for each open file. The file description is assigned a number for each user who opens it, and the user can henceforth access it by this number (as long as he has it open).

Under UNIX, the Shell opens three files at the time a program is started:

Number	File
0	Standard input
1	Standard output
2	Standard error output

These files are equivalent to the corresponding files in the standard library and can be redirected from the keyboard or screen to files using the Shell command language.

12.2 Opening and creating files

Syntax of the open function:

open(<filename>, <mode>)

The file name must be specified as a string, and the mode as an 'int' value as follows:

Mode	Opening mode
0	Open for read access
1	Open for write access
2	Open for read and write access

The return value of the open function is the value of the file descriptor ('int' value) if the open was successful, and −1 otherwise.

Example:

```
...
int fdr;     /* File descriptor */
fdr = open("/usr/cs/list.lis", 0);
...
```

The file /usr/cs/list.lis is opened for read access. Henceforth, the file is accessed under its file descriptor fdr.

Syntax of the create function:

creat(<filename>, <access-code>)

While an error message will be produced if you attempt to open a non-existent file, it is permitted to create an already existing file. This will cause the contents of the existing file to be cleared.

Creating a file opens it for write access. The access code is represented by a three-digit octal number. Read, write and execution permission are (in that order) granted by set bits of the first octal digit to the owner, by set bits of the second octal digit to the owner's group, and by set bits of the third octal digits to other users.

Example:

```
dbn = creat("/usr/cs/test.dat", 0644);
```

Creating the file /usr/cs/test.dat with read and write permission for the owner and read permission for his group and other users.

12.2.1 Further processing with standard library functions

Since all input/output functions in the standard library require a file pointer to identify the accessed file, there exists (in the standard library) a function that assigns a file pointer to the file descriptor of an open file:

fdopen(<file-descriptor>, <mode>)

The return value of this function is a file pointer to the open file.

Important:

You must specify the access mode in the same way as when opening a file with fopen (cf. "Input/output" section), and the mode must, of course, be compatible with the mode specified at open time.

12.2.2 Converting a file pointer to a file descriptor

Conversely, if you want to use the operating system interfaces to process a file that was opened by the fopen function from the standard library, you must use the number of the associated file descriptor that matches your file pointer. You can obtain the number with the following function:

fileno(<file-pointer>)

The function returns the value of the file descriptor as an 'int' number.

12.3 Reading and writing

Syntax:

read(<file-descriptor>, <string>, <number>**)**

write(<file-descriptor>, <string>, <number>**)**

<File-descriptor> is the value of the file descriptor obtained when opening or creating the file. <String> is the input or output buffer and must be specified as a 'char' array name or address. <Number> is the number of characters to be transferred from the file to the buffer or vice versa.

The file is read sequentially with the input/output system remembering internally up to which point a file was read with the last read function. Writing is also performed sequentially, usually at the end of file.

The function values returned by both functions are the number of characters actually transferred. When reading, this number may be less than the specified <number>. The value 0 indicates that you are at the end of file, and −1 is an error indication. When writing, however, the value should always be equal to <number>, and any value less than that indicates an error.

The following simple example shows how the standard library function getchar might be simulated by the UNIX input function read.

Example:

```
getchar()   /* Single-character input */
{
    char c;

    return((read(0, &c, 1) > 0) ? c & 0377 : EOF);
}
```

The value of the character read is returned by the getchar function as an 'int' value ('int' is an executable type) although only one character (= 1 byte) was actually read.

12.4 Random access

The read and write functions presented in the preceding subsection operate sequentially, but it is possible to set explicitly the file pointer maintained in the process. This provides you with the ability to access files randomly.

Syntax:

lseek(<file-descriptor>, <offset>, <reference-point>**)**

<File-descriptor> is the value of the file descriptor returned when opening or creating the file. The <offset> is specified in number of bytes relative to the <reference-point>. The <reference-point> can be specified in one of the following ways:

<Reference-point>	Meaning
0	Beginning of file
1	Current file pointer
2	End of file

The <file-descriptor> and <reference-point> must be of type 'int', while the <offset> must be of type 'long int'.

On successful translation of the pointer, the function returns the pointer's distance from the beginning of file (as a 'long int'). If an attempt is made to move to a position before the beginning of the file, the pointer remains unchanged and the function value is −1.

Example:

```
...
if((offset = lseek(fdr, 0L, 2)) == -1)    /* End of file */
...
```

Positioning the file pointer to the end of file.

12.5 Support of input/output from/to terminal files

Since normal data files are treated in the same way as special (device) files, you can, when you open a file, specify the path name of a terminal (/dev/ttyxx) just as well as the name of a normal data file.

The C standard library provides two functions enabling you to use the file descriptor to find out if the open file is a terminal and which name it has.

isatty(<file-descriptor>**)**

ttyname(<file-descriptor>**)**

Isatty returns (the 'int' value) 1 if the <file-descriptor> stands for an open terminal, and 0 otherwise.

The function "ttyname" returns a 'char' pointer to a string that contains the path name of the terminal that has been opened with <file-descriptor>. If <file-descriptor> does not stand for an open terminal, ttyname returns a null pointer.

12.6 Closing and unlinking files

Syntax (close function):

close(<file-descriptor>**)**;

"Close" cuts the connection between the file descriptor and the file. Depending on the operating system used, you can have open approximately 15 to 20 files for one program at any one time. Remember that the numbers 0, 1 and 2 are always used for standard files (see above). Once a file has been closed, its old file descriptor can be re-used to open another file.

Note:

At the end of a program, all open files are automatically closed.

Syntax (unlink function):

unlink(<filename>);

This function removes from the UNIX file system the file whose name is specified in the parameter as a string variable or constant.

12.7 Terminating a process with exit

To understand the exit statement you must be aware of the fact that, under UNIX, each program is started as a child process of another process. Exit immediately terminates a program and closes all open files. Exit has one parameter of type 'int' in which the termination code is returned to the parent process. The following convention applies to the return code:

Parameter	Meaning
0	Successful completion
1 – 255	Errors have occurred

13 Programming style

C programs can be designed so compactly and, on the other hand, provide so many variations that the strictest discipline in writing the code is required of the programmer to obtain readable and easily maintainable programs.

Various software teams have given a number of programming recommendations that should help you in coding easy-to-read and maintainable programs. The most important recommendations are presented in this chapter. However, it is up to you to decide whether you want to adhere strictly to the recommendations or develop your own C programming style by slightly modifying these rules here and there.

- Never let a source file, and even more so a function definition, become bigger than 500 lines. Subdivide the source file and modularize your program.
- Common definitions and type declarations should be written in a separate file and included with #include.
- The following order is recommended for the local declarations and definitions: extern declarations, register variables, auto variables and, finally, static variables.
- Each variable, type or function definition should be followed by an explanatory comment.
- Definitions should be as local (static) as possible.
- User-defined names occurring in #define or a typedef statement should be written in capital letters.
- External variables and functions should always (and not only when required) be explicitly declared with extern.
- Extern declarations should begin in the first column of a line.
- The local declarations and definitions within a function should be separated from the statements by a blank line.

- There should be neither blank lines between the statements nor multiple spaces within a line.
- If a statement (such as if, for etc.) controls one or more statements, you should indent the latter by a certain number of columns (at least three). Use the C beautifier cb.
- The same applies to statements or expressions that do not fit on a single line.
- You should never use the goto statement.
- Each binary operator should be enclosed in spaces, whereas each unary operator should be followed immediately by the operand.
- Although the logical operators &&, || etc. operate on 'int' expressions, you should not use them to form numeric values.
- The guidelines for converting a program segment into a separately defined function are as follows: if you need to indent more than five levels because of nested control or your function is much longer than a (printer) page, transfer the program segment into a separate function.

14 Operating the programming environment

The C programming environment under the UNIX operating system is described in this chapter. The easy-to-use, but all the same powerful, preprocessor of the C compiler is also discussed.

14.1 Creating source files

Since UNIX has only a single file type for data, you can enter a source file in the same way as texts or other data using the normal editor.

If no program is active on your terminal, you are in the Shell command mode. The command for calling the editor is

```
e <filename>.c
```

(The name of your system's editor may be different, for example ed, re, vi, edit etc.; ask your system manager.)

The <filename> can be chosen at will and can consist of letters, digits and special characters. The name is usually followed by a dot and a qualifier that gives a clue to the use of the file. Although the qualifier can normally be chosen arbitrarily, the C compiler expects the qualifier ".c" for C source files.

The file name including the qualifier must be no longer than 14 characters.

If the source file existed before, it is displayed on the screen (depending on the working mode of the editor) and can be edited. If it does not yet exist, the file is created and can then be edited.

When the editor session is complete, the file is saved on disk. The file can then be referenced at any time by its file name.

14.1.1 Beautifying source programs

Syntax:

cb [-s] [-j] [-l <length>] [<filename]^{n}

Cb reads C source code from the standard input or from the specified files and outputs it to the standard output with the spaces and indentations that reflect the structure of the C code. When specified without options, cb will not change the arrangement of the lines as entered by the user.

Options:

-s	Output in the style specified by Kernighan and Ritchie ("The C Programming Language").
-j	Broken lines are joined.
-l <length>	Lines exceeding the specified length are broken.

14.1.2 Cross-reference lists of source programs

Syntax:

cxref [-cs] [-**w** <width>] [-**o** <filename>] {<filename>}^{n}

Cxref analyzes the specified C source files and produces on standard output a cross-reference list of all occurring symbols. When specified without the -c option, a separate list is produced for each file, whereas with the -c option, a combined list for all files is produced. The position at which a symbol is defined is marked with an *.

Cxref includes a preprocessor run.

Options:

-c	Combined list for all files.
-w[<width>]	Change the output width (default: 80, even if you specify <width> at less than 51).
-o <filename>	Redirect the output to the file <filename>.
-s	Suppress output of the source file names.
-t	Output width is 80 characters.

14.1.3 Control flow charts of source programs

Syntax:

cflow [-r] [-ix] [-i_] [-d <depth>] {<filename>}^{n}

Cflow attempts to depict graphically the external references of the specified files to reflect the static control flow of the program. The chart is output to standard output. Cflow processes files ending with .y, .l, .c, .i, .a and .o.

Each output line begins with a line number, followed by the variable or function name of the external reference (default: functions only). Depending on the nesting depth of the reference, the name is indented by that many tab positions. The name is followed by a colon and the type (e.g. char *() is a function returning a 'char' pointer) and, finally, the source file and the line, enclosed in angle brackets, in which this element is defined.

For definitions that occur in linked object files, the location counters in the object file are specified rather than line numbers.

If the same reference occurs many times, only the line number of the first occurrence is specified in the angle brackets for the subsequent occurrances. If a reference is undefined, the angle brackets contain nothing.

Example:

File test.c:

```
main()
{
        f();
        g();
        f();
}

f()
{
        h();
}
```

The command "cflow test.c" produces the following output:

```
1       main: int(), <test.c 1>
2               f: int(), <test.c 8>
3                       h: < >
4               g: < >
5               f: <2>
```

Options:

-r — The relationship between calling and called function is reversed with a sorted list of the callers of each function being produced.

-ix — The list contains not only functions, but also external and static data.

-i_ — Functions whose names begin with _ (usually system functions) are included.

-d<depth> — Specification of the maximum nesting depth for references that are to be output (standard: very large).

14.2 Compiling and linking C programs

To compile and link a C program, you need a single Shell command, cc. You must know which source files make up your program because you must append the file names as parameters to the cc command.

```
cc <filename-1>.c ... <filename-n>.c
```

The source files are now separately translated into object files. This includes, for each file, the preprocessor run and the following four steps:

- Translation into preliminary assembler code,
- optimizer,
- translation into final-form assembler code,
- translation into object code.

The object files are then called

```
<filename-i>.o
```

These files are then linked to produced an executable program. If the source program consists of a single file only, the object file (ending with .o) will be cleared after the link run.

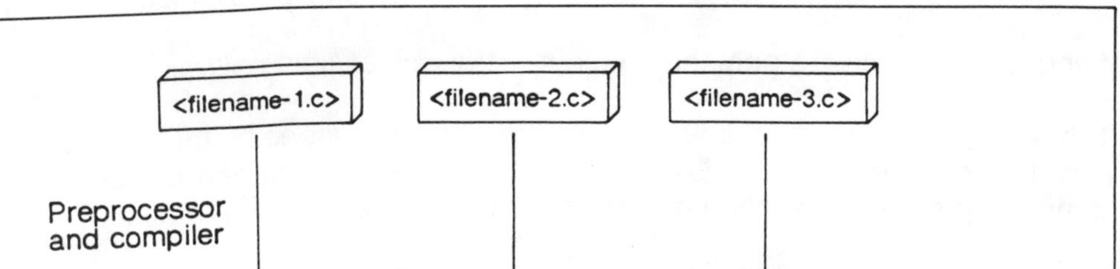

If the terminal displays no error message, the program has been successfully compiled and linked and can be executed. Note that the linker always assigns the same name "a.out" to the executable file. If you want to prevent this file from being overwritten during the next compiler run, you should rename it now using

mv a.out <name>

Rather than renaming the executable file subsequently, you can specify an option in the cc command assigning a name to the executable file:

cc -o <name> <source-file>

This will prevent the old file a.out from being destroyed.

If a program is compiled several times and you change, say, only one of the multiple source files, you only need to re-compile that source file. You then specify the source file name (.c) only of this file in the cc command, while for the other files you specify the object file names (.o).

Example:

```
cc hanu.o ctex.o erf.c form.o
```

Only erf.c will be re-compiled and then linked with the other object files.

If you want to activate only the linker, specify object files only.

If, however, you want to invoke only the compiler, specify the "-c" option with the cc command. In this case the source files do not need to contain a complete program, and the smallest possible unit is the source file.

Example:

The above example is equivalent to:

```
cc -c erf.c
cc hanu.o ctex.o erf.o form.o
```

Following is the complete description of the cc and ld command from the "System V Interface Description":

NAME

cc - C compiler

SYNOPSIS

cc [options] file ...

DESCRIPTION

The cc command is the interface to the C compilation system. The system conceptually consists of a preprocessor, compiler, optimizer, assembler, and link-editor. The cc command processes the supplied options and then executes the various tools with the appropriate arguments.

The suffix of a file-name argument indicates how the file is to be treated. Files whose names end with .c are taken to be C source programs, and may be preprocessed, compiled, optimized, and link-edited. The compilation process may be stopped after the completion of any pass if the appropriate options are supplied. If the compilation process is allowed to complete the assembly phase, then an object program is produced; the object program for a source file called xyz.c is created in a file called xyz.o. However, the .o file is normally deleted if a single C program is compiled and loaded all at one go.

In the same way, arguments whose names end with .s are taken to be assembly source programs, and may be assembled and link-edited. Files with names ending in .i are taken to be preprocessed C source programs and may be compiled, optimized, assembled, and link-edited. Files whose names do not end in .c, .s, or .i are handed to the link-editor.

By default, if an executable file is produced (i.e., the link-edit phase is allowed to complete), the file is called a.out. This default name can be changed with the -o option (see below).

The following options are interpreted by cc:

-c Suppress the link edit phase of the compilation, and do not remove any object files that are produced.

-f Include floating-point support for systems without an automatically included floating-point implementation. This option is ignored on systems that do not need it.

-g Cause the compiler to generate additional information needed for the use of sdb.

-o outfile
Use the name outfile, instead of the default a.out, for the executable file produced. This is a link-editor option.

-p Arrange for the compiler to produce code that counts the number of times each routine is called; also, if link editing takes place, a profiled version of the standard C library is linked, and monitor [see MONITOR(SD_LIB)] is automatically called. A mon.out file will then be produce at normal termination of execution of the program. An execution profile can then be generated by use of prof.

-q This option is reserved for specification of implementation specific profiling directives.

-E Run only cpp on the named C programs and send the result to the standard output.

-F This option is reserved for implementation specific optimization directives.

-O Do compilation phase optimization. This option will not affect files.

-P Run only cpp on the named C programs and leave the result on corresponding files suffixed .i. This option is passed to cpp.

-S Compile and do not assemble the named C programs, and leave the assembler-language output on corresponding files suffixed .s.

-Wc,arg1[,arg2...]
Hand off the argument[s] argi to phase c where c is one of [p02al] indicating preprocessor, compiler, optimizer, assembler, or link editor, respectively. For example, -Wa, -m passes -m to the assembler phase.

The cc command also recognizes the options -C, -D, -I, and -U, and passes them (and their associated arguments) directly to the preprocessor without using the -W option. Similarly, the loader options -a, -l, -o, -r, -s, -u, -L, and -V are recognized and passed directly to the loader. See CPP(SD_CMD) and LD(SD_CMD) for descriptions of these options.

Other arguments are taken to be C-compatible object programs, typically produced by an earlier cc or pcc run, or perhaps libraries of C-compatible routines, and are passed directly to the links-editor. These programs, together with the results of any compilations specified, are linked (in the order given) to produce an executable program with the name a.out (unless the -o link-editor option is used).

The standard C library is automatically available to the C program. Other libraries (including the math library) must be specified explicitly using the -l option with cc; see LD(SD_CMD) for details.

FILES

file.c	input file
file.i	preprocessed C source file
file.o	object file
file.s	assembly language file
a.out	link-edited (executable) output

SEE ALSO

CPP(SD_CMD), LD(SD_CMD), PROF(SD_CMD), SDB(SD_CMD), EXIT(BA_OS), MONITOR(SD_LIB).

USAGE

Since the cc command usually creates files in the current directory during the compilation process, it is typically necessary to run the cc command in a directory in which a file can be created.

NAME

ld - link editor for object files

SYNOPSIS

ld [options] file ...

DESCRIPTION

The ld command combines several object files into one, performs relocation, resolves external symbols, and supports symbol table information for symbolic debugging. In the simplest case, the names of several object programs are given, and ld combines them, producing an object module that can either be executed or, if the -r option is specified, used as input for a subsequent ld run. The output of ld is left in a.out. By default this file is executable if no errors occurred during the load. If any input file, filename, is not an object file, ld assumes it is an archive library.

If any argument is a library, it is searched at the point it is encountered in the argument list. Only those routines defining an unresolved external reference are loaded. The library (archive) symbol table is searched to resolve external references which can be satisfied by library members. The ordering of library members is unimportant, unless there exist multiple library members defining the same external symbol.

The following options are recognized by ld:

-e epsym
: Set the default entry point address for the output file to be that of the symbol epsym.

-lx Search the library which has the abbreviation x (e.g., -lm to search the math library). A library is searched when its name is encountered, so the placement of a -l is significant.

-o outfile
: Produce an output object file by the name outfile. The name of the default object file is a.out.

-r Retain relocation entries in the output object file. Relocation entries must be saved if the output file is to become an input file in a subsequent ld run. The link editor will not complain about unresolved references, and the output file will not be made executable.

-s Strip all symbolic information from the output object file.

-u symname
: Enter symname as an undefined symbol in the symbol table. This is useful for loading entirely from a library, since initially the symbol table is empty and an unresolved reference is needed to force the loading of the first routine.

-L dir
: Change the algorithm of searching for the library to look in dir before looking in the default library directories. This option is effective only if it precedes the -l option on the command line.

-V Output a message giving information about the version of ld being used.

FILES
: a.out output file

SEE ALSO
: AR(BU_CMD), CC(SD_CMD), STRIP(SD_CMD).

USAGE

When the link editor is called through cc, a startup routine is linked with the user's program. This routine calls exit() after execution of the main program. If the user calls the links editor directly, then the user must ensure that the program always calls exit() rather than falling through the end of the entry routine.

The symbols etext, edata, and end are reserved and are defined by the link editor. It is erroneous for a user program to redefine them.

14.3 Executing programs and listing files

Programs are executed in the Shell command mode by entering the name of the executable file.

If the program was compiled and linked without options, the call is

a.out

If, on the other hand, the program was after compilation renamed with "mv" or if the option "-o <objfile>" was specified with the cc command, you must specify that name when calling the program.

If parameters are to be passed to the program, append them to the name or a.out, separated by spaces:

<name> <parameter-1> <parameter-2> ... <parameter-n>

If you want to list all the files you already have (source files, object files, executable files, data files, ...) enter the command

ls

and a list of your files will appear on the screen.

To remove one or more of your files, use the following command:

rm { <filename> }^{n}

14.4 The C compiler preprocessor

The C compiler is provided with an easy-to-use yet powerful preprocessor capable of performing the following tasks:

- Substitution of names
- Parameter-controlled substitution of names
- Including other source files
- Conditional compilation
- Changing of line numbers

The preprocessor is called by the character "#" and a keyword at the **beginning** of a source file line. Preprocessor calls are permitted at any position within a program. If a preprocessor statement does not fit on a line, end the line with a "\" and continue it on the next line.

If you want to examine what the preprocessor does to your source file, you can execute only the preprocessor on your files by calling it with

cc -P {<filename>.c}^{n}

The result is stored in a file with the same name and the ending qualifier **.i**.

14.4.1 Substitution of names (constant definition)

Syntax:

#define <name> [<def-string>]

The domain of this definition is from the point where it occurs to the end of the source file. Wherever the <name> appears after that point in the source file, it will be substituted by the specified string. Only if <name> appears within a string constant or a comment will it not be substituted.

The <def-string>, unlike a string constant, is not enclosed in double quotes but begins with the first character other than space after <name> and ends at the end of the line.

If there is no <def-string>, then <name> is considered as defined (see subsection 14.4.3, "Conditional compilation of program segments") but does not have a value and, if it occurs in the program, is replaced by nothing, i.e. deleted.

The main advantage of this construct is that a constant is defined only at one point in a source file and can then be inserted any number of times. If you want to change the constant, you need to do this only in the definition, and you can be sure that the change is reflected everywhere in the source file.

Example:

```
#define EOL   10
#define JA    1
#define TEXT  "Enter your password!\n"
```

A subsequent program line of the following form

```
if (ist_static == JA) ...
```

will look as follows after the preprocessor run:

```
if (ist_static == 1) ...
```

Although the #define statement is normally used to define constants, names can also be replaced by other names or symbols.

Example:

If you want to use "begin ... end" instead of braces (similar to Pascal), the following definitions are required:

```
#define begin   {
#define end    ;}
```

Name substitution can be parameterized with the actual parameter at substitution time being inserted in the define line instead of the formal parameter. This is also called a "macro definition".
Be sure not to insert a space between <name> and the parameter list enclosed in parentheses.

Example:

```
#define max(A, B)    ((A) > (B) ? (A) : (B))
```

If the program contains the line

```
len = max(len1, len2);
```

it will look as follows after the preprocessor run:

```
len = ((len1) > (len2) ? (len1) : (len2));
```

The parentheses in the #define statement are required in case you have the following expression in your program:

```
max(p+q, r+s)
```

Although a macro call in the program cannot syntactically be distinguished from a function call, there are some important differences:

- Macros cannot be addressed;
- There is no type for the result of a macro, so the results may be unpredictable;
- The use of increment/decrement or compound assignment operators for the parameters of macros may have undesired side-effects.

If you want to disable a definition or a macro before the end of the source file, use

#undef <name>

From this point on, <name> will no longer be substituted.

The defining and undefining of names can also be performed by an option in the compiler and, hence, the preprocessor call.

Example:

```
cc -DMAX=20 ctex.c
```

has the same effect as

```
#define MAX 20
```

at the beginning of the file ctex.c.

All the possibilities of name substitution and macro definition are demonstrated once again by the following example in which a program is shown before and after the preprocessor run:

```
#define begin          {
#define end            ;}
#define LIM            25
#define intest(i,s)    (s[i] = getchar()) != '\n' && \
                       s[i] != '\0' && i < LIM - 1
main()  /* Constant definition, name substitution */
begin
   int i = 0;
   char string[LIM];

   printf("begin \n");
   printf("Please enter your name: ");
   while(intest(i, string)) i++;
   string[i] = '\0';
   printf("%s\n", string)
end
```

After the preprocessor run the program looks as follows:

```
main()
{
   int i = 0;
   char string[25];

   printf("begin \n");
   printf("Please enter your name: ");
   while((string[i] = getchar()) != '\n' &&
      string[i] != '\0' && i < 25 - 1) i++;
   string[i] = '\0';
   printf("%s\n", string)
;}
```

14.4.2 Including other source programs

Syntax:

#include "<filename>"

This line is replaced by the entire file named <filename>. The <filename> here includes the ending qualifier (".c" or ".h").

It is good programming practice to write definitions, declarations and macros that are identically used in several source files or even several programs into a separate file (header file) and include them in each source file using #include.

Files included with #include may in turn contain #include's.

If the file name is not enclosed in double quotes but in angle brackets, the preprocessor will assume that this is a system file from the directory /usr/include, whereas double quotes cause it to assume a private file for which the complete or the relative path name to the current directory should be specified (cf. the option "-I<directory>" of the cc command).

14.4.3 Conditional compilation of program segments

Syntax:

#if <const-expression>

.
.   /* Program segment A */
.
[**#else**
.
.   /* Program segment B */
.]
#endif

If the constant expression is greater than 0 (true), the statements of program segment A will be compiled only, while program segment B will be ignored. If the expression is 0 (false), the statements (including preprocessor statements) between #if and #else or, if #else is not present, between #if and #endif, will be ignored (#else is optional in program segment B).

Instead of #if <const-expression>, the following preprocessor statement may be used:

#ifdef <name>

This condition is considered to be met if, at this point, <name> has occurred in a #define statement and has not yet been disabled by #undef <name>.

Conversely, you can also test to see whether the name is undefined:

#ifndef <name>

This condition is considered to be met if <name> has not yet occurred in a #define statement or has been disabled by #undef <name>.

Conditional compilation is mainly used for two reasons:

- To make programs portable that use certain system-specific functions by compiling the function calls associated with that system;
- To include test and debugging statements for the test version or eliminate them from the final version.

14.4.4 Line number control

Syntax:

#line <constant> [<name>]

This preprocessor statement sets the line number of the next statement to the value of a constant.

If a name is specified in addition, the name of the current source file for the C compiler is changed to <name>.

This preprocessor statement is used for error diagnosis, although it is relatively seldom used. #line is used by various tools generating the C code.

15 Testing C programs

The test is the execution of a program with the intention of finding mistakes.

Glenford J. Myers

The first test your C program has to stand is when it is being compiled. If mistakes are found in this process, this is not so serious because the location of the bugs are precisely indicated by the source file name and line number where the error was detected. The actual cause of the bug may, of course, be further upstream in the program, but in most cases you will find that the bug is exactly on the specified line or on the line above.

In addition, the compiler issues an error message that gives you a clue to the possible cause of the error. For more information refer to the chapter "Error messages".

Errors detected by the compiler obviously do not require debugging because both the error location and the cause of the error are rather precisely indicated. Debugging goes somewhat further and means the isolation of errors detected during testing and the determination of the error cause.

A more difficult situation arises when the compiler has not detected any bugs in your program but you find in the logic test that your program does not do what you expected.

Another preliminary verification you should perform in this case is provided by the "lint" utility which will check your C program not only from the syntactic but also from a programming technique point of view. For its functionality refer to the next subsection.

Lint also gives you information about the error location and cause.

The further testing procedure is the more effective the better you have modularized your program. The C language with its strong modularization features and the ability to compile each module separately is ideally suited for writing modular, easy-to-test programs. Already in the program design stage you should design-in testability. Besides subdividing your program into small modules, you should have an eye on clean interfaces between the modules with as few global data items as possible, instead having transfer of values by function parameters and returned function values.

There are two strategies for further testing from which you can choose:

- top-down test
- bottom-up test (module test)

When you use the top-down test, you first verify the main function. Functions called by the main functions are provisionally included as dummy functions or, if required, as functions with a constant function value.

Examples:

```
reverse(s)
char *s;
{ }

weekday(day, month, year)
int day, month, year;
{ return(1); }
```

If your program passes the test, you gradually include the real functions instead of the dummy functions. When the first error occurs, then it can only have been caused by the last included function or the way it is integrated. If your program failed the first test, there must be a bug in the main function.

When you have in this way determined the program module containing the bug, you must isolate the error or the cause of the error. This is what we call debugging. Frequently you will have a suspicion as to the cause of the error, but if not, you must systematically "walk through" the module.

The capabilities the C preprocessor makes available to you in this case without requiring different test and production versions of your program to be created are discussed in detail in the "Debugging" subsection.

When using the bottom-up test, you begin exactly at the opposite end: you test a module that does not call any other functions by simulating its environment. Once tested, you can include the module in other functions that call it and then perform another module test.

This procedure will also enable you to isolate the module containing a bug.

For a more detailed description of module testing refer to subsection 15.2, "Module testing".

Once the bottom-up or top-down test has shown that all modules are free from bugs you are justified in believing that your program is actually error-free and can release it.

The production use of your program is the last test it will have to stand. Unfortunately, no one has so far been able to simulate this final test. You should take comfort from the fact that even brilliant programmers and debugging experts very often fail to notice bugs that are later detected by some user, perhaps in a quite simple application.

15.1 Verifying C programs using lint

Syntax:

lint [-abchnpuvx] [<filename>]^n

Lint attempts to detect features of the C program files that are likely to be bugs, non-portable, or wasteful. It also checks type usage more strictly than the compilers. Among the things that are currently detected are unreachable statements, loops not entered at the top, automatic variables declared and not used, and logical expressions whose value is constant. Moreover, the usage of functions is checked to find functions that return values in some places and not in others, functions called with varying numbers or types of arguments, and functions whose values are not used or whose values are used but none returned.

Arguments whose names end with .c are taken to be C source files. Arguments whose names end with .ln are taken to be the result of an earlier invocation of lint with either the -c or the -o option used. The .ln files are analogous to .o (object) files that are produced by the cc command when given a .c file as input. Files with other suffixes are warned about and ignored.

Lint will take all the .c, .ln, and llib-l<lib>.ln (specified by -l<lib>) files and process them in their command line order. By default, lint appends the standard C lint library (llib-lc.ln) to the end of the list of files. However, if the -p option is used, the portable C lint library (llib-port.ln) is appended instead. When the -c option is not used, the second pass of lint checks this list of files for mutual compatibility. When the -c option is used, the .ln and the llib-l<lib>.ln files are ignored.

Any number of lint options may be used, in any order, intermixed with file-name arguments. The following options are used to suppress certain kinds of complaints:

-a Suppress complaints about assignments of long values to variables that are not long.

-b Suppress complaints about break statements that cannot be reached. (Programs produced by lex or yacc will often result in many such complaints).

-h Do not apply heuristic tests that attempt to intuit bugs, improve style, and reduce waste.

-u Suppress complaints about functions and external variables used and not defined, or defined and not used. (This option is suitable for running lint on a subset of files of a larger program).

-v Suppress complaints about unused arguments in functions.

-x Do not report variables referred to by external declarations but never used.

The following arguments alter lint's behavior:

-l<lib> Include additional lint library llib-l<lib>.ln. For example, you can include a lint version of the Math Library llib-lm.ln by inserting -lm on the command line. This argument does not suppress the default use of llib-lc.ln. These lint libraries must be in the assumed directory. This option can be used to reference local lint libraries and is useful in the development of multi-file projects.

-n Do not check compatibility against either the standard or the portable lint library.

-p Attempt to check portability to other dialects (IBM and GCOS) of C. Along with stricter checking, this option causes all non-external names to be truncated to eight characters and all external names to be truncated to six characters and one case.

-c Cause lint to produce a .ln file for every .c file on the command line. These .ln files are the product of lint's first pass only, and are not checked for inter-function compatibility.

-o<lib> Cause lint to create a lint library with the name llib-l<lib>.ln. The -c option nullifies any use of the -o option. The lint library produced is the input that is given to lint's second pass. The -o option simply causes this file to be saved in the named lint library. To produce a llib-l<lib>.ln without extraneous messages, use of the -x option is suggested. The -v option is useful if the source file(s) for the lint library are just external interfaces (for example, the way the file llib-lc is written). These option settings are also available through the use of "lint comments" (see below).

The -D, -U, and -I options of cpp and the -g and -O options of cc are also recognized as separate arguments. The -g and -O options are ignored, but, by recognizing these options, lint's behavior is closer to that of the cc command. Other options are warned about and ignored. The pre-processor symbol "lint" is defined to allow certain questionable code to be altered or removed for lint. Therefore, the symbol "lint" should be thought of as a reserved word for all code that is planned to be checked by lint.

Certain conventional comments in the C source will change the behavior of lint:

/*NOTREACHED*/ at appropriate points stops comments about unreachable code. (This comment is typically placed just after calls to functions like exit).

/*VARARGSn*/ suppresses the usual checking for variable numbers of arguments in the following function declaration. The data types of the first n arguments are checked; a missing n is taken to be 0.

/*ARGSUSED*/ turns on the -v option for the next function.

/*LINTLIBRARY*/ at the beginning of a file shuts off complaints about unused functions and function arguments in this file. This is equivalent to using the -v and -x options.

Lint produces its first output on a per-source-file basis. Complaints regarding included files are collected and printed after all source files have been processed. Finally, if the -c option is not used, information gathered from all input files is collected and checked for consistency. At this point, if it is not clear whether a complaint stems from a given source file or from one of its included files, the source file name will be printed followed by a question mark.

The behavior of the -c and the -o options allows for incremental use of lint on a set of C source files. Generally, one invokes lint once for each source file with the -c option. Each of these invocations produces a .ln file which corresponds to the .c file, and prints all messages that are about just that source file. After all the source files have been separately run through lint, it is invoked once more (without the -c option), listing all the .ln files with the needed -l<lib> options. This will print all the inter-file inconsistencies. This scheme works well with make; it allows make to lint only the source files that have been modified since the last time the set of source files were linted.

15.2 Module testing

As was mentioned above, the module under test should be a clearly defined program segment with precisely defined interfaces. In the C language this will usually be a function.

The ideal size of a module does not only depend on the number of program lines, but also on its complexity and its relationships with other modules.

It will certainly be easy for you to determine the correct results, obtained from specific inputs, of program modules of this type. Enter the inputs and desired results in separate files. You should begin with a function that does not call any other (user-defined) function. Functions that call other functions should not be tested before the called functions (bottom-up procedure).

Then extract the module to be tested from the overall program by placing it in a separate source file. Any global data or include file that may be used must be declared/included separately in the source file.

This procedure leaves the overall program unchanged. This meets an important requirement to be fulfilled by an efficient testing procedure, namely that the test version of a program and of each individual module must be identical with the production version.

For the extracted module under test you write a so-called test driver. In C, you use a special main function whose sole purpose is to supply the module under test with inputs (and possibly to initialize global data) and then to output its results.

Example:

The module under test is the following function "weekday()".

```
weekday(day, month, year)    /* Weekday, Monday = 1 */
int day, month, year;
{
    int yearfac;

    yearfac = year + (month - 14)/12;
    return(((13 * (month + 10 - (month + 10)/13*12) - 1)/5
        + day + 77 + 5 * (yearfac % 100)/4
        + yearfac / 400 - yearfac / 100 * 2) % 7 + 1);
}
```

The test driver might look as follows:

```
#include <stdio.h>
main()    /* Test driver for weekday() */
{
    int day, mon, year;

    for(;;)
    {
        scanf("%d.%d.%d", &day, &mon, &year);
        if(day == 0) exit(0);
        printf("%d\n", weekday(day, mon, year));
    }
}
```

Input of a 0 terminates the test. Now you should create a file into which you should input some dates of which you know the weekday. Create another file with the corresponding weekdays.

```
File1:              File2:
31.8.84             5
1.9.84              6
20.8.84             1
20.8.49             6
26.11.54            5
2.12.81             3
0
```

Next you compile the module under test together with the test driver and execute the test program with standard input redirected from the file with the input dates. Redirect the standard output to a file.

Now you just need to compare the output data with the desired values you have entered. This can also be done automatically because UNIX provides various tools for this purpose (cmp, diff, ...).

If the comparison is successful with the module under test supplying the correct results you should check closely to see if your test values are scattered widely enough and borderline cases are covered. The test may have to be repeated several times with different data.

Be sure to retain the input and output data because this will save you time when the test has to be repeated after program modifications.

15.3 Debugging

A powerful tool for isolating errors and detecting the causes is the C preprocessor. It enables you to create a test version and a production version of a program with totally identical source code, but with the test version being provided with versatile test features by simply specifying a compiler option.

The standard debugging aids include:

- Interrupt execution
- Interrupt execution on specific events
- Output variable values on program interruption and possibly change them
- Skip statements
- Check program loops
- Check nested function calls

The test version is produced when you specify a defined name (e.g. DEBUG) at compile time:

```
cc -DDEBUG <source-files>
```

and in addition include test statements at interesting points in the source program that are enclosed in the following processor statements:

```
#ifdef DEBUG
...
...        /* Test statements      */
...
#endif
```

Following the last successful test, the program is compiled without the "-D" option:

```
cc <source-files>
```

This compiler run ignores the test statements. There is no reason why you should not leave the test statements in your source program, and doing so will simplify debugging after slight modifications.

15.3.1 Interrupting execution

The simplest way of interrupting program execution is to request a keybord entry.

Example:

```
main()
{
#ifdef DEBUG
   char din[10];
#endif
   ...
   ...
   ...      /* Stop after this statement */
#ifdef DEBUG
   printf("Continue (Y/N)? ");
   scanf("%s", din);
   if(strcmp(din, "N")) exit(255);
#endif
...
```

Program execution stops after the commented statement. When "N" is entered, the program will be aborted, and any other entry causes it to continue.

15.3.2 Conditional interruption of execution

You can, of course, make the interruption of program execution conditional on the occurrence of certain events, e.g. the calling of a specific function or the modification of a specific variable.

Example:

You want to find out whether a string checking function strchk modifies the string passed to it.

```
   ...
#ifdef DEBUG
   strcpy(strold, str);
#endif
   sstat = strchk(str);
#ifdef DEBUG
   if(!strcmp(str, strold))
   {
      printf("str before: %s after: %s\n", strold, str);
      printf("Continue (Y/N)? ");
      scanf("%s", din);
      if(strcmp(din, "N")) exit();
   }
#endif
   ...
```

The variable strold must previously also be defined in a #ifdef section.

15.3.3 Displaying and changing values

Displaying and changing of variable contents is easily done by printf and scanf statements within the #ifdef section.

If you only want to output values, these can easily be redirected to a file using the Shell and compared with later versions.

Example:

```
   ...
#ifdef DEBUG
   printf("Number: %d, Price: %.2f\n", number, price);
   printf("Change (Y/N)? ");
   scanf("%s", din);
   if(strcmp(din, "Y")) scanf("%d %f", &number, &price);
#endif
   ...
```

If the program only runs satisfactorily with manually changed values, it obviously requires modification.

15.3.4 Skipping statements

Statements that are to be skipped during the debug run should be bracketed as follows:

```
    ...
#ifndef DEBUG
    ...   /* Statements, to be */
    ...   /* skipped during    */
    ...   /* debugging         */
#endif
    ...
```

15.3.5 Displaying loop calls with loop counters

If you want to check a program loop, you should define a counter variable that is incremented and output on each loop execution.

Example:

```
powsum(lim) /* Next higher sum of powers */
int lim;
{
#ifdef DEBUG
   int count = 0;
#endif
   int ps = 0, i = 0;

   while(ps < lim)
   {
#ifdef DEBUG
      printf("%d. Loop execution\n", ++count);
#endif
      ps += i++;
   }
   return(ps);
}
```

15.3.6 Displaying the calling sequence of functions

To check the calling sequence of functions, you insert an output function in each function after the opening brace. If you want to count in addition how often a function is called, you define a counter as a static variable, increment it on each call and output it.

Example:

```
main()
{
#ifdef DEBUG
   printf("Entry main\n");
#endif
   reverse();
#ifdef DEBUG
   printf("Exit main\n");
#endif
}

reverse()    /* Read string and output in reverse */
{
   char c;
#ifdef DEBUG
   static int count;
   printf("Entry reverse, %d. times\n", ++count);
#endif
   if ((c = getchar()) != '\n')
   {
      reverse();
      putchar(c);
   }
#ifdef DEBUG
   printf("Exit reverse\n");
#endif
}
```

The output indicates how often the function reverse() is called directly and recursively.

16 Error messages

Error messages issued during the compilation of your C program are written to the standard error output. Each error message produces an output line of the following form:

<compiler-step>:<source-file>:<line-no>:<error-text>

The majority of error messages are issued by the syntax verifier and converter "c0". <Source-file> and <line-no> indicate in which of the source files specified in your cc command and on which line the error has occurred. Incidentally, the output of these two fields can be manipulated with the preprocessor statement #line.

The error texts of the syntax verifier and converter "c0" are listed in the following in alphabetic order. The possible cause and the recommended action is given in each case. Note that the error texts are not identical for each C implementation but depend on your specific compiler.

<Error text>	Possible cause/recommended action
<Name> multiply defined	<Name> is multiply defined; use different names
<Name> redeclared	<Name> is defined/declared multiple times; use different names
<Name> undefined	A name that is used is not defined; include definition
<Name1> undefined, func. <name2>	<Name1> is not defined in function <name2>; include definition
0-length row: <name>	Array dimensioning error
Ambiguous structure reference for <name>	Ambiguous structure reference for <name>; use a different name
Bad func. storage class	Incorrect storage class for function declaration (only extern or auto are allowed)
Bad structure/union/enum name	Use different name

<Error text>	Possible cause/recommended action
Bad type for field	Incorrect type in field definition (only 'unsigned' and 'int' are allowed)
Botch in outcode	Abort during compilation
Break/continue error	Break not in a loop or a switch statement, continue not in a loop
c0 internal error: tree not active	Compiler error; inform system specialist
Call of non-function	Call of a function that is not declared or defined as a function
Can't create temp	Disk management error; inform system specialist
Can't find source <name>	Source file <name> does not exist or is read-protected
Case not in switch	Switch statement contains no case and no default
Compiler error (length)	Compiler error; inform system specialist
Compound statement required	Switch statement without braces
Conflict in storage class	Storage classes are incompatible
Declaration syntax	Syntax error in definition/declaration statement including initialization
Default not in switch	Default label occurs outside switch statement
Disallowed conversion	Illegal conversion; check conversion rules
Divide check	Illegal division; division by 0?
Expression overflow	Expression nested too deeply
Expression syntax	Illegal number of operands for an operator; operator incorrect or missing
External definition syntax	Syntax error in a definition outside functions

<Error text>	Possible cause/recommended action
Field too long	Field element exceeds the length of an 'int' variable
Illegal #	# is illegal (preprocessor statement does not begin in the first column)
Illegal condition	Condition is illegal (if outside functions)
Illegal conversion	Conversion is illegal; check conversion rules
Illegal enum constant for <name>	Illegal name in an enumeration type definition
Illegal enumeration <name>	Incorrect enumeration type constant
Illegal indirection	Illegal address formation
Illegal lvalue	Syntax element to the left of the "=" sign is illegal
Illegal storage class	Specification of a storage class before a function definition; incorrect name of a storage class, illegal storage class
Illegal structure ref	Illegal use of structures
Illegal type of operand	Use a different operand type
Illegal use of type	Use a different type
Illegal use of type name	Type name used illegally
Illegal use of void object	'Void' function used illegally used
Inappropriate 'else'	"Else" has no matching "if"
Inappropriate parameters	Parameters (types) are inappropriate
Incompatible structures	Structures are incompatible
Int constant required, 0 assumed	Constant missing for case in the switch statement
Long character constant	'Char' constant too long (longer than 'int' variable)

<Error text>	Possible cause/recommended action
Lvalue required	The only elements permitted here include a simple variable, a pointer, an array element, a structure element, a union element or a field element
Misplaced 'long'	'Long' must not appear in this place
Misplaced 'unsigned'	'Unsigned' must not appear in this place
Missing '}'	Number of '{' is different from number of '}'
More than 1 'default'	Only one default label is allowed in a switch statement
Names <name1> and <name2> conflict	Conflict of names; change a name
Negative field width	Field element length is negative
No auto. aggregate initialization	No aggregate initialization is allowed for the storage class auto (allowed only for global definitions and storage class static)
No field initialization	Fields must never be initialized
No strings in automatic	String initialization only allowed for global definitions and storage class static
Non-terminated comment	Comment terminator '*/' missing
Non-terminated string	String constant terminator " missing
Not an argument: <name>	<Name> cannot be used as an actual parameter
Null dimension	Dimension is null
Number syntax	Incorrect notation of a number constant
Out of space	Compiler error; inform system specialist
Stackframe larger than 32k	Stack is too large; optimize or divide program
Statement syntax	Illegal syntax in a statement

<Error text>	Possible cause/recommended action
Structure redeclaration	A structure is multiply defined; use different names
Switch table overflow	Too many different cases in a switch statement
Too many initializers: <name>	Too many element values were specified in the aggregate initialization of <name>
Too many structure initializers	Too many element values were specified in the aggregate initialization of the structure <name>
Too many structure members	Structure has too many elements; divide or form arrays
Too many }'s	Number of '{' is not equal to number of '}'
Too much declaring in an expression	Too many implicit declarations in an expression; declare explicitly
Type clash	Incompatible types
Type is too complicated	Type is nested too deeply
Undefined structure	Structure is not defined
Undefined structure initialization	Initialization of an undefined structure
Unexpected EOF	Number of '{' is > number of '}'
Unknown character	Character is unknown
Unknown keyword	A reserved word is unknown
Unknown option	An option is unknown (on a compiler call)
Warning: <name> not member of cited struct/ union	<Name> is not the name of a structure element
Warning: <op>= operator assumed	Either the old notation of the compound assignment operator was used or a space is missing between the = and the unary operator

<Error text>	Possible cause/recommended action
Warning: conversion from short to pointer	Type conflict: the value may be corrupted (forgot function declaration?)
Warning: illegal conversion from unsigned long to float	Type conflict: the value may be corrupted
Warning: very large data structure	Compound data structure too large; divide
Write error on temp	Compiler error; inform system specialist

A1 Appendix 1: Summary of statements

```
<variable> = <expression>;
```

To the variable is assigned the value of the expression

```
<variable> <op> = <expression>;
```

Manipulation of a variable by <expression> with operator <op>

```
<variable>++;
<variable>--;
++<variable>;
--<variable>;
```

Incrementation/decrementation of a variable

```
{
      [ <definitions> ]
      <statement-1>
      <statement-2>
      .
      .
      .
      <statement-n>
}
```

Compound statement

```
if ( <expression> ) <statement1>

[ else <statement2> ]
```

Conditional execution of statements

```
if ( <expression-1> )
    <statement-1>
else if ( <expression-2> )
    <statement-2>
    .
    .
else if ( <expression-n> )
    <statement-n>
[ else
    <statement-n+1> ]
```

Conditional execution of statements with distinction between several cases

```
switch ( <expression> ) {
    case <const-expr-1>: ...
    case <const-expr-2>: ...
    .
    .
    .
    case <const-expr-n>: ...
    [ default          : ... ]
    }
```

Multi-way branch

```
break;
```

Exit a loop or switch statement

```
while ( <expression> )
    <statement>
```

Loop with test of continuation condition before execution

```
for ( <expression1>; <expression2>; <expression3> )
    <statement>
```

Loop with test of continuation condition before execution

```
do
    <statement>
while ( <expression> );
```

Loop with test of continuation condition after execution

```
continue;
```

Start another loop execution

```
return [<expression>];
```

Exit a function with possible passing of the function value

```
<fnct-name>( [<actual-parameter-list>] );
```

Function call

```
goto <label>;

<label> ::= <name>:
```

Branch to a symbolically named location

```
;
```

Dummy statement

A2 Appendix 2: Summary of operators

The operators within a group have the same precedence level. They are processed in the order they are specified. The operators in a higher group have a higher precedence level.

The table columns have the following meanings:

- Operator
- Meaning
- Processing sequence in the same group

()	Function and expression parentheses	L→R
[]	Array brackets	
->	Structure element via structure pointer	
.	Structure element via structure name	
!	Negation (not operation)	R→L
~	Bitwise complement	
++	Increment	
--	Decrement	
-	Minus sign	
(<type>)	Explicit type conversion	
*	Reference operator	
&	Address of	
sizeof	Size in bytes	
*	Multiplication	L→R
/	Division	
%	Modulo	
+	Addition	L→R
-	Subtraction	
<<	Left shift	L→R
>>	Right shift	
<	Less than	L→R
<=	Less than or equal to	
>	Greater than	
>=	Greater than or equal to	
==	Equal to	L→R
!=	Unequal to	
&	Bitwise AND	L→R
^	Bitwise XOR	L→R
|	Bitwise OR	L→R
&&	Logical AND	L→R
||	Inclusive OR	L→R
?:	Conditional expression operator	R→L
=	Assignment operator	R→L
+=	Compound assignment operators	
-=		
etc.		
,	Separator for expression lists	L→R

A3 Appendix 3: Summary of functions

A3.1 Input/output functions

Unless otherwise stated, your source file must contain the preprocessor statement

#include <stdio.h>

You then do not need to declare the function.

void clearerr(<file-pointer>) Clear error and EOF status
int fclose(<file-pointer>) Close file
int feof(<file-pointer>) Test EOF status of input/output function
int ferror(<file-pointer>) Test error status of input/output function
int fflush(<file-pointer>) Flush the output buffer
int fgetc(<file-pointer>) Read the next character from a file

char *fgets(<string>, <max-number>, <file-pointer>**)**

Read a line from a file

FILE *fopen(<file-name>, <mode>**)**

Open a file

int fprintf(<file-pointer>,<control-string>,<expression-list>**)**

Formatted output to a file

int fputc(<char>, <file-pointer>**)**

Append a character to a file

int fputs(<string>, <file-pointer>**)**

Append a string to a file

int fread(<address>,<field-length>,<no-of-fields>, <file-pointer>**)**

Read fields from a file to the location address points to

FILE *freopen(<filename>,<mode>,<file-pointer>**)**

Close a file and open another file

int fscanf(<file-pointer>,<control-string>,<pointer-list>**)**

Formatted input from a file

int fseek(<file-pointer>,<offset>,<reference-point>**)**

Modify a file pointer

long ftell(<file-pointer>)

Test the file pointer

---

int fwrite(<address>,<field-length>,<no-of-fields>,
<file-pointer>)

Write fields located at address to a file

---

int getc(<file-pointer>)

Read the next character from a file

---

int getchar()

Read a character from standard input

---

char *gets(<string>)

Read a line from standard input

---

int getw(<file-pointer>)

Read the next memory word from a file

---

char *mktemp(<string2>)

Find file name for temporary file (must be declared)

---

int pclose(<file-pointer>)

Close a pipe and pass process status

FILE *popen(<command-string>, <mode>)

Open a pipe and start a second process

int printf(<control-string>, <expression-1>,...,<expression-n>)

Formatted output of values to standard output

int putc(<char>, <file-pointer>)

Output a character to a file

int putchar(<char>)

Output a character to standard output

int puts(<string>)

Output a string to standard output

int putw(<int-value>, <file-pointer>)

Output a memory word to a file

void rewind(<file-pointer>)

Set file pointer to beginning of file

int scanf(<control-string>,<pointer-1>,...,<pointer-n>)

Formatted read via standard input

void setbuf(<file-pointer>,<pointer>)

Define input/output buffer; determine unbuffered input/output

int sprintf(<output-string>,<control-string>,<expression-list>**)** Format values into a string
int sscanf(<input-string>,<control-string>,<pointer-list>**)** Formatted read from a string
char *tempnam(<directory>, <prefix>**)** Find file name for temporary file
FILE *tmpfile() Open temporary file
char *tmpnam(<pointer>**)** Find file name for temporary file
int ungetc(<char>, <file-pointer>**)** Cancel the last read character

A3.2 Other standard library functions

Character classification and conversion #include <ctype.h>
int isalpha(<char>) Check for letters
int isupper(<char>) Check for uppercase letters
int islower(<char>) Check for lowercase letters
int isdigit(<char>) Check for decimal digits
int isxdigit(<char>) Check for hexadecimal digits
int isalnum(<char>) Check for letters and digits
int isspace(<char>) Check for space, tab, CR, newline, vertical tab and form feed characters
int iscntrl(<char>) Check for control characters (codes: 00 through 037, 0177)

Character classification and conversion #include <ctype.h>
int ispunct(<char>) Check for punctuation characters (other than letters/digits or control characters)
int isprint(<char>) Check for printable characters (codes: 040 through 0176)
int isgraph(<char>) Check for graphic characters (= printable characters except space)
int isascii(<char>) Check for ASCII characters (codes: 00 through 0177)
int toupper(<char>) Convert to uppercase letters
int _toupper(<char>) Convert to uppercase letters (macro)
int tolower(<char>) Convert to lowercase letters
int _tolower(<char>) Convert to lowercase letters (macro)
int toascii(<int-value>) Convert to ASCII characters

Other conversion functions

int atoi(<string>**)**

Convert a number from a string to an 'int' value

long atol(<string>**)**

Convert a number from a string to a 'long' value

double atof(<string>**)**

Convert a number from a string to a 'double' value

long strtol(<string>,<pointer>,<base>**)**

Convert a number from a string to a 'long' value with different bases

double strtod(<string>, <char-ptr-ptr>**)**

Convert a number from a string to a 'double' value

char *ecvt(<double-value>,<no-of-digits>,<dec-point>,<sign>**)**

Convert a 'double' value to a string

char *gcvt(<double-value>,<no-of-digits>,<string>**)**

Convert a 'double' value to a string

String handling functions

#include<string.h>

char *strcat(<string1>, <string2>)

Concatenate string 2 to string 1

char *strncat(<string1>, <string2>, <number>)

Concatenate string 2 to string 1 (max. number of characters)

int strcmp(<string1>, <string2>)

Compare strings

int *strncmp(<string1>, <string2>, <number>)

Compare strings (max. number of characters)

char *strcpy(<string1>, <string2>)

Copy string 2 to string 1

char *strncpy(<string1>, <string2>, <number>)

Copy string 2 to string 1 (max. number of characters)

int strlen(<string>)

Current string length

char *strchr(<string>, <char>)

First occurrence of a character in the string

char *strrchr(<string>, <char>)

Last occurrence of a character in the string

String handling functions

#include <string.h>

char *strpbrk(<string1>, <string2>**)**

First occurrence of a character from string 2 in string 1

int strspn(<string1>, <string2>**)**

Length of that part of string 1 that consists of characters from string 2

int strcspn(<string1>, <string2>**)**

Length of that part of string 1 that does not consist of characters from string 2

char *strtok(<string1>, <string2>**)**

Retrieve fields from string 1 that are separated by characters from string 2

Mathematical functions

#include <math.h>

int abs(<int-value>**)**

Absolute value of an 'int' value

double fabs(<double-value>**)**

Absolute value of a 'double' value

double fmod(<double-value1>, <double-value2>**)**

Modulo function for 'double' values

Mathematical functions

#include <math.h>

double gamma(<double-value>)

Gamma function

double j0(<double-value>)

Bessel function of first type, zeroth order

double j1(<double-value>)

Bessel function of first type, first order

double jn(<int-value>,<double-value>)

Bessel function of first type, n-th order

double y0(<double-value>)

Bessel function of second type, zeroth order

double y1(<double-value>)

Bessel function of second type, first order

double yn(<int-value>, <double-value>)

Bessel function of second type, n-th order

double exp(<double-value>)

Exponential function

double log(<double-value>)

Logarithm to the base e

Mathematical functions

#include <math.h>

double log10(<double-value>)

Logarithm to the base 10

double pow(<double-value1>, <double-value2>)

First value to the power of second value

double sqrt(<double-value>)

Square root

double floor(<double-value>)

Next lower integer

double ceil(<double-value>)

Next higher integer

double sin(<double-value>)

Sine

double cos(<double-value>)

Cosine

double tan(<double-value>)

Tangent

double asin(<double-value>)

Arc sine

Mathematical functions

#include <math.h>

double acos(<double-value>)

Arc cosine

double atan(<double-value>)

Arc tangent

double atan2(<double-value1>, <double-value2>)

Arc tangent of the quotient of the two values

double hypot(<double-value1>, <double-value2>)

Euclidian distance

double sinh(<double-value>)

double cosh(<double-value>)

double tanh(<double-value>)

Hyperbolic sine/cosine/tangent

Random number generation

int rand()

Random numbers between 0 and $2^{15} - 1$

void srand(<unsigned-value>)

Set starting point for the rand function

Floating-point number manipulation

double frexp(<double-value>, <int-pointer>**)**

Mantissa and exponent (in 'int' pointer) of a floating-point number

double ldexp(<double-value>, <int-value>**)**

<double-value> * pow(2, <int-value>)

double modf(<double-value>, <double-pointer>**)**

Fractional and integer portion (in 'double' pointer)

Management of data structures

char *lsearch((char *)<pointer>, (char *)<base>,<no-of-pointers>,
<length>, <comp-fnct>)

Search/enter an element in a list

char *lfind((char *)<pointer>, (char *)<base>,<no-of-pointers>,
<length>, <comp-fnct>)

Find an element in a list

void qsort((char *)<base>,<number>,<length>,<comp-fnct>)

Quicksort of a list

char *bsearch((char *)<pointer>, (char *)<base>,<number>,
<length>, <comp-fnct>)

Binary search of a sorted list for an element

Management of data structures
#include <search.h>
int hcreate(<number>**)** Create hash table
ENTRY *hsearch(<entry>, <action>**)** Search/enter in the hash table with actions FIND/ENTER
void hdestroy() Destroy hash table
char *tsearch((char *)<data-ptr>, **(char **)**<root-ptr>, <comp-fnct>**)** Search/enter a node in a binary tree
char *tfind((char *)<data-ptr>, **(char **)**<root-ptr>,<comp-fnct>**)** Search for a node in a binary tree
char *tdelete((char *)<data-ptr>, **(char **)**<root-ptr>, <comp-fnct>**)** Delete a node in a binary tree
void twalk((char *)<root-ptr>, <action>**)** Walk through a binary tree, initiating a function at each node

Memory management

char *malloc(<size>**)**

Allocate contiguous memory area

char *calloc(<number>, <el-length>**)**

Allocate and initialize contiguous memory area

char *realloc(<pointer>, <length>**)**

Expand/contract memory area

void free(<pointer>**)**

Free memory area

System interfaces

int abort()

IOT signal to the current process

void monitor(<address1>, <address2>, <'short'-array>, <size>, <number>**)**

Prepare execution profile between address 1 and 2

char *crypt(<string>, <modifier>**)**

Password encryption function

void perror(<string>**)**

Issue system error message with user comment

System interfaces
int assert(<expression>**)** Check assertions
unsigned sleep(<seconds>**)** Cause program to sleep
int system(<string>**)** Execute Shell command
void swab(<char-ptr1>, <char-ptr2>, <number>**);** Copy and swap bytes
int ftw(<string>, <fnct-ptr>, <depth>**);** Walk through directory tree, initiating functions

Miscellaneous information
long clock() CPU time
char *cuserid(<'char'-pointer>**)** **char *getlogin()** **char *logname()** Login name (owner of the process / from /etc/utmp / from Shell environment)

Miscellaneous information

char *ctermid(<'char'-pointer>)

Path name of the terminal

---

int ttyslot()

Index of your own entry in /etc/utmp

#include <string>
char *ctime(<'long'-pointer>)

Convert date and time to string

#include <time.h>
struct tm *localtime(<'long'-pointer>)

struct tm *gmtime(<'long'-pointer>)

Date and local/Greenwich Mean Time in a structure

#include <time.h>
char *asctime(<'struct tm'-pointer>)

Convert date/time structure to a string

#include <time.h>
void tzset()

Set the external variables timezone and tzname

---

char *getcwd(<'char'-pointer>, <size>)

Current directory

---

char *getenv(<name>)

Value of a parameter from the user environment

Miscellaneous information
#include <grp.h> **struct group *getgrent()** Next entry from /etc/group as a structure
#include <grp.h> **struct group *getgrgid(**<group-id>**)** Entry from /etc/group with specified group identifier as a structure
#include <grp.h> **struct group *getgrnam(**<name>**)** Entry from /etc/group with specified group name as a structure
#include <grp.h> **void setgrent()** Set file pointer to the beginning of /etc/group
#include <grp.h> **void endgrent()** Close /etc/group
#include <pwd.h> **struct passwd *getpwent()** Next entry from /etc/passwd as a structure
#include <pwd.h> **struct passwd *getpwuid(**<user-id>**)** Entry from /etc/passwd with specified user identifier as a structure

Miscellaneous information

#include <pwd.h>
struct passwd *getpwnam(<name>)

Entry from /etc/passwd with specified user name as a structure

---

#include <pwd.h>
void setpwent()

Set file pointer to the beginning of /etc/passwd

#include <pwd.h>
void endpwent()

Close /etc/passwd

---

int getpw(<user-id>, <string>)

Entry from /etc/passwd with specified user identifier as a string

---

int getopt(argc, argv, <opt-string>)

Check the next option from the Shell command line

#include <a.out.h>
int nlist(<filename>, <'struct nlist'-array-pointer>)

List of names of an executable program

Protecting programs by passwords

char *getpass(<prompt>)

Input a blanked password

A4 Appendix 4: ASCII code table

Representation of the ASCII characters:

- decimal
- octal
- hexadecimal
- graphic character

0	000	00	NUL	32	040	20	Blank	64	100	40	@	96	140	60	'
1	001	01	SOH	33	041	21	!	65	101	41	A	97	141	61	a
2	002	02	STX	34	042	22	"	66	102	42	B	98	142	62	b
3	003	03	ETX	35	043	23	#	67	103	43	C	99	143	63	c
4	004	04	EOT	36	044	24	$	68	104	44	D	100	144	64	d
5	005	05	ENQ	37	045	25	%	69	105	45	E	101	145	65	e
6	006	06	ACK	38	046	26	&	70	106	46	F	102	146	66	f
7	007	07	BEL	39	047	27	'	71	107	47	G	103	147	67	g
8	010	08	BS	40	050	28	(	72	110	48	H	104	150	68	h
9	011	09	HT	41	051	29	)	73	111	49	I	105	151	69	i
10	012	0A	LF	42	052	2A	*	74	112	4A	J	106	152	6A	j
11	013	0B	VT	43	053	2B	+	75	113	4B	K	107	153	6B	k
12	014	0C	FF	44	054	2C	,	76	114	4C	L	108	154	6C	l
13	015	0D	CR	45	055	2D	—	77	115	4D	M	109	155	6D	m
14	016	0E	SO	46	056	2E	.	78	116	4E	N	110	156	6E	n
15	017	0F	SI	47	057	2F	/	79	117	4F	O	111	157	6F	o
16	020	10	DLE	48	060	30	0	80	120	50	P	112	160	70	p
17	021	11	DC1	49	061	31	1	81	121	51	Q	113	161	71	q
18	022	12	DC2	50	062	32	2	82	122	52	R	114	162	72	r
19	023	13	DC3	51	063	33	3	83	123	53	S	115	163	73	s
20	024	14	DC4	52	064	34	4	84	124	54	T	116	164	74	t
21	025	15	NAK	53	065	35	5	85	125	55	U	117	165	75	u
22	026	16	SYN	54	066	36	6	86	126	56	V	118	166	76	v
23	027	17	ETB	55	067	37	7	87	127	57	W	119	167	77	w
24	030	18	CAN	56	070	38	8	88	130	58	X	120	170	78	x
25	031	19	EM	57	071	39	9	89	131	59	Y	121	171	79	y
26	032	1A	SUB	58	072	3A	:	90	132	5A	Z	122	172	7A	z
27	033	1B	ESC	59	073	3B	;	91	133	5B	[	123	173	7B	{
28	034	1C	FS	60	074	3C	<	92	134	5C	\	124	174	7C		
29	035	1D	GS	61	075	3D	=	93	135	5D	]	125	175	7D	}
30	036	1E	RS	62	076	3E	>	94	136	5E	^	126	176	7E	~
31	037	1F	US	63	077	3F	?	95	137	5F	_	127	177	7F	DEL

Index